Recent Progress in MANY-BODY THEORIES

VOLUME 2

International Advisory Committee for the Sixth Conference

A. J. Kallio (Oulu, Finland)—Chairman
R. F. Bishop (Manchester, U. K.)—Secretary
C. E. Campbell (Minneapolis, U.S.A.)—Treasurer
N. Ashcroft (Cornell, U.S.A.)
Y. Avishai (Beer Sheva, Israel)
J. W. Clark (St. Louis, U.S.A.)
S. Fantoni (Pisa, Italy)
I. M. Khalatnikov (Moscow, U.S.S.R.)
H. G. Kümmel (Bochum, F.R.G.)
M. L. Ristig (Koln, F.R.G.)
D. Yoshioka (Fukuoka, Japan)

Local Organising Committee for the Sixth Conference

Y. Avishai (Beer Sheva, Israel)—Chairman
A. Lonke (Beer Sheva, Israel)—Secretary (deceased)
O. Entin Wohlman (Tel Aviv, Israel)
H. Gutfreund (Jerusalem, Israel)
Y. Imry (Rehovot, Israel)
M. Kaveh (Bar Ilan, Israel)
M. Revsen (Haifa, Israel)

Recent Progress in MANY-BODY THEORIES

VOLUME 2

Edited by

Y. Avishai
Ben Gurion University
Beer Sheva, Israel

Plenum Press • New York and London

Proceedings of the Sixth International Conference on Recent Progress in Many-Body Theories, held November 5–10, 1989, in Arad, Israel

Library of Congress Catalog Card Number 88-645051

ISBN 0-306-43705-8

A Division of Plenum Publishing Corporation
233 Spring Street, New York, N.Y. 10013

Printed in the United States of America

PREFACE

The present volume contains the texts of the invited talks delivered at the Sixth International Conference on Recent Progress in Many-Body Theories held in Arad, Israel during the period November 5-10 1989. The host institute was the Physics Department at the Ben Gurion University of the Negev. Beside the invited talks there have been also two poster sessions.

The general format and style of the meeting followed closely those which had preceded it and evolved from the earlier conferences in the series: Trieste 1978, Oxtapec 1981, Altenberg 1983, San Francisco 1985 and Oulu 1987. The conferences in this series are intended to cover the broad spectrum of modern physics related to the many body problem. It starts on the smallest length scales in dealing with quark gluon plasmas and ends on the largest length scales in tackling many body problems in astrophysics. In between, it is concerned with nuclear physics, solid state physics, statistical mechanics, field theory, molecular dynamics, hydrodynamics, chaos, etc.

Like in the Oulu conference, special emphasis has been put on recent developments in high T_c superconductivity. Beside the half day session devoted to it we have also organized an evening round table discussion on the future of the whole subject. Furthermore, a new, and in the opinion of the editor a very refreshing element which was absent in earlier conferences was the presentation of a few talks by experimentalists.

Another aspect which the local organizing committee found worth presenting is concerned with many body aspects which are not directly related to traditional many body theories. Thus, we have dedicated sessions to localization which included also propagation of waves in disordered media, and a session to the rapidly developing field of mesoscopic systems.

We have decided to organize the lectures in groups in such a way that each group is related to a special topic. This classification is somewhat artificial, but we found it very useful, especially for those who are interested in a certain part of the topics which have been presented. The main subjects covered in this conference and which are presented in this book are High T Superconductivity, Localization, Nuclear Physics, Mesoscopic Systems, Strongly Correlated Systems and Microscopic Approaches (including Quantum Fluids, Surface Structure, Deep Inelastic Scattering,and other topics).

A special event which took place in this conference is the presentation of the Feenberg Medal Award to Prof. M. Kalos of Courant Institute and Cornell

University. This medal is awarded each conference (starting from the San Francisco Conference at 1985) for outstanding achievements in the physics of many body problems. Prof. Kalos is then joining Prof. D. Pines from Urbana (recipient of the medal in 1985) and Prof. J. Clark of Washington University (recipient of the medal in 1987).

At the height of our work for the preparation of the conference we were faced by the very sad event of the death of our conference secretary and colleague Dr. Aharon Lonke. The first session was preceded by a few words commemorating his life and scientific activity.

I would like to thank all those who helped me organize this conference. They include the international advisory board - [N. Ashcroft, R. Bishop, C. Campbell, J. W. Clark, S. Fantoni, I. M. Khalatnikov, A. Kallio, H. Kummel. M. Ristig and D. Yoshioka.], the local organizing committee [O. Entin Wohlman, H. Gutfreund, Y. Imry, M. Kaveh and M. Revsen.] and our sponsors [The U.S Navy, the Technion, Bar Ilan University, Tel Aviv University, Ben Gurion University, the Weizmann Institute of Science, the Israeli Academy of Science, the Israeli Ministry of Research and Development and the Israeli Ministry of Tourism. Special thanks are directed to Mr. G. Neuman for suggesting and drawing the conference symbol. However, the warmest thanks must go to all the participants and the speakers without whom the conference and this volume would not have been established.

Y. Avishai
Ben Gurion University of the Negev
April 1990

CONTENTS

SECTION I: HIGH T_c SUPERCONDUCTIVITY

SECTION II: LOCALIZATION

SECTION III: NUCLEAR PHYSICS

SECTION IV: MICROSCOPIC APPROACHES I

SECTION V: MESOSCOPIC SYSTEMS

SECTION VI: STRONGLY CORRELATED SYSTEMS

SECTION VII MICROSCOPIC APPROACHES II

SECTION VIII: FEENBERG MEDAL PRESENTATION AND SUMMARY

Flux Phases in the $t-J$ Model

T. M. Rice

Theoretische Physik
ETH-Hönggerberg
8093 Zürich, Switzerland

After a short review of the case for the $t-J$ model as a suitable low energy model Hamiltonian for CuO_2 planes in the high-T$_c$ superconductors, the renormalized mean field approximation to this model is discussed. At half-filling the reduction to a Hilbert space containing only spin degrees of freedom leads to a wide class of degeneracies in the fermion meanfield description. One particular class of states which has generated much interest in the past year, is that built from one electron Hofstadter states which are eigenstates in the presence of a uniform magnetic flux. The mean field theory of such states in a $t-J$ model is reviewed followed by a discussion of their unusual properties. In conclusion, some comments on the relationship to anyon models and other open questions are given.

1. Introduction

The discovery of high-T$_c$ superconductors has greatly stimulated interest in the properties of strongly correlated electrons. The simplest model that one can use to describe such a system has a single tight-binding band made from a single orbital on each site [1]. The strong correlation condition forbids double occupancy of this orbital. At an electron density of exactly 1 electron/site this restriction limits the Hilbert space to spin degrees of freedom only which are coupled with a Heisenberg interaction (J). This simple model on a square lattice is a good description of the low energy spectrum [2] of the insulating CuO_2 planes which have a formal valence Cu^{2+}. The question of how to describe the removal or addition of electrons (i.e. hole or electron doping) has been the subject of a great deal of discussion. The simplest description was proposed initially by Anderson [1] namely that the extra charge would be carried by formal valence Cu^{3+} or Cu^{+} ions which would be spin singlets. The Hilbert space that results for electron count less than 1 per site say, has 3 configurations on each site namely spin up or down and empty. Electrons can transfer on the neighboring empty sites with a matrix element, t. These two terms make up the $t-J$ Hamiltonian which will be the subject of this talk.

Recent Progress in Many-Body Theories, Volume 2
Edited by Y. Avishai
Plenum Press, New York, 1990

The adequacy of the simplest $t-J$ model to describe doped CuO_2 planes is an interesting question which cannot be covered here in detail. The key question is whether the starting Hamiltonian containing $Cu-d$ and $O-p$ orbitals can be reduced to the much smaller Hilbert space with 3 configurations per Cu site. Given that added holes go mainly into O-orbitals, the $t-J$ model requires that the formal Cu^{3+} ion that carries the charge is a tightly bound singlet between the central Cu spin and the hole with spin residing on the four neighboring $O2p$-orbitals, as discussed by Zhang and Rice [3]. Unfortunately there is no small parameter which can be used to formally justify the reduction procedure. In addition there are many terms in the starting Hamiltonian such as direct $O-O$ overlap or the Coulomb repulsion on O-orbitals which complicate the description. There are also questions concerning the values of parameters in the starting Hamiltonian. Recently Hybertsen, Stechel, Schluter and Jennison [4] have reported a detailed set of a priori calculations based on the local density functional method to estimate the parameters. They then examined the low energy effective Hamiltonian by comparing to the low energy eigenvalues of the full Hamiltonian obtained by exact diagonalization on a small cluster [5]. The set of parameter values they obtain agree quite well with the obtained in other calculations and empirically [6]. The $t-J$ model extended through a small n.n.n. hopping term t' is found to give a good representation for both hole and electron doping. The Heisenberg coupling J estimated [4] at $\approx 130\ meV$ agrees very well with experimental values [2] on La_2CuO_4. The n.n. hopping matrix element $t \approx 420\ meV$ is essentially the same for both electron and hole doping [4] leading to a value for the ratio $J/t \approx 0.3$. The ratio $t'/t \approx -0.2$ is rather small. Other estimates [7] gave a rather larger value for $|t'|$. Nonetheless it seems that the $t-J$ model is the simplest reduced Hamiltonian that can be considered reasonable for the CuO_2 planes.

2. Renormalized Mean Field Theory

This talk will concentrate on only one of the many methods being used currently to tackle the $t-J$ model. The restriction to a reduced Hilbert space complicates the problem considerably, even in meanfield theory as we shall discuss.

Thus a general Hartree-Fock-Bogoliubov wavefunction $|\Psi_{HFB}\rangle$ needs to be projected on to the reduced Hilbert space so that the actual mean field wavefunction is

$$|\Psi_{MF}\rangle = P_d\,|\Psi_{HFB}\rangle\ ; \quad P_d = \prod_i (1-n_{i\uparrow}n_{i\downarrow})\,. \tag{1}$$

One approximation scheme which is reasonably accurate when compared to numerical calculations and used by Zhang et al [8], replaces the projection operator by a renormalization factor calculated according to Gutzwiller's prescription i.e. one replaces

$$\begin{aligned} H_{t-J} &= P_d\{-t\sum_{<ij>\sigma} d^+_{i\sigma}d_{j\sigma} + h.c. + J\sum_{<ij>}\vec{S}_i\cdot\vec{S}_j\}P_d \\ &= P_d\{H_t + H_J\}P_d \end{aligned} \tag{2}$$

by the simpler form

$$\tilde{H}_{t-J} = g_t \, H_t + g_J \, H_J , \tag{3}$$

where g_t, g_J are numerical factors which depend on the average densities of $\uparrow$ and $\downarrow$-spin electrons. For paramagnetic translationally invariant states [8]

$$g_t = 4\delta(1+2\delta)^{-1} \quad g_J = 4(1+2\delta)^{-2} \tag{4}$$

where $n_\uparrow = n_\downarrow = 1/2 - \delta$. ($2\delta$ is the hole concentration). This approximation has been shown by Vollhardt and coworkers [9] to be exact in infinite dimensions and was shown by Kotliar and Ruckenstein [10] to be expressible as a slave boson meanfield theory.

At half-filling ($\delta = 0$) the $t-J$ model reduces to a Heisenberg model and it now generally agreed that this has a antiferromagnetically ordered ground state with a sublattice magnetization reduced to 60% of the Neel value due to quantum fluctuation. The purely spin-problem is best described by the spin operators $\{\vec{S}_i\}$ which operate in a Hilbert space containing 2^L configurations, (L: Number of sites). But we are interested in representing it in the much larger fermion Hilbert space which applies for $\delta \neq 0$. This introduces a tremendous redundancy since the unprojected $|\Psi_{HFB}\rangle$ are vectors in 4^L Hilbert space. Anderson, Affleck and coworkers [1,11] were the first to point out that this results in local SU(2) gauge invariance. Thus if we on any set of sites $\{i\}$ make a transformation

$$\begin{aligned} d^+_{i\sigma} &\longrightarrow \alpha_i \, d^+_{i\uparrow} + \beta_i \, d_{i\downarrow} \\ d_{i-\sigma} &\longrightarrow -\beta^*_i \, d_{i\uparrow} + \alpha^*_i \, d_{i\downarrow} \, ; \qquad [| \alpha_i |^2 + | \beta_i |^2 = 1] \end{aligned} \tag{5}$$

we change the unprojected wavefunction $|\Psi_{HFB}\rangle$ drastically but after projection of wavefunction which is now purely a spin wavefunction is the same. In the fermion mean field theory this redundancy has the consequence that a large class of meanfield solutions are degenerate at $\delta = 0$, even though they ostensibly represent fermion states with quite different characteristics. For example two especially interesting states are the d-wave superconducting state (or d-wave RVB state) and the Affleck-Marston [12] state. These are represented by quite different diagonal and off-diagonal expectation values in the unprojected states i.e. ($\tau = x \; or \; y$)

$$\begin{aligned} \tilde{\Delta}_\tau &= \langle d^+_{i\uparrow} \, d^+_{i+\tau\downarrow} - d^+_{i\downarrow} \, d^+_{i+\tau\uparrow} \rangle \\ \chi_{i,i+\tau} &= \sum_\sigma \langle d^+_{i,\sigma} \, d_{i+\tau,\sigma} \rangle \end{aligned}$$

take values

$$\tilde{\Delta}_x = -\tilde{\Delta}_y = \chi_{i,i+x} = \chi_{i,i+y} = 0.339 \tag{6}$$

$$\tilde{\Delta}_x = \tilde{\Delta}_y = 0; \; \chi_{i,i+\tau} = 0.479 \; exp(\pm i\pi/4) \tag{7}$$

in the d-wave RVB and Affleck-Marston states respectively, yet when projected onto the reduced Hilbert space these two states are identical.

These two states are of special interest because within this class of mean field solutions they give the best expectation value of H_t when a small number of holes are introduced. To leading order in δ they give the same value [8] for the kinetic energy per added hole. These states do not have long range AF order and since they are written as fermion wavefunction are easily generalized to finite doping. At $\delta = 0$ Gros [13] found numerically an energy of -0.64 J/site which is several percent higher than current estimates [14] of an AF ordered spin state -0.67 J/site. Lee and Feng [15] have shown that by introducing a long range AF into these states that their energy can be lowered to values much closer to that obtained with spin wavefunctions. Introducing holes suppresses the AF long range order and they find the generalizations of the d-wave RVB state are stable against AF ordering beyond 5% doping. Therefore we will concentrate on the generalizations of the paramagnetic wavefunctions at finite doping and ignore the AF ordered states.

3. Flux States

The name flux states has been applied to the Affleck-Marston (AM) state [12] and generalizations thereof built from wavefunctions that describe electrons in an external magnetic field or flux. The relevance of these wavefunctions was clarified recently by the following conjecture [16]. Consider a system of non-interacting electrons (density ν electrons/site) moving in a square lattice described by a n.n. tight-binding matrix element $(-t)$ but we introduce a phase factor $e^{i\varphi_{ij}}$ for each n.n. bond (ij). If we seek a solution with uniform density and choose the phases to derive from a vector potential corresponding to a uniform magnetic flux (in units of the flux quantum) $\Phi = (2\pi)^{-1} \sum_{\square} \varphi_{ij}$ in each lattice square, then according to this conjecture the ground state is not given by $\Phi = 0$ (i.e. Bloch states) but by the choice $\Phi = \nu$. The eigenvalues of the uniform flux problem on a square lattice were studied extensively by Hofstadter [17] who found a complex spectrum $\{\varepsilon_l\}$ with many bands which depend on the commensurability of the flux Φ. The choice $\Phi = \nu$ corresponds to placing the Fermi level in the large gap between the 1st and 2nd Landau-Hofstadter bands. Clearly energy is gained by having an energy gap at the Fermi energy. However the Hofstadter spectrum is compressed overall relative to the Bloch spectrum so that the question of whether

$$E_G(\nu, \Phi) \ < \ E_G(\nu, \Phi = 0)\,; \ E_G(\nu, \Phi) \ = \sum_{l}^{occ} \varepsilon_l \tag{8}$$

depends on which of these two factors is dominant. Hasegawa et al [16] and Montambeaux [18] and Nori [19] have made a series of numerical calculations which support this conjecture. $E_G(\nu, \Phi)$ has a cusp minimum as a function of Φ at $\Phi = \nu$. Note $E_G(\nu, \Phi = \nu)$ seems to be a continuous function of ν without obvious structure at commensurate values of ν [18].

Earlier this year Anderson, Shastry and Hristopolous [20] proposed generalizations of the AM-state and in particular a variational state which can be written as

$$|\,\Psi_{ASH}\rangle \ = \ P_d \prod_{l,l'}^{occ} d_{l\uparrow}^{+}\, d_{l'\downarrow}^{+}\,|\,\mathrm{vac}\rangle\,, \tag{9}$$

where $d^+_{l\sigma}$ creates a Hofstadter state $\{l\sigma\}$. Lederer, Poilblanc and Rice [21] (LPR) showed that this state can be used to minimize the energy of the exchange term H_J when the flux in the Hofstadter problem is chosen to be commensurate with density of $\uparrow$ (or $\downarrow$) spin electrons i.e. $\Phi = n_\uparrow = n_\downarrow$. Consider just one term in the Heisenberg interaction, then we may write

$$\langle \Psi_{ASH} \mid S_i^+ S_j^- \mid \Psi_{ASH} \rangle = \langle \Psi_{ASH} \mid e^{i\varphi_{ij}} d^+_{i\uparrow} d_{j\uparrow} \cdot e^{-i\varphi_{ij}} d^+_{j\downarrow} d_{i\downarrow} \mid \Psi_{ASH} \rangle \qquad (10)$$

and the expectation value factors into $\uparrow$ and $\downarrow$ expectation values. Introducing the phase factor simplifies the problem when we choose the phase φ_{ij} to be the same as in the one-elctron Hamiltonian used to generate the Hofstadter eigenvalues. This choice allows us to express each expectation value simply as a sum over occupied Hofstadter eigenvalues and using the singlet character of the wavefunction and the renormalization approximation LPR get

$$\langle \Psi_{ASH} \mid H_J \mid \Psi_{ASH} \rangle = -\frac{3J}{16} g_J \Big(\frac{1}{L} \sum_l^{occ} \varepsilon_l\Big)^2 . \qquad (11)$$

This energy is then minimized by the choice of the commensurate flux $\Phi = n_\uparrow$, leading to the name commensurate flux phase (CFP) for this choice. The fact that H_J describes an exchange process whereby $\uparrow$ and $\downarrow$-spin electrons interchange positions allowed us to introduce a phase factor to the hop for $\uparrow$ and $\downarrow$-spin separately and this in turn allowed us to introduce a fictitious flux for each species determined by the other species. Note since the Hofstadter eigenvalues depend only on the flux and not on gauge in which this flux is represented, this property carries over to the H_J-term.

In the presence of holes we have to examine also the kinetic energy, H_t. Introducing the phase factor again into the expectation value gives again a sum over Hofstadter eigenvalues but now the phase factor is not compensated so that

$$\langle \Psi_{ASH} \mid H_t \mid \Psi_{ASH} \rangle = 2t\, g_t\, L^{-1} \sum_l^{occ} \varepsilon_l \cdot \langle cos\ \varphi_{ij} \rangle_{\text{ave}} \qquad (12)$$

but now the average over all bonds depends on the gauge in which the flux is represented. This degree of freedom can be used to minimize the energy. LPR defined a function $K(\Phi) = max \langle cos\, \varphi_{ij} \rangle_{\text{ave}}$ where the maximum is taken w.r.t. the choice of gauge and also there is a discrete degree of freedom since a unit flux may be added or subtracted from a square in the magnetic supercell. This gauge dependence shows up clearly in the case of $\Phi = 1/2$ (AM-state) where a Landau gauge give $1/2$ but a choice with alternating fluxes $\pm 1/2$ on neighboring squares and $\mid \varphi_{ij} \mid = \pi/4$ gives $K(\frac{1}{2}) = 1/\sqrt{2}$. Also $K(\Phi \to 0) = 1$ but in between, $0 < \Phi < 1/2$, $K(\Phi)$ varies with a lower bound of $2/\pi$. This function shows up also in the study of Josephson junction arrays and it has been extensively studied [22] in that context. It is clear that these values of $K(\Phi)$ represent a considerable cost in kinetic energy in the CFP-states.

This kinetic energy cost shows up as a limitation on the stability of the CFP for $J/t < 1$. This in the renormalized mean field theory it has the consequence that for $2\delta = 1/8$ the state with $\Phi = 7/16$ requires a value of $J/t \gtrsim 1.8$ to be stabilized. This unfavorable cost in kinetic energy is not just a consequence of the renormalized mean field theory. Poilblanc, Hasegawa and Rice [23] examined a 4×4 cluster with 2 holes (i.e. $2\delta = 1/8$) by exact evaluation of the ASH-states and by exact numerical diagonalization. Again the choice $\Phi = 7/16$ is not favored for $J/t < 1$ but smaller commensurate flux values (e.g. $\Phi = 1/16$) can occur. The overall comparison to the exact energies is not bad. There are many interesting aspects to this comparison for small clusters which cannot be covered here such as the role of the broken translation symmetry and the introduction of hole-hole repulsion.

Liang and Trivedi [24] have reported a series of exact numerical evaluations on quite large clusters (20×20) which agree in their overall conclusion that relative large values of J/t are required to stabilize the CFP.

There is another generalization of the AM-state namely a staggered flux phase (SFP) with alternate values $\pm \Phi$ on neighboring squares. Such states were proposed by Harris, Lubensky and Mele [25] and their energies were considered in some detail by Poilblanc and Hasegawa [26]. Such states do not have a gap at the Fermi energy but they gain energy through the splitting of the van Hove singularity at the Bloch band center. The analogous $K_S(\Phi)$-function which appears in the kinetic energy H_t-term is however much larger since a continuous choice $|\varphi_{ij}| = \pi\Phi/2$ is now possible so that $K_S(\Phi) \geq K(\Phi)$. Thus the overall energy is improved for $J/t < 1$, relative to CFP. Poilblanc and Hasegawa [26] find in their exact studies on small clusters that the kinetic energy term favors SFP over CFP in this region. There remains however the possibility that introducing further correlations in the wavefunctions or additional terms in the Hamiltonian might change this.

4. Properties of Flux States and Relation to Anyons

In this part the CFP will be emphasized although SFP are clearly of interest too. The first property that is clearcut is the chiral spin order in these states. Following Wen, Wilczek and Zee [27] the chiral spin order expectation value on a square can be expressed as

$$\begin{aligned}\langle E_{123} + E_{134} + E_{124} - E_{234}\rangle = & -4\mathcal{I}m\{\langle\chi_{12}\chi_{23}\chi_{34}\chi_{41}\rangle - \\ & -\langle\chi_{14}\chi_{43}\chi_{32}\chi_{21}\rangle\}\end{aligned} \qquad (13)$$

with $E_{123} \equiv \vec{\sigma}_1 \cdot (\vec{\sigma}_2 \times \vec{\sigma}_3)$. The r.h.s. depends directly on the value of $sin(2\pi\Phi)$ so that for $\Phi \neq 1/2$ (i.e. except AM-state), it is nonzero and such states have a chiral spin order.

A second property concerns orbital currents. Since the current on an individual bond varies as $\sum_\sigma \langle d^+_{i\sigma} d_{i-x,\sigma} - d^+_{i-x,\sigma} d_{i\sigma}\rangle$ it will be proportional to $sin(\varphi_{ij})$ so that there will in general be a current pattern in the magnetic supecell when $\delta \neq 0$. The detailed current pattern will depend on the flux Φ and the gauge. Thus these states are in the category of orbital magnets discussed in the classification of diagonal long range order by Halperin and Rice [28]. The CFP-states are characterized by a particular orbital current pattern which in turn will generate spatially varying magnetic fields in the sample. Note these fields are not to be confused with the flux generated selfconsistently through the meanfield treatment of the correlated motion. The real magnetic fields are much smaller in magnitude since they are proportional to the density of holes through the g_t-factor. However they should still be quite measurable [29] in NQR, NMR, μSR and other experiments which are sensitive to small internal magnetic fields.

The most interesting properties of the CFP are their superconducting properties discussed by LPR. Since the expectation value of the kinetic energy, (H_t-term) depends directly on the gauge in which the flux is represented, these states have a collective gauge coordinate. This was used by LPR [21] to construct a Ginzburg-Landau-like expansion for the energy as a function of an applied e.m. field. The coupling between the gauges for $\uparrow$ and $\downarrow$-spin electron implied by the factorization in Eq. (10) leads to a gauge coupling similar to that in usual BCS-states and to flux quantization with a charge $2e$. The cusp property of exchange energy, (H_J-term) as a function of the flux, Φ leads to a rigidity of the wavefunction w.r.t. a small external e.m. flux so that this will be screened out by a surface current i.e. a Meissner effect. The question of the compressibility of this state is trickier since the function $K(\Phi)$ is not differentiable [22] w.r.t. to Φ at all rational values of Φ leading to an incompressible state. However as discussed out by Lederer and Rice [30], various perturbations such as lattice imperfections, impurities or interlayer coupling etc. may smooth out the cusps in $K(\Phi)$.

The CFP described above have a number of properties in common with the anyon gas model. This model for the superconductivity in the cuprates is based on Laughlin and Wiegmann's proposal that the charge carrying holes are particles which obey 1/2-statistics [31]. This proposal has led to many papers on the properties of the free anyon gas [27,37,33]. Such a gas may be represented as a system of spinless fermions which carry a fraction of a flux quantum leading to long range interactions due to the resulting vector potential. If these are treated in a mean field approximation the parallel to the mean field CFP is quite close. In the anyon gas model there are two species of holes labelled by isospin which as free particles in a fictitious flux such that each species exactly fills its lowest Landau level [33]. Such an anyon gas has been shown to exhibit a Meissner effect and to have a Goldstone mode characteristic of superfluidity [37,33]. Further Halperin et al [33] show that a net orbital moment results on each CuO_2 plane. In addition to this diagonal long range order they show that there is an off diagonal long range order defined by adding and removing two nearby holes with the accompanying flux at large distances.

There is a close parallel between the CFP and the anyon gas as regards the diagonal long range order but although it is plausible that CFP also have ODLRO this remains to be shown explicitly. One important difference is the role of the underlying lattice. In the anyon gas model, it determines only the effective mass of the anyons and none of the commensurability effects associated with cusps in $K(\Phi)$ appear.

5. Conclusions

The past year has seen a lot of progress in the understanding of flux phases and the closely related anyon models and their fascinating properties. Clearly there remains many open questions in this subject especially regarding the relative stability of the commensurate and staggered flux phases and the d-wave RVB state at finite doping levels. In this connection, Zhang [34] has recently reported that the staggered flux phase when treated in the Gutzwiller approximation is unstable towards d-wave pairing. The interrelationship between the mean field theory for the $t - J$ model discussed above and the anyon gas models needs to be clarified further. A word of caution however is in order since very recent reports [35] of μSR experiments stimulated by the prediction of Halperin et al [33] give a preliminary limit on the magnetic field at the muon sites which is much below values estimated from an orbital magnetic moment. The definitive results on this point will be a crucial test of the diagonal long range order in the CFP or anyon gas models.

Acknowledgements

This work is the result of a very close collaboration with P. Lederer, D. Poilblanc, Y. Hasegawa and P. Wiegmann and exchanges with many others especially G. Blatter, P.A. Lee and M. Roos.

References

1. *P.W. Anderson, Science 235, 1196 (1987) and in Proc. Int. School E. Fermi, Course CIV, eds. R.A. Broglia and J.R. Schrieffer, North Holland p. 1, 1987.*
2. *For an account of the magnetic properties see T.R. Thurston et al., Phys. Rev. B 40, 4585 (1989);*
   K. Yamada et al., ibid. 4557;
   J.M. Tranquada et al., ibid 4503 and references therein.
3. *F.C. Zhang and T.M. Rice, Phys. Rev. B 37, 3759 (1988).*
4. *M.S. Hybertsen, E.B. Stechel, M. Schlüter and D.R. Jennison, preprint.*
5. *See also A. Ramsak and P. Prelovsek, Phys. Rev. B 40, 2239 (1989);*
   C.H. Chen, H.B. Schüttler and A.J. Fedro, preprint;
   W.H. Stephan, W. von der Linden and P. Horsch, Phys. Rev. B 39, 2924 (1989);
   M. Ogata and H. Shiba, preprint.
6. *F. Mila, Phys. Rev. B 38, 11358 (1988);*
   A.K. Mc Mahon, R.M. Martin and S. Satpathy, Phys. Rev. B 38, 6650 (1989);
   H. Eskes and G.A. Sawatzky, Phys. Rev. Lett. 61, 1413 (1988).
7. *H. Eskes, L.H. Tjeng and G.A. Sawatzky, Springer Series in Mater. Science, Vol. 11, 20 (1989).*
8. *F.C. Zhang, C. Gros, T.M. Rice and H. Shiba, Super. Sci. and Tech. 1, 36 (1988).*
9. *W. Metzner and D. Vollhardt, Phys. Rev. Lett. 62, 324 (1989);*
   D. Vollhardt preprint, E. Müller-Hartmann preprint, M. Roos preprint.

10. G. Kotliar and A.E. Ruckenstein, Phys. Rev. Lett. 57, 1362 (1986).
11. P.W. Anderson in Ref. 1;
I. Affleck, Z. Zou, T. Hsu and P.W. Anderson, Phys. Rev. B 38, 745 (1988).
12. I. Affleck and J.B. Marston, Phys. Rev. B 37, 3774 (1988);
G. Kotliar, Phys. Rev. B 37, 3664 (1988).
13. C. Gros, Phys, Rev. B 38, 931 (1988).
14. see for example N. Trivedi and D.M. Ceperley, Phys. Rev. B 40, 2737 (1989) and preprint.
15. T.K. Lee and S. Feng, Phys. Rev. B 38, 11809 (1988).
16. Y. Hasegawa, P. Lederer, T.M. Rice and P. Wiegmann, Phys. Rev. Lett. 63, 907 (1989).
17. D.R. Hofstadter, Phys. Rev. B 14, 2239 (1976).
18. G. Montambeaux, Phys. Rev. Lett. 63, 1657 (1989).
19. F. Nori, private communication.
20. P.W. Anderson, B.S. Shastry and D. Hristopolous, preprint.
21. P. Lederer, D. Poilblanc and T.M. Rice, Phys. Rev. Lett. 63, 1519 (1989).
22. S. Teitel and C. Jayaprakesh, Phys. Rev. Lett. 51, 1999 (1983);
T.C. Halsey, Phys. Rev. B 31, 5728 (1985).
23. D. Poilblanc, Y. Hasegawa and T.M. Rice, preprint.
24. S. Liang and N. Trivedi, preprint.
25. A.B. Harris, T.C. Lubensky and F.J. Mele, Phys. Rev. B 40, 2631 (1989).
26. D. Poilblanc and Y. Hasegawa, preprint.
27. X.G. Wen, F. Wilczek and A. Zee, Phys. Rev. B 39, 11413 (1989);
X.G. Wen and A. Zee, preprint.
28. B.I. Halperin and T.M. Rice, Solid State Phys. 21, 115 (1968) (ed. F. Seitz, D. Turnbull and H. Ehrenreich, Acad. Press, N.Y.).
29. P. Lederer, D. Poilblanc and T.M. Rice, to be published.
30. P. Lederer and T.M. Rice, to be published.
31. V. Kalmeyer and R.B. Laughlin, Phys. Rev. Lett. 59, 2095 (1987).
R.B. Laughlin, Phys. Rev. Lett. 60, 2677 (1988) and Science 242, 525 (1988);
P.B. Wiegmann, Phys. Rev. Lett. 60, 821 (1988).
32. Y.H. Chen, F. Wilczek, E. Witten and B.I. Halperin, preprint.
A.L. Fetter, C.B. Hanna and R.B. Laughlin, Phys. Rev. B 39, 9679 (1989);
G.S. Canright, S.M. Girvin and A. Brass, preprint.
33. B.I. Halperin, J. March-Russell and F. Wilczek, preprint.
34. F.C. Zhang, preprint.
35. G.M. Luke, private communication.

TWO DIMENSIONAL PHENOMENA IN HIGH TEMPERATURE SUPERCONDUCTORS

Baruch Horovitz

Department of Physics
Ben-Gurion University
Beer-Sheva 84105
Israel

The presence of Kosterlitz-Thouless type resistive fluctuations and of non-linear current-voltage relation in some of the high temperature superconductors suggests the presence of two-dimensional (2D) superconductivity. It is shown here that an array of 2D Josephson junctions can account for this behavior. The junctions exhibit an additional phase transition above which the layers are uncorrelated. The sesitivity of the latter transition to magnetic fields and the temerature dependence of the non-linear current-voltage relation are consistent with experimental data.

INTRODUCTION

Two-dimensional (2D) superconductivity, as studied on thin films made of conventional superconductors[1], exhibits a Kosterlitz-Thouless type phase transition[2] at a temperature T_c. This phase transition has two peculiar features: a) Above T_c the resistivity ρ drops to zero as $T \to T_c$ in the form

$$\rho(T) \sim \exp[-b/(T-T_c)^{1/2}] \qquad (1)$$

where b is a constant. b) Below T_c, the current (I) - voltage (V) relation has the form

$$V \sim I^{a(T)} \qquad (2)$$

with $a(T)>1$ a temperature dependent exponent. Just below T_c $a(T)=3$ (seen only in high quality films) and increases linearly with T upon cooling. Eq. (2) implies zero resistivity, but unlike three dimensional superconductors, there is no threshold current for the onset of a voltage.

A number of experimental groups[3-6] have observed the behavior of Eqs. (1,2) in the high temperature superconductor $Y_1Ba_2Cu_3O_7$ (YBCO). The form (2) was seen in both polycrystalline samples[3] and in single crystals[4-6] in a 1-3°K range below $T_c \cong 80$-90°K. In one case[3] the exponent seems to diverge at a temperature T_{c2} with $T_c - T_{c2} \cong 1$°K. Furthermore, T_{c2} seems to be extremely sensitive to a magnetic field H, dropping by ~1°K with H=50G; the sharp dependence $T_{c2}(H)$ has a cusp shape. Other data[4-5] shows a more gradual increase in a(T), but still faster than that expected in a 2D system, reaching[4] a remarkable $a(79°K) \cong 8$ with $T_c \cong 82$°K.

Recent Progress in Many-Body Theories, Volume 2
Edited by Y. Avishai
Plenum Press, New York, 1990

In a strong magnetic field (a few KG) the temperature range over which Eq. (2) is observed seems to increase. Eq. (2) was fitted[6] with an H dependent exponent at 77°K. A possibly related phenomena is the temperature $T_{irr}(H)$ for the onset of irreversibility in the magnetization curve[7] with $T_{irr}(H)$ also exhibiting a cusp shape.

The behavior of Eq. (1) was also seen in the high temperature superconductor $Bi_2Sr_2CaCu_2O_8$ (BSCCO) with $T_c \cong 85°K$.The related behavior of Eq. (2) was, however, not reported so far. Thus BSCCO is a second candidate, not fully confirmed, to exhibit 2D superconductivity.

In the next section we develop the theoretical framework[8] in which 2D superconductivity appears as a result of thermall fluctuations of 2D Josephson junctions. A second transition, which corresponds to an ordering of the junctions, then allows the formation of a three dimensional superconductor. Finally, the physical origin of these junctions is discussed--either as twin boundaries[8] or as resulting from non-conventional elctrodynamics.

2D JOSEPHSON JUNCTIONS

In this section Josephson's derivation[9] of the free energy of a junction is reconsidered and then the effects of thermal fluctuations are studied. I consider a uniform 2D junction separating two bulk superconductors. In each superconductor the free energy has the form

$$F = \int d^3r\, [\mathbf{h}^2/8\pi + (\hbar^2/2m)n_s(\nabla\phi - 2e\mathbf{A}/\hbar c)^2] \qquad (3)$$

where $\mathbf{h}(\mathbf{r})$ is the internal magnetic field, n_s is the density of superconducting electrons, ϕ is their phase and $\mathbf{A}$ is the electromagnetic vector potential. The equation for the current $\mathbf{J}$ follows from $\delta F/\delta \mathbf{A}=0$, i.e.

$$\mathbf{J}=(c/4\pi)\nabla x\mathbf{h}=(e\hbar/2m)n_s(\nabla\phi - 2e\mathbf{A}/\hbar c) \qquad (4)$$

To find the free energy of the junction, I follow Josephson[9] and first find an equation of motion for the Josephson phase $\phi(x,y)$,

$$\phi(x,y)=\phi(x,y,\infty) - \phi(x,y,-\infty) \qquad (5)$$

where z=0 is the plane of the junction. The boundary conditions require that all currents vanish far from the junction, hence

$$\nabla\phi(x,y,z)=2e\mathbf{A}/\hbar c \qquad z\to\pm\infty \qquad (6)$$

Consider an area perpendicular to y with width dx in the x direction and extending to $\pm\infty$ in the z direction. Choosing a gauge $A_z=0$, the flux through this area can be written in the form

$$h_y^{av}(x,y)dx \equiv dx \int_{-\infty}^{\infty} h_y(x,y,z)dz=$$

$$=\oint \mathbf{A}\cdot d\mathbf{l}=[A_x(x,y,\infty) - A_x(x,y,-\infty)]dx=dx(\hbar c/2e)\partial\phi(x,y)/\partial x \qquad (7)$$

A similiar equation follows for $h_x^{av}(x,y)$, the z integrated $h_x(x,y,z)$. Thus the z dependence can be eliminated by using the integrated 2D fields $\mathbf{h}^{av}=(h_x^{av},h_y^{av})$ for which

$$\mathbf{h}^{av}(x,y)=(\hbar c/2e)\hat{z}\times\nabla\phi(x,y) \tag{8}$$

where $\hat{z}$ is a unit vector normal to the junction.

The z component of the equation of motion, Eq. (4), becomes

$$(c/4\pi)[-\partial_y h_x(x,y,z)+\partial_x h_y(x,y,z)]=(e\hbar/2m)n_s\partial\phi(x,y,z)/\partial z \tag{9}$$

Integrating z yields the Josephson phase $\phi(x,y)$ on the right hand side. The junction effect[9] is to replace ϕ by $\sim\sin\phi$, so that observables are periodic in ϕ and spacial variations in ϕ matter only where $\phi\neq 2\pi$. Substituting (8) into the z integrated (9) finally yields

$$\nabla^2\phi(x,y)=(1/\lambda_J)^2\sin\phi(x,y) \tag{10}$$

where λ_J is the Josephson penetration length. Note that no assumption on an exponential decay of $\mathbf{h}(x,y,z)$ in the z direction (as in usual derivations[9]) has been used.

Eq. (10) allows to construct a free energy such that Eq. (10) is its minimum condition, i.e.

$$F\sim\int dxdy\{I_J[(1-\cos\phi)+\tfrac{1}{2}\lambda_J^2(\nabla\phi)^2]-2\mathbf{H}\cdot\hat{z}\times\nabla\phi\} \tag{11}$$

The last term in (11) is a surface term and does not affect the equation of motion. It represents the coupling of an external magnetic field **H** (assumed parallel to the junction) to the internal field Eq. (8).

The free energy of the form (11) has been extensively studied in the context of commensurate-incommensurate phase transitions in 2D[10-12]. Considering first **H**=0, the partition function of Eq. (11) has a Kostelitz-Thouless type phase transition at a temperature T_0. At $T>T_0$ thermal fluctuations destroy the Josephson coupling, i.e. $I_J\rightarrow 0$ in a renormalization group sense; hence $\langle\cos\phi\rangle=0$ and the two superconductors are uncorrelated.

The above scenario may seem peculiar: ordering in the bulk which does not enforce ordering on the surface. This possibility is specific to the superconducting order parameter which is not just a complex one, but has also a related vector potential as a gauge field. Thus, as the phase fluctuates in the bulk, the vector potential **A** can also fluctuate so that the current (4) vanishes far from the junction; the resulting energy fluctuations are localized near the junction and describe an effective 2D free energy.

At $T<T_0$ the junction becomes coherent and $\langle\cos\phi\rangle\neq 0$. In the renormalization group approach I_J scales to ~1 while the length scale λ_J becomes ξ, where

$$\xi\sim\exp[b/(T_0-T)^{1/2}] \tag{12}$$

with b a constant.

A line in the (x,y) plane of the junction, across which $\phi(x,y)$ varies by 2π defines a fluxon line; it carries a unit of magnetic flux (Eq.8) along the line. The last term in (11) shows that an external field **H** couples to the fluxon density favoring a finite density, while the other terms in (11) favor zero density. Thus there is a critical field H_J for flux penetration, which is proportional to the (H=0) fluxon energy. The latter energy at T=0 is $\sim\lambda_J$, hence thermal fluctuations which renormalize λ_J to ξ yield the critical line

$$H_J \sim \exp[b/(T_0-T)^{1/2}] \tag{13}$$

The fluxon density ρ vanishes at this line[10-12] as $\rho\sim(H-H_J)^{1/2}$; the theory also predicts[12] that at $T>T_0$ when $H_J=0$ the fluxon density is nonanalytic in H due to thermally excited fluxon loops with the result

$$\rho \sim H-cH^{\eta(T)-3} \tag{14}$$

where $\eta(T)>4$ and increases with T.

This completes the study of the thermodynamics of a single junction and I turn now to examine its correlation with the data on high T_c superconductors.

DISCUSSION

Experiments on the high T_c compounds[3-6] dealt with bulk three dimensional samples. Thus one has to conjecture who are the 2D structures and why do they appear to be uncorrelated in a small temperature interval. As argued first, the latter feature can be understood in terms of an array of junctions, and is largely independent of the identification of these junctions.

Consider an array of parallel junctions which define a system of parallel layers between these junctions. At T_c the layers become superconducting while the junctions are thermally disordered, as shown in the previos section. Thus each layer can fluctuate independently and exhibit 2D superconductivity. As T is lowered, the junctions order at T_0 and three dimensional correlation is achieved. Thus T_0 is identified with T_{c2} of Ref. (3), i.e. the temperature at which a(T) diverges; the observed cusp[3] in $T_{c2}(H)$ is accounted for quatitatively by Eq. (13). If there is a distribution of junction parameters, the transtion at T_{c2} would appear smeared, as seen in the other experiments[4-6]. Furthermore, the exponent a(T) in Eq.(2) is proportional to the width of a superconducting layer[2]; thus as more junctions order the effective width of the layers increases and so does a(T). This provides a qualitative explanation for the rapid increase of a(T) upon cooling.

We face now the problem of identifying the 2D structures. The obvious candidates are the Cu-O planes; the difficulty with this possibility is that their separation is too small, ~15Å even for BSCCO. The distance between the 2D layers must be large enough to allow for fluctuations of vortex-antivortex pairs, responsible for Eq. (2). The latter fluctuations carry magnetic flux perpendicular to the layers which should not penetrate the adjacent layers; this is possible if the layer separation is large, roughly[13] as the penetration depth $\lambda \cong 10^3$Å.

YBCO has in fact suitable candidates for well separated junctions: these are twin boundaries--lattice boundary structures which are remarkably smooth and clearly seen by electron microscopy[14]. These twin boundaries are oriented in the (110) direction and separated by $\sim 10^3$ Å; thus they can account for the 2D behavior. Note also that the nonlinearity (2) was seen in both the (a,b) plane and in the c direction[5]. I note in passing that twin boundaries which are weakly pinned (probably not the case in YBCO) have unusual dynamics and may enhance superconductivity[15]; the latter was in fact seen in some low T_c elements[16].

BSCCO has no twin boundaries; thus *if* there are no other well separated junctions (not yet identified), and *if* the nonlinearity (2) is also observed, one must reconsider the Cu-O planes, but then with non-conventional electrodynamics for which the "magnetic" length is much shorter. This may be the case with "anyonic superconductivity"[17], though the coupling between layers in this theory has not yet been studied.

I conclude that more experiments on the high T_c compounds are essential for identifying the 2D structures. This identification can indicate the type of electrodynamics effective in this materials, information which may be a clue for the underlying mechanism of superconductivity at high temperatures.

REFERENCES

1. A. F. Hebard and A. T. Fiory, Physica B **109&110**, 1637 (1982); A., M. Kadin K. Epstein and A. M. Goldman, Phys, Rev. B**27**, 6691 (1983); A. T. Fiory, A. F. Hebard and W. I. Glaberson, Phys. Rev. B**28**, 5075 (1983); J. C. Garland and H. J. Lee, Phys. Rev. B**36**, 3638 (1987).
2. B. I. Halperin and D.R. Nelson, J. Low Temp. Phys. **36**, 599 (1979).
3. M. A. Dubson, S. T. Herbert, J. J. Calabrese, D. C. Harris, B. R. Patton and J. C. Garland, Phys. Rev. Lett. **60**, 1061 (1988).
4. P. C. E. Stamp, L. Forro and C. Ayache, Phys. Rev. B**38**, 2847 (1988).
5. N. C. Yeh and C. C. Tsuei, Phys. Rev. B**39**, 9708 (1989).
6. T. Onogi, T. Ichiguchi and T. Aida, (to be published).
7. Y. Yeshurun et al., Cryogenics **29**, 258 (1989).
8. D. Browne and B. Horovitz, Phys. Rev. Lett. **61**, 1259 (1988).
9. B. D. Josephson, Adv. Phys. **14**, 419 (1965); M. Tinkham, *Introduction to Superconductivity* (McGraw-Hill, New York, 1975), Ch. 6.
10. H. J. Schulz, Phys. Rev. Lett. **46**, 1685 (1981).
11. F. D. M. Haldane, P. Bak and T. Bohr, Phys. Rev. B**28**, 2743 (1983).
12. B. Horovitz, T. Bohr, J. M. Kosterlitz and H. J. Schulz, Phys. Rev. B **28**, 6596 (1983).
13. P. C. E. Stamp, Phys, Rev. Lett. **63**, 582 (1989).
14. For reviews see J. Electron Microsc. Tech. **8**, No. 3 (1988).
15. B. Horovitz, G.R. Barsch and J. A. Krumhansl, Phys. Rev. **36**, 8895 (1987).
16. I. N. Khlyustikov and A. I. Buzdin, Adv. Phys. **36**, 271 (1987).
17. Y. Chen, F. Wilczek, E. Witten and B. I. Halperin, Int. J. Mod. Phys. B**3**, 1001 (1989).

NOVEL SINGULAR SOLUTIONS OF BCS THEORY

M. Weger and R. Englman*

Racah Institute of Physics
Hebrew University, Jerusalem
*Soreq Nuclear Research Center
Yavne 70600, Israel

Abstract

The Jahn-Teller Effect gives rise to a negative McMillan coupling constant λ. This negative λ gives rise to oscillatory solutions of the BCS gap function $\Delta(E)$. The period of oscillations is given (approximately) by $(4/3)\,\omega$, where ω is the phonon or exciton frequency. These oscillations require an Einstein spectrum, or at least small dispersion. The oscillatory solutions are insensitive to the Coulomb repulsion μ. They are obtained from the Dyson equation which serves as a starting point of Eliashberg theory, but not from the Eliashberg equations per se. We suggest that the singular nature of these solutions requires a minor modification of Eliashberg's derivation. These solutions are particularly favorable for excitonic superconductivity, for which the range of the interaction $\eta=(3\pi/2)v_F/\omega$ is one, or a few, lattice constants, and the Cooper pairs have some "bosonic" character.

1. INTRODUCTION

In spite of intensive efforts, no theory for the superconductivity of the cuprates has as yet proved to be

Recent Progress in Many-Body Theories, Volume 2
Edited by Y. Avishai
Plenum Press, New York, 1990

correct. The large number of theories that have been proposed can roughly be divided into two classes: non-BCS and BCS-type theories. Non-BCS type theories have been discussed in some detail at this conference, in the talks of Rice, Wiegman, Schmeltzer and others. A BCS-type theory has just been discussed by Aharony. We wish to discuss here a BCS-type theory of Englman, Halperin, and Weger. BCS-type theories [1] assume a Cooper pairing, which can be s-wave or "exotic" (such as p or d wave), and a dissolution of this pairing at T_C, or closely above T_C; there is a fluctuating behavior of the resistivity when the coherence length ξ is small [2] and in the diamagnetic susceptibility even if ξ is not small [3]; the fluctuating diamagnetism is particularly large in 2-D systems [4,5]. The interaction between the electrons forming the Cooper pair can be mediated by phonons, as in the "classical" BCS theory, by non-magnetic electronic excitations, such as excitons, plasmons, etc. or by magnetic excitations, such as paramagnons, or spin-flips, as suggested by Aharony. The electrons are assumed to behave as a Fermi liquid at least in the vicinity of E_F. Ample experimental evidence that the electrons are indeed described by band theory, at least within 0.3 eV of E_F, has been provided recently by Arko's group [6]. The theory presented here depends on these experimental results in an essential way, and moreover requires the electrons to be described by band theory even some distance away from E_F.

2.BCS-ELIASHBERG THEORY

A rigorous mathematical formulation of the conceptual BCS theory has been given by Eliashberg [7,8] and the Eliashberg equations have been solved by McMillan [9] for moderate values of λ ($\lambda \lesssim 2$), and by Bergmann & Rainer [10] and Allen & Dynes [11] for larger values of λ .

According to McMillan,

$$T_C \simeq \omega \exp\{-(1+\lambda)/[\lambda-\mu^*(1+0.6\,\lambda)]\} \tag{1a}$$

where the electron-phonon coupling constant $\lambda = 2g^2n(E_F)/\hbar\omega$ where g is the electron-phonon coupling constant in the

Frohlich Hamiltonian:

$$\mathcal{H} = \sum g\, c_{k+q}^{*}\, c_k\, (a_q + a_{-q}^{*}) , \qquad \text{and}$$

$$\mu^{*} = \mu/(1+\mu \ln(\Gamma/\omega)) \qquad \text{where} \quad \mu = U\, n(E_F)$$

is the Coulomb coupling and Γ is the band edge, of order E_F. For large values of λ:

$$T_c = 0.18\omega \sqrt{\lambda/(1+2.6\,\mu^{*})} \qquad (1b).$$

The Eliashberg equations suggest a severe limitation to the value of T_c. In principle, T_c can approach, and even exceed ω if λ is extremely large (say, of order 10). However, in this case instabilities are to be expected, and indeed observed experimentally every time one attempts to obtain values of λ exceeding about 2.

Alternatively, one might attempt to increase T_c by increasing ω, either by making use of very high frequency phonons, or electronic excitations, such as excitons, as proposed by Ginzburg [12], Little [13], and Allender, Bray and Bardeen [14]. However, increasing ω reduces the factor $\ln(\Gamma/\omega)$ thus increasing μ^{*}. The increased value of μ^{*} requires a still larger value of λ to obtain a high T_c, thus the need to obtain a large value of λ, which is liable to cause instabilities, is not avoided.

Gutfreund & Little investigated this problem in great detail [15]. They showed that excitonic superconductivity is in principle possible, but only under exceedingly restrictive conditions, that probably cannot be met in practice.

This problem (of the large μ^{*}) is particularly severe in the cuprates, because of the large value of U, both of copper ($\sim$8 eV) and of oxygen ($\sim$4-5 eV).

One way to overcome the difficulty of the large μ^{*} is to consider molecular, or cluster states instead of the ordinary atomic states. When two electrons are placed on a CuO_n cluster, the repulsion between them is much smaller than when they are placed on a Cu (or O) atom, because of the larger separation, and the screening that becomes effective over such distances. Plasmons are observed at 1eV and around 20 eV [16], the first being probably due to the cluster U. The cluster U gives rise to a small value of μ^{*}.

An ab-initio band calculation [17] showed the presence of valence fluctuations, which are an indication of a small value of the cluster U. However, when we talk of cluster electronic states, the phonons interacting with them are "external" modes, i.e. rigid motion of the cluster (such as translations and rotations), in contrast with "internal" modes of the cluster. Since only a small fraction of the phonon modes (i.e. the external ones) interact with the electronic cluster modes, the associated λ is rather small, and not sufficient to account for the high T_c. This situation is analogous to that in the organic metals, where the molecular states give rise to a small value of μ^*, as evidenced by the plasma frequency of 1 eV, but there is only weak coupling with the phonons [18], in contrast, internal modes have a strong electron-phonon coupling [19], but because of the large "atomic" μ^*, these modes do not contribute much to superconductivity (as evidenced by the large pressure dependence of T_c) [20].Thus, the limitation of the value of T_c to about 30 K or 40 K seems to be inherent in Eliashberg theory, for physically reasonable values of the parameters λ and μ. In this work, we suggest that within the framework of BCS theory, it is possible that a minor modification of Eliashberg theory may overcome the problem of the large μ^* (for large values of ω), and thus facilitate the high value of T_c observed in the cuprates.

3. THE JAHN-TELLER EFFECT, NEGATIVE λ, AND OSCILLATORY SOLUTIONS OF THE FROHLICH HAMILTONIAN

The work of Alex Muller on the cuprates was motivated by the large Jahn-Teller effect of the cuprate ion. This implies a large matrix element g for the electron-phonon coupling and consequently a large value of $\lambda = 2g^2\, n(E_F)/\hbar\omega$. In addition to this consideration, the JTE means that two (or more) energy levels are degenerate, and the phonon removes this degeneracy, causing one energy level to rise and the other to fall linearly as function of the distortion Q (Fig. 1).

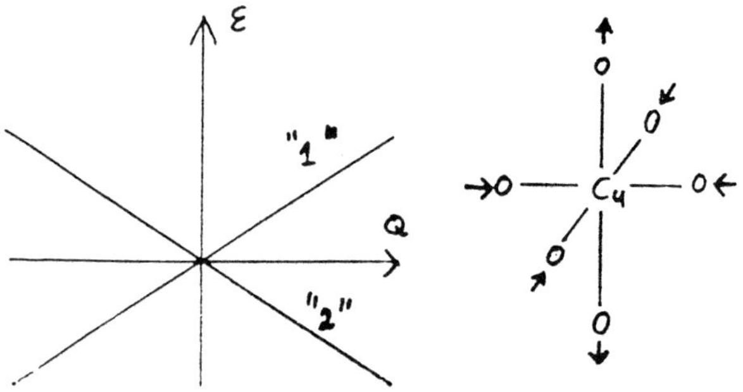

Figure 1

The Jahn-Teller Effect.

Variation of energies of states "1" and "2" caused by lattice distortion Q .

The opposite sign of $\partial\varepsilon_1/\partial Q$ and $\partial\varepsilon_2/\partial Q$ implies opposite signs of the matrix elements g_1 and g_2, and as a result the product $g_1 g_2$ is negative [21].

A negative value of λ indicates a repulsive interaction; however, when we consider the Frohlich Hamiltonian

$$V_{kk'}^F = 2g_1g_2 / [(\varepsilon_k - \varepsilon_{k'})^2 - \omega^2] \tag{2}$$

the interaction becomes attractive when $|\varepsilon_k - \varepsilon_{k'}| > \omega$ (Fig.2).

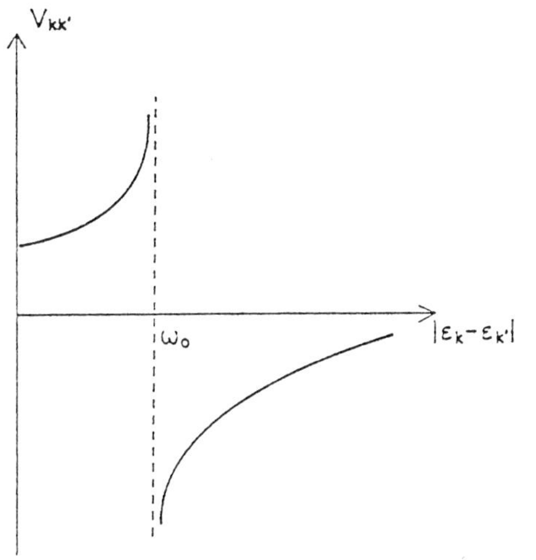

Figure 2

The effective electron-electron potential $V_{kk'}$ as function of the half energy transfer $|\varepsilon_k - \varepsilon_{k'}|$.

This attractive interaction gives rise to a bound state, even if $|\lambda|$ is small. The solution of the BCS gap equation

$$\Delta_k = -\tfrac{1}{2} \sum_{k'} V_{kk'}^F \; \Delta_{k'}/\sqrt{\varepsilon_{k'}^2 + \Delta_{k'}^2} \tag{3}$$

is shown in Fig. 3 for both negative and positive values of λ as function of energy ε_k where $\varepsilon_k = \hbar^2 k^2/2m$. It is seen that for negative λ, Δ oscillates as function of ε, somewhat like $\cos(3\pi\varepsilon/2\omega)$. The shape of the curve depends on the value of λ, but the oscillation period is given by $(4/3)\omega$. We proved this for large $|\lambda|$. For very small $|\lambda|$, the behavior is shown in Fig. 7. For positive λ, Δ changes sign at (or very near to) $\varepsilon = \omega$, as is well known; i.e. the period is 4ω. For the Einstein spectrum that we consider, the oscillatory solution is obtained for negative λ even for infinitesimally small values of $|\lambda|$. However, when dispersion is present, the oscillatory solution disappears very rapidly, in particular for small values of $|\lambda|$ (Fig.4) [22].

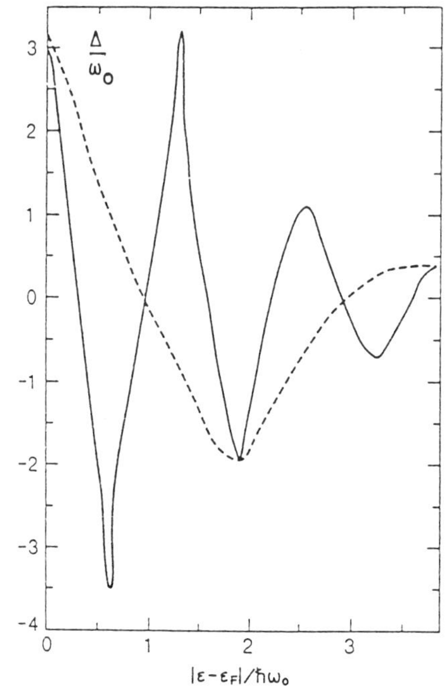

Figure 3

The gap parameter Δ (in units of ω_0) versus energy.
Full line: A discretized version (100 intervals) of Eq. 3 is solved iteratively for near-zero temperature and moderate coupling strength ($\lambda = 1.6$) to a 10^{-3} accuracy in each Δ for a band width of $8\omega_0$.
Broken line: BCS case; similar solution of Eq. (3) after changing the sign of $V_{kk'}$.

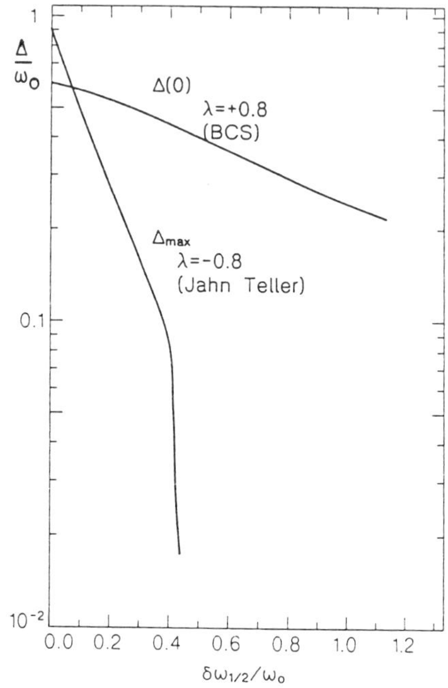

Figure 4

Effect of dispersion of the phonon (or exciton) spectrum on the gap parameter $\Delta(\varepsilon)$. $\Delta(0)$ is plotted for the positive-λ case, and $\varepsilon \cong 1.33\ \omega_0$ gives a maximum value Δ_{max} to $\Delta(\varepsilon)$ for the negative-λ case. Here strong coupling effects ($Z(\varepsilon) \neq 1$) are included.

Because of the rapid oscillation in $\Delta(E)$, the Coulomb contribution

$$\sum_{k'} U_{kk'}\, \Delta_{k'}/\sqrt{\varepsilon_{k'}{}^2 + \Delta_{k'}{}^2} \tag{4}$$

is very small, and essentially vanishes ($U_{kk'}$ is the Coulomb term, which is a smooth function of energy, and can be assumed to be constant up to $|\varepsilon_k|$, $|\varepsilon_{k'}| = \Gamma$). Thus the oscillatory solutions are insensitive to μ , and are not significantly affected even for values of μ that are large enough to induce a Mott metal-to-insulator transition.

Super-Retardation

The effective elimination of the Coulomb interaction by the negative- λ solutions can be understood as follows: The "normal" interaction described by the Frohlich Hamiltonian with positive λ is retarded; however, the interaction via the negative λ interaction is even more retarded [23]. Electron 1 causes a polarization at a given site R_0 ; this polarization repels electron 2 (Fig.5b). After half a period of the phonon

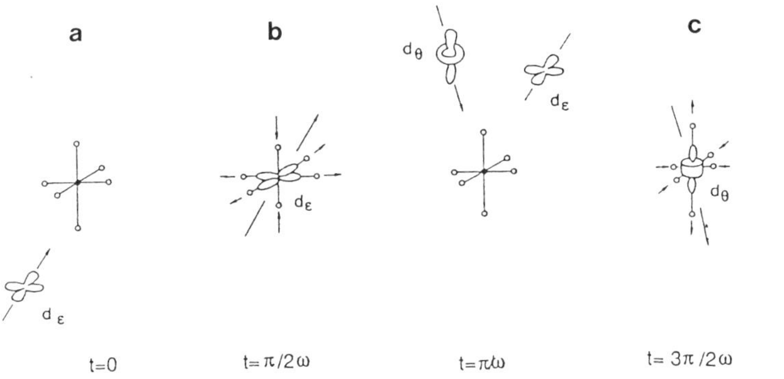

Figure 5

Attractive interaction between $d_{x^2-y^2}$ and d_{z^2} states mediated by breathing mode.

oscillation the phase of the polarization is reversed; now the reversed polarization attracts electron 2 (Fig.5d).

Thus, while for the "normal" BCS interaction the retardation is by $0<t<\pi/\omega$, for the JTE the retardation is by $\pi/\omega<t<2\pi/\omega$. Because the retardation is larger, the distance between the two electrons is larger; specifically, since the relative velocity is $2v_F$, and the retardation is $3\pi/2\omega$, the distance is given by: $\eta = 3\pi v_F/\omega$, which is rather large and the Coulomb repulsion is effectively screened out at this distance. We call this "Super Retardation". Obviously, Super-Retardation requires a spectrum with small dispersion (Fig. 4), otherwise the "undershoot" at time $3\pi/2\omega$ is strongly reduced (Fig. 6).

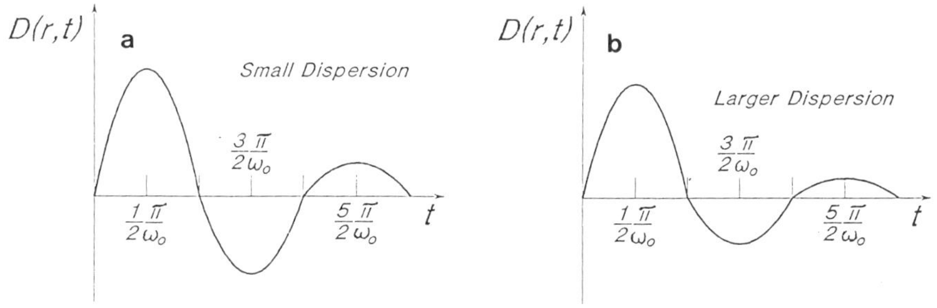

Figure 6

The use of the Frohlich Hamiltonian in the present context is questionable. The Frohlich Hamiltonian is an effective Hamiltonian, whose form depends on the initial and final states; it specifically assumes energy-conservation in the phonon-exchange process. In BCS theory, the phonon-exchange processes are virtual and there is no energy-conservation in the virtual process [24]. The question whether the interaction has features like those of the Frohlich Hamiltonian, and specifically, a resonant character, turns out to be very difficult. Therefore, before looking into it, it may be worthwhile to see what is the nature of the solutions of the BCS gap equation when a Frohlich potential is employed; and this turns out to be an "easy" exercise.

4. THE ELIASHBERG EQUATIONS

The rigorous formulation of BCS theory was given by Eliashberg [7,8]. The Eliashberg equations for the gap function Δ can be written for T=0 and $T=T_c$ as [25]:

$$\Delta(E) = [\lambda\omega/Z(E)]\int_{\Delta_0}^{\Gamma} dE' \; \mathrm{Re}[\Delta(E')/(E'^2-\Delta^2(E'))^{\frac{1}{2}}] \; [D(E'+E)+D(E'-E)] \qquad (T=0)$$

$$\Delta(E) = [\lambda\omega/Z(E)] \int_0^{\Gamma} dE' \; \Delta(E')/E' \; [D(E'+E)+D(E'-E)] \; \tanh(E'/2T_c) \qquad (T=T_c)$$

$$[1-Z(E)] \; E = \lambda\omega \int_0^{\Gamma} dE' \; [D(E'+E)-D(E'-E)] \qquad (5)$$

where: $D(E)=(E+\omega-i\delta)^{-1}$.

We assume $T_c \ll \omega$, where ω is the phonon frequency.

These equations are obtained by solving the Dyson equation, making use of Migdal's theorem that applies when $\omega \ll \Gamma$. Recently, a more general solution was derived for the case of an Einstein spectrum for the case when $\omega \simeq \Gamma$ and Migdal's theorem no longer applies [26] .

The best-known feature of the Eliashberg equations is

the rigorous treatment of strong coupling by means of the renormalization function Z(E). $Z(E) \simeq 1+\lambda$ [9] and therefore $\Delta(E)$ and T_c are smaller than the values obtained when Z(E) is assumed to be 1 (weak coupling theory). Z(E) depends only on the absolute value of λ, therefore this reduction of Δ and T_c takes place for negative λ as well. The solution with renormalization is shown in Fig. 7; Δ is seen to oscillate with E with the same period as when renormalization is ignored (Fig. 3).

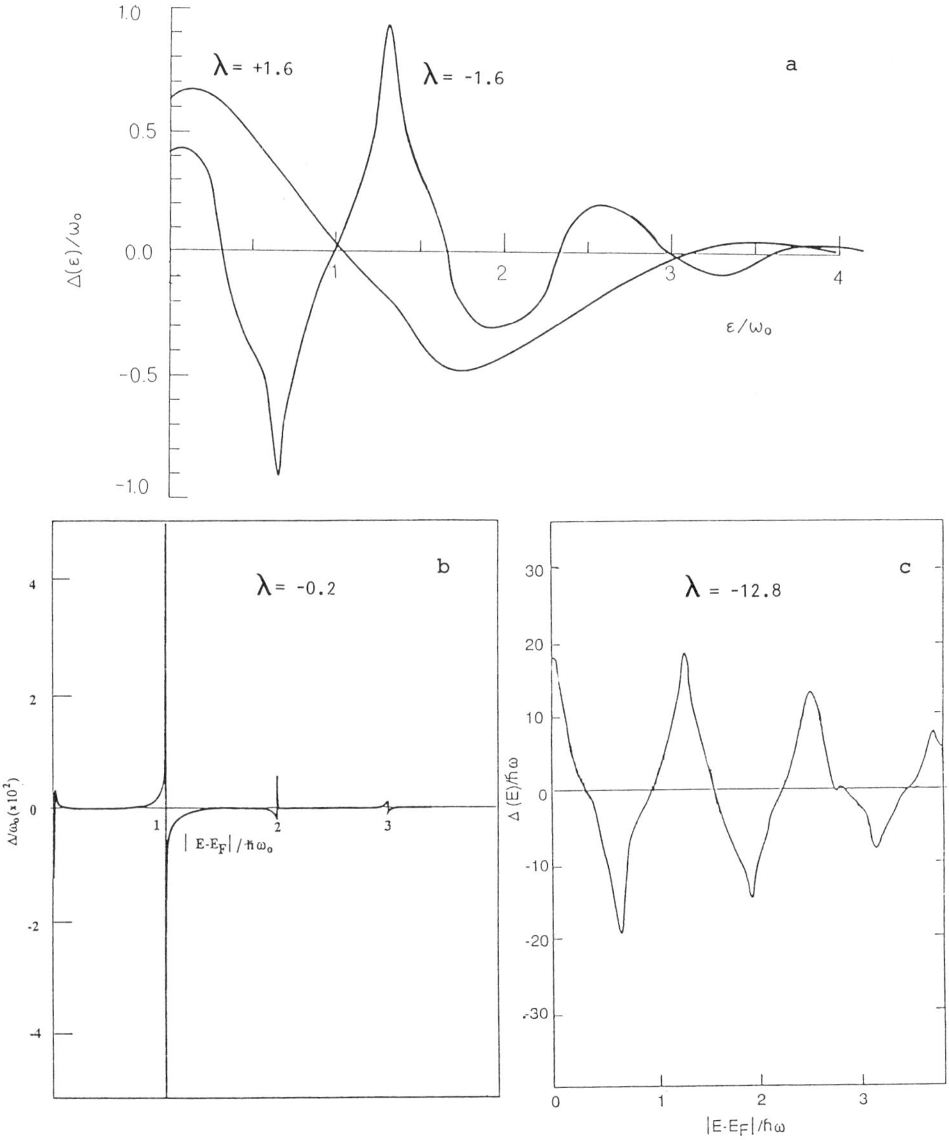

Figure 7

Δ as function of E for medium (a), weak (b), and strong (c) coupling.

A more subtle, yet vital aspect of the Eliashberg equation is the removal of the resonance at $|\varepsilon_k - \varepsilon_{k'}| = \omega$. This removal of the resonance (which was already noted by Bogoliubov [27] and is treated in detail by KMK [28]) is obtained mathematically by a deformation of the integration path in the complex plane. The Dyson equation for the gap function, simplified for the case of weak coupling, is given by:

$$\Delta(E) = \frac{i}{(2\pi)^3} \int_0^{q_m} \alpha_q^{\,2} q\, dq \int_L dz\, D(q,z-E) \quad \Delta(z) \int_{\infty}^{\infty} \frac{d\xi}{k_o(\xi)} \; \frac{1}{\xi^2 - z^2 + \Delta^2(z)} \qquad (6)$$

where $\xi = v_F|\vec{k}-\vec{k}'|$ (p.700 of [7]).

Here, $D(q,z) = 2\,\omega_q/(z^2 - \omega_q^{\,2} + i\delta)$ and $k_0(\xi)$ is essentially k_F. Now, Eliashberg deforms the path of integration in the complex z-plane from L to L_1 (for the first pole) and L_2 (for the second pole). (Fig. 8). This removes the resonance.

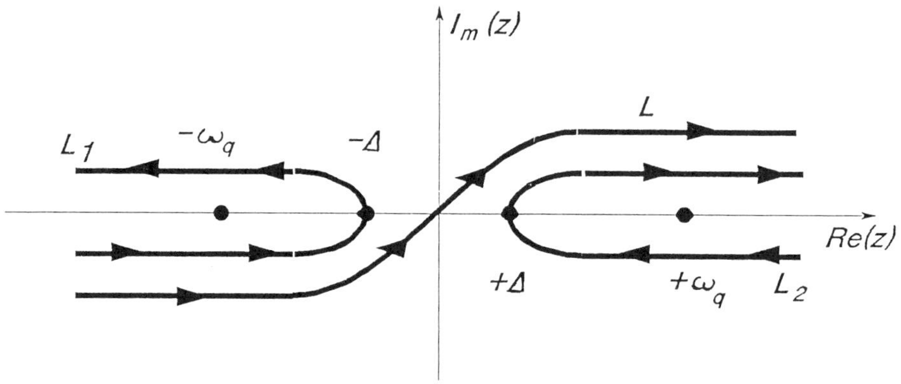

Figure 8

Deformation of the path of integration in the complex plane.

Then, a transformation $z \to -z$ turns L_1 into L_2. Subsequently, the integration over ξ is performed, and the Eliashberg equations obtained.

The effective Eliashberg potential, the Cooper potential, and the Frohlich potential (possessing the resonance) are illustrated in Fig. 9.

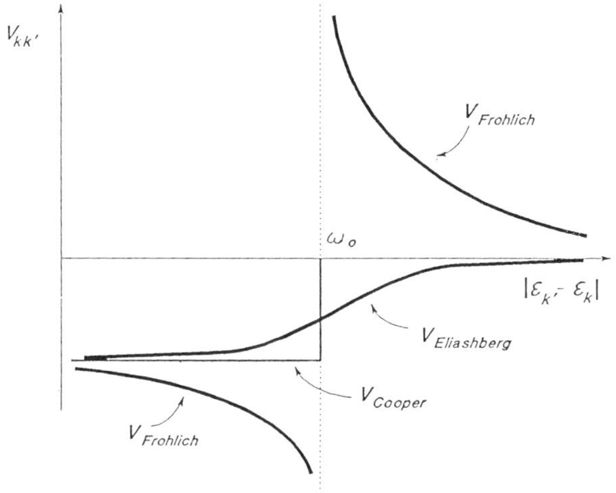

Figure 9

The resonant Frohlich (Bardeen-Pines) potential, the postulated Cooper potential, and the calculated potential of Bogoliubov; Eliashberg; Liu; and Kirzhnitz, Maximov, & Khomskii.

The deformation of the path of integration in ref. [7] assumes that the function $\Delta(z)/\sqrt{[z^2 - \Delta^2(z)]}$ is regular in the upper and lower half planes (it has a branch-cut along the real axis). If $\Delta(z)$ is not regular, we cannot deform the path of integration . However, assuming that eq. (6) is valid, we can avoid the deformation of the path of integration by integrating first over ξ, and then over z; In this way we obtain the equation for $\Delta(E)$ [29]:

$$\Delta(E) = \frac{1}{4\pi^2} \int \alpha_q^2 \, q \, dq \int_L dz \, D(q,z-E) \frac{\Delta(z)}{[z^2-\Delta^2(z)]^{\frac{1}{2}}} \frac{1}{k_o(z)} \tag{7}$$

Here, the resonance at $|\varepsilon_k - \varepsilon_{k'}| = \omega$ is preserved, and the equation is equivalent to the BCS gap equation with a Frohlich potential.

5. DIFFICULTIES OF THE PROPOSED THEORY

The present theory depends critically on the suggestion that the Dyson equation of BCS theory has in addition to the regular solutions, derived by Eliashberg, also singular solutions. For negative λ, there are no regular solutions (except the trivial $\Delta = 0$ one), while singular solutions exist. The singular solutions exist only for a spectrum with a small dispersion (Fig. 4). The singular solutions oscillate sharply with energy. They behave approximately like:

$\exp[i(3\pi/2)\, E/\omega]$, i.e. $\mathrm{Re}[\Delta(E)] \propto \cos[(3\pi/2)E/\omega]$ and

$Im[\Delta(E)] \propto \sin[(3\pi/2)|E|/\omega]$. Along the imaginary axis,

$$\Delta(E) \propto \exp(-3\pi/2\ |E|/\omega). \tag{8}$$

We derived these solutions using the causal, i.e. T=0, Green's function. The free-electron causal Green's function $G^{(0)}(E)$ has poles above the real axis for $Re(E)<0$ (hole states) and below the real axis for $Re(E)>0$ (electron states). Thus, if the interacting Green's function G(E) has the same analytical properties as $G^{(0)}(E)$, as Gorkov stated [30], the deformation of the path of integration is legitimate. However, an essential feature of BCS theory is that there are (excited) hole states at positive energy, and electron states at negative energy. Taking this property at face value, we may expect singularities above the real axis for positive E, and below it for negative E, and the deformation of the path of integration is no longer justified.

The conventional procedure for handling this problem is to use the finite-temperature (Matzubara) Green's function, which even in the limit $T \to 0$ does not have any poles in the upper and lower half planes, but only a branch-cut along the real axis. Therefore, the deformation of the path of integration is justified. We were not yet able to derive our singular solutions by making use of the finite-temperature Green's function. Therefore, our theory is as yet incomplete for the $T \neq 0$ and $T \to 0$ situations.

6. THE LENGTH SCALE η

We can relate energy and momentum by: $E-E_F = \hbar(k-k_F)v_F$. By this ansatz, we can express $\Delta(E)$ as function of k, and perform a Fourier transform, to obtain $\Delta(r)$. The function obtained in this way possesses a very sharp peak at $r = \eta = (3\pi/2)(v_F/\omega)$ (for negative λ; Fig.10). For positive λ the range is given by : $r = \eta = (\pi/2)(v_F/\omega)$, but there is no sharp peak in $\Delta(r)$ at this distance. Thus, for negative λ the two electrons forming the Cooper pair can be considered to be a distance η apart. Because of this distinct separation, the Coulomb interaction is weak, possibly even weaker than in the normal metallic state, where the electrons do not possess this separation. The length scale η usually doesn't play a role in BCS theory; it occurs implicitly when we consider the conditions of validity of Migdal's theorem; Migdal's theorem requires that η be large compared with the average distance between the electrons (or between the Cooper pairs). Since for normal superconductivity $\omega \ll E_F$, $\eta \gg k_F^{-1}$, and the condition for the validity of Migdal's theorem is fulfilled.

For excitonic superconductivity this is not the case, since $\omega \simeq E_F$, and consequently $\eta k_F \simeq 1$. η characterizes the range of the interaction leading to superconductivity which normally is large so that many Cooper pairs overlap and the BCS mean-field approach is valid. If $\eta k_F \ll 1$, the Cooper pairs should be considered to be small bosons, each electron of a Cooper pair interacting mainly with its mate, and the mean-field approach becomes questionable; the bosons are expected to be present even at temperatures T considerably higher than T_c.

The relation between η and ξ_0 is given by:

$$\eta = (3\pi^2/2)\ \xi_0\ \Delta_0/\omega \ , \tag{9}$$

i.e. even if ω is an order of magnitude larger than Δ_0, η is not much smaller than ξ_0. For ω of order 1/2 to 1 eV, η is about 1 to 3 lattice constants (for negative λ). Unlike $\xi(T) = \xi_0 \sqrt{T_c/(T_c-T)}$, η does not diverge at $T = T_c$, but is essentially temperature independent.

Since ηk_F is not large, Migdal's theorem is not necessarily valid, but according to [26] , this doesn't seem to present serious problems.

A general description of boson theories has been given by Friedberg & Lee [31]; they consider the s-channel (Fig. 11b) rather than the normal t-channel (Fig. 11a). This approach was motivated by the small value of ξ_0 [32]. We suggest here that the length scale η may play a vital role here; and the "bosonic" character may be a general feature of excitonic superconductivity.

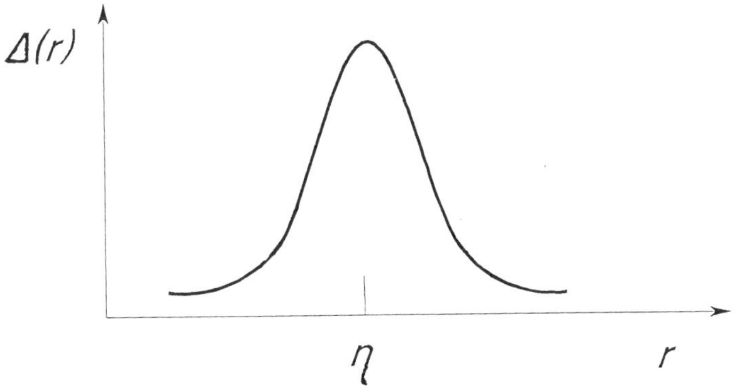

Figure 10

The distance between the two electrons forming the Cooper pair is sharply peaked at $r=\eta$, for negative λ .

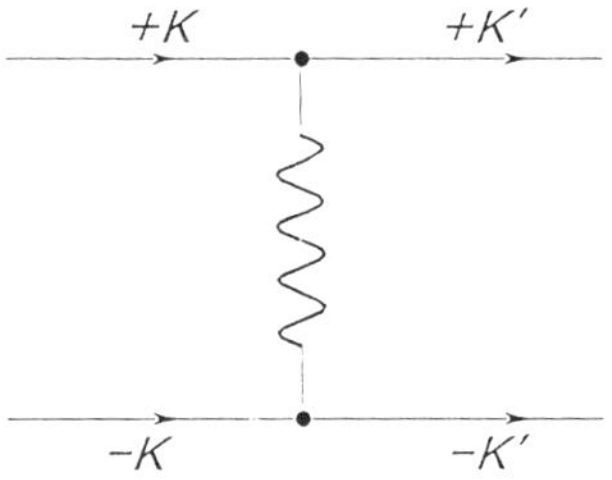

a

Figure 11

(a) The t-channel pairing, characteristic of the normal BCS theory.

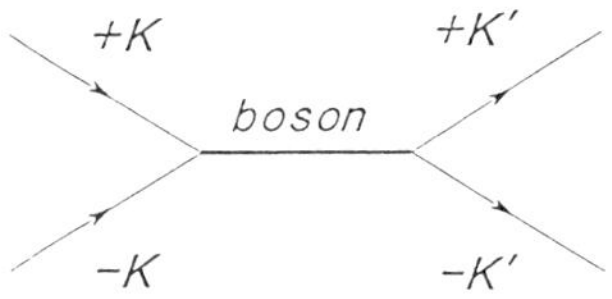

b

(b) s-channel pairing, characteristic of the "bosonic" theory of T.D. Lee.

7. REFERENCES

1) J. Bardeen, L.N. Cooper, J.R. Schrieffer, Phys. Rev. 108 (1957)1175.

2) L.G. Azlamazov & A.I. Larkin, Sov. Phys. Solid State 10 (1968)875.

3) J.P. Gollub, M.R. Beasley, R.S. Newbower, M. Tinkham, Phys. Rev. Lett. 22(1969)1288.

4) A. Schmid, Phys. Rev. 180(1969)527.

5) T. Sugiyama M. Weger, J. Phys. C21(1988)951.

6) B.W. Veal, J.Z. Liu, A.P. Paulikas, K. Vandervoost, H. Claus, J.C. Campuzano, C. Olson, A.B. Yang, R. Liu, C. Gu, R.S. List, A.J. Arko, R. Bartlett, Physica C158, (1989)2706 and subsequent publications.

7) G.M. Eliashberg, Sov. Phys. JETP 11(1960)696.

8) G.M. Eliashberg, Sov. Phys. JETP 12(1961)1000.

9) W.L. McMillan, Phys. Rev. 167(1968)331.

10) G. Bergmann & D. Rainer, Z. Phys. 263(1973)59.

11) P. Allen & R. Dynes, Phys. Rev. B12(1975)905. See also V. Kresin, H. Gutfreund, W.A. Little, Solid State Commun. 51(1984)339.

12) V.L. Ginzburg, Sov. Phys. JETP 20(1965)1549.

13) W.A. Little, Phys. Rev. 134(1964)A1416.

14) D. Allender, J. Bray, J. Bardeen, Phys. Rev. B7(1973) 1020; ibid B8(1973)4433.

15) H. Gutfreund & W.A. Little, in "Highly Conducting One Dimensional Solids", J.T. Devreese, E.V. Evrard, V.E. Van Doren Edts. Plenum Press 1979, p. 305.

16) H. Rietchel, J. Fink, E. Gering, F. Gompf, N. Nucker, L. Pintchovius, B. Renker, W. Reichardt, Physica C153-155, (1988)1067.

17) R.A. de Groot, H. Gutfreund, M. Weger, Solid State Commun. 63(1987)451.

18) M. Weger, M. Kaveh, H. Gutfreund, Solid State Commun. 37 (1981)421.

19) B. Horovitz, H. Gutfreund, M. Weger, J. Phys. C 7(1974) 383; M.J. Rice & N.O. Lipari, Phys. Rev. Lett. 38(1977) 437.

20) M. Weger, J. Kubler, D. Schweitzer in "Novel Superconductivity", S.A. Wolf V.. Kresin Edts. Plenum Press 1987, p. 149.

21) R. Englman, B. Halperin, M. Weger, Solid State Commun. 70(1989)57.

22) M. Weger, R. Englman, B. Halperin, Solid State Commun. 71(1989)17.

23) M. Weger, R. Englman, B. Halperin Proc. NATO ASI Bad Windsheim Aug. 1989, R. Kossowsky & S. Methfessel Edts.

24) S.H. Liu, Phys. Rev. 125(1962)1244.

25) J.R. Schrieffer, D.J. Scalapino, J.W. Wilkins, Phys. Rev. Lett. 10(1963)336.

26) J. Cai, X.L. Lei, L.M. Xei, Phys. Rev. B39(1989)11618.

27) N.N. Bogoliubov, V.V. Tolmachev, D.. Shirkov, "A New Method in the Theory of Superconductivity", Consultants Bureau NY 1959.

28) D.A. Kirzhnitz, E.G. Maximov, D.I. Khomskii, J. Low Temp. Phys. 10(1973)79.

29) R. Englman, M. Weger, B. Halperin, Proc. Stanford. Conf. on HTSC July 1989.

30) L.P. Gorkov, Sov. Phys. JETP 7(1958)505.

31) R. Friedberg & T. D. Lee, Phys. Lett. A138 (1989) 423.

32) G. Deutscher & K.A. Muller, Phys. Rev Lett. 39(1987)1745.

This work is supported by BSF grant No. 87-00275.

MAGNETISM AND SUPERCONDUCTIVITY IN DOPED LAMELLAR COPPER OXIDE SYSTEMS

Amnon Aharony

Raymond and Beverly Sackler Faculty of Exact Sciences
School of Physics and Astronomy
Tel Aviv University, Tel Aviv 69978, Israel

EXTENDED ABSTRACT

Most of the doped lamellar CuO_2 based high temperature superconductors exhibit very rich phase diagrams. The behavior varies strongly as function of the dopant concentration, which is directly related to the concentration of the electronic holes. In the present paper we emphasize the interplay among the various phases which appear in the temperature - concentration phase diagram.

Without doping, these systems are insulating antiferromagnets. The antiferromagnetic exchange interactions between the localized spin-1/2 copper ions are strong in the CuO_2 planes, and weak among the planes. The quantum spin fluctuations can be renormalized, and the behavior in each plane is excellently described by an effective classical two-dimensional Heisenberg antiferromagnetic model.[1] Recently,[2] we have shown that the weak coupling between the planes is very well described by a mean field theory in which the two dimensional fluctuations are calculated exactly. In La_2CuO_4, the orthorhombic rotation of the CuO_6 octahedra generates an additional antisymmetric Dzyaloshinskii-Moriya interaction between the Cu spins in the planes. This results in a bilinear coupling between the staggered magnetization perpendicular to the plane, yielding a canting of the magnetic moments out of the planes. At zero magnetic field, the ferromagnetic moments of neighboring planes order antiparallel to each other. However, external magnetic fields cause a variety of spin flip transitions.[2,3] The bilinear coupling between the staggered and the uniform magnetizations implies an indirect coupling between the external uniform field and the staggered magnetization, explaining the unusual sharp peak in the uniform susceptibility at the Neél point.[2] Our mean field theory for weakly coupled planar Heisenberg models fits the susceptibility data, with practically no adjsutable parameters.[2] Similar theories should work for all other properties of these weakly coupled planar systems, e.g. in the superconducting phase.

Doping introduces quenched randomness, as well as deviations from stoichiometry, usually adding electronic holes, which reside mainly on the oxygen ions in the planes. At low concentration, these holes are localized, exhibiting variable range hopping conductivity.[4] The holes always remain localized in the planes, with a small perpendicular localization length. Within the plane, the localization length at low doping is of order 10Å (about 2.5 lattice constants). In La_2CuO_4, the hole conductivity is very sensitive to the external magnetic field, changing by about a factor 2 at the spin flip transition.[2,3] This shows a strong coupling between the hole states and the magnetic ordering.

In order to discuss the coupling of the hole to the underlying magnetic ordering of the copper spins, we considered[5] a single hole, localized at an oxygen site. There is a strong

Recent Progress in Many-Body Theories, Volume 2
Edited by Y. Avishai
Plenum Press, New York, 1990

superexchange coupling of the spin of this hole to those of the two neighboring copper ions, resulting in an effective ferromagnetic exchange interaction between these copper spins. The exchange interaction between the hole spin and the copper spins will be ferromagnetic (antiferromagnetic) if the hole sits mainly on a p_π (p_σ) oxygen state, with the wave function mainly perpendicular (parallel) to the bond.[6] The resulting three-spin state will have total spin 3/2 (or 1/2). It is not yet completely clear which state wins. However, both situations create effective ferromagnetic Cu-Cu interactions, and frustrate the antiferromagnetic copper state. We believe that this frustration is responsible for the fast decrease of the Neél temperature with doping. We also predicted[5] that at higher doping, the antiferromagnetic phase should be replaced by a spin glass phase, as indeed confirmed by many experiments. Alternative pictures, which couple the hole into a singlet state with one copper ion, do not contain the frustration necessary for the spin glass.

The frustration model implies that the hole spin prefers to be perpendicular to the copper spins, and may help in understanding the sensitivity of the hole mobility to the spin ordering. In addition, the disturbance to the copper antiferromagnetic ordering is smaller if two holes sit on neighboring or next nearest neighboring (nnn) oxygens. Since the former seems to be excluded by Coulomb repulsion, we considered the attractive pairing potential between two holes on nnn oxygens.[6] This potential decays with increasing doping, as the antiferromagnetic correlation length decreases. Assuming that the holes move freely in the p-band, we used a BCS theory with the cutoff set at the Fermi energy (proportional to the number of mobile holes).[6] Adjusting only the temperature scale, we were able to reproduce the concentration dependence of the superconducting transition temperature of $La_{2-x}Sr_xCuO_4$, including the peak at around $x \simeq 0.15$ and the decay for larger x.[7] More recently,[8] we also considered the tight binding band structure for the motion of the Cu-O-Cu "polaron", and the results are qualitatively similar.

Work supported by grants from the U.S.-Israel Binational Science Foundation and from IBM. Parts of this work were done at M.I.T., in collaboration with R. J. Birgeneau and M. A. Kastner.

REFERENCES

1. S. Chakrvarty et al, Phys. Rev. Lett. 60, 1057 (1988).
2. T. Thio et al, Phys. Rev. B38, 905 (1988).
3. T. Thio et al, to be published.
4. M. A. Kastner et al, Phys. Rev. B37, 111 (1988);
   N. W. Preyer et al,, Phys. Rev. B39, 11563 (1989).
5. A. Aharony et al, Phys. Rev. Lett. 60, 1330 (1988).
6. R. J. Birgeneau et al, Z. Phys. B71, 57 (1988).
7. A. Aharony et al, IBM J. Res. Develop. 33, 287 (1989).
8. L. Klein and A. Aharony, unpublished.

A VORTEX LATTICE OF ANYONS IN STRONGLY CORRELATED SYSTEMS

D. Schmeltzer

Department of Physics, City College of the City University of New York, New York, NY 10031, U.S.A.*
and
Department of Physics, Technion-Israel Institute of Technology, Haifa 32000, ISRAEL

1. INTRODUCTION

The CuO new high-temperature superconductors[1] are Mott insulators spin 1/2 antiferromagnets[2]. Doping the CuO layers by substituting Sr or La destroys the magnetic order leading to a spin liquid with a metal conductivity at zero temperature. In the undoped system the strong Coulombic repulsion leads to a correlation gap. After doping we obtain a metal (this is similar to the semiconductors with the difference that the gap is due to strong Coulombic repulsion). The magnetism is destroyed by doping. One obtains a disordered magnetic phase which is different from the paramagnetic phase. This new phase is called quantum spin liquid. A known example of such a spin liquid is the Heisenberg spin 1/2 in one dimension. The excitations are chargeless spin 1/2 particles with a linear energy momentum relation.

As a consequence of these facts we conclude that the ordinary fermi liquid theory is not applicable to strongly correlated systems. Instead we have separated charge and spin excitations. A microscopic model which describes the situation in strongly correlated systems, can be postulated in terms of few parameters t,J,V (these parameters have to be calculated from an explicit microscopic model).

$$H=-t \sum_{x,\mu} \tilde{C}^{\dagger}_{\sigma}(x)\tilde{C}_{\sigma}(x+\mu)+H.C.+J \sum_{x,\mu} \vec{S}(x)\cdot\vec{S}(x+\mu)+V \sum_{x,\mu} \tilde{n}(x)\tilde{n}(x+\mu) \qquad (1a)$$

This hamiltonian is described on a constrained subspace $n(x)\neq 2$ (double occupancy is excluded): $\tilde{C}^{\dagger}_{\sigma}=|\sigma\rangle\langle o|$, $\tilde{C}_{\sigma}=|0\rangle\langle\sigma|$, $\tilde{n}=\sum_{\sigma=1}^{N}|\sigma\rangle\langle\sigma|$ $\sigma=1,\ldots N$,

*New address beginning January 1990.

Recent Progress in Many-Body Theories, Volume 2
Edited by Y. Avishai
Plenum Press, New York, 1990

$\vec{S}=\tilde{C}^{\dagger}_{\alpha}\ \vec{\tau}_{\alpha\beta}\ \tilde{C}_{\beta}$, for spin 1/2 N=2 and $\vec{\tau}$ is the Pauli matrix. We will consider here the case V=0. For t>>J we have a fermi liquid description[3] with an effective hopping $t_{eff} \simeq t\delta$, $\delta=1-\langle n\rangle$ (δ is the hole concentration).

In the opposite limit $t \lesssim J$ one obtains in two dimensions the following phases: a spin Peierls phase for $t/J \lesssim 0.1$ and $\delta \lesssim 0.2$. For $t/J>0.1$ a new phase called flux phase is possible.[4] This phase has a complex order parameter $\chi_{\mu}(x)=\exp[iA_{\mu}(x)]$. $A_{\mu}(x)$ is a gauge field on the bond $(x,x+\mu)$. Due to this gauge field the tight binding part in Eq. 1a is identical to a tight band model in an external magnetic field. In agreement with Hofstatder and Wannier[5] the minimum of the energy is obtained for $\oint A_{\mu}(x)=2\pi f$ ($\oint$ is the sum around the plaquette). For a two dimensional square lattice and exactly at the half filling f=1/2, $\oint A_{\mu}(x)=\pi$. This solution suggests the existence of anyons[6] in two dimensions $\oint A_{\mu}(x)=\theta^{-1}\rho(x)$ ($\theta^{-1} = \frac{2\pi}{n}$, n=1,2,...) where $\rho(x)$ is the charge density. (The case $\theta^{-1} = \frac{2\pi}{2}$ corresponds in the mean field approximation to the flux phase condition if $\langle\rho(x)\rangle=1$.) The response of a charged gas to an external magnetic field is identical to the Meisner effect in a superconductor with a charge of ne.

For n=2 we map the problem on a vortex lattice. Using the charge neutrality condition we find that the electric charge can have two opposite fractional vortices ±1/2. As a result two vortices bind forming a boson of charge 2e. The critical temperature for binding is the Kosterlitz Thoules one. An anisotropic s wave solution for the superconducting phase is obtained.

2. THE t-J MODEL IN THE BOSON-FERMION COHERENT STATE REPRESENTATION

Using the formalism introduced in Ref. 7 we obtain an effective action for the t-J model. The natural variables for the constrained space ($n\neq 2$) is to use for each particle two fields - one for spinless charges and the second one for chargeless spins. We choose[7] $\tilde{C}^{\dagger}_{\sigma}=bf^{\dagger}_{\sigma}$, $\tilde{C}^{\dagger}_{\sigma}=b^{\dagger}f_{\sigma}$, $b^{\dagger}(x)b(x)+\sum_{\sigma=1}^{2} f^{\dagger}_{\sigma}(x)f_{\sigma}(x)=1$. b is a charged core boson operator and f_{σ} is a spin fermion operator (an opposite choice is also possible). In the path integral formulation the operators are replaced by their eigenvalues in the coherent state representation. We have two complex fields $b(x,\tau)$, $b^{*}(x,\tau)$ and 4 Grassman fields $f_{\sigma}(x,\tau)$, $f^{*}_{\sigma}(x,\tau)$ at each space-time point. Using the fact that for any scalar η we have $Z_{\sigma}=\eta f_{\sigma}$, $dZ_{\sigma}=\eta^{-1}df_{\sigma}$ and not $dZ_{\sigma}=\eta df_{\sigma}$ where Z_{σ}, Z^{*}_{σ} are the new 4 Grassman fields. The constraint condition is solved by choosing $\sum_{\sigma=1}^{2} Z^{*}(x,\tau)Z_{\sigma}(x,\tau)=1$,

$f_\sigma(x,\tau)=[1-b^*(x,\tau)b(x,\tau)]^{1/2}Z_\sigma(x,\tau)$, $b^2=(b^*)^2=0$. The partition function of the t-J model in the coherent state representation takes the form

$$Z = \int Db^*DbDZ^*_\sigma DZ_\sigma \exp\{-\int_0^\beta d\tau S(\tau)\} \tag{1b}$$

$$S(\tau) = \sum_x b^*(x,\tau)[\partial_\tau+\mu-\sum_{\sigma=1} Z^*_\sigma(x,\tau)\partial_\mu Z_\sigma(x,\tau)]b(x,\tau) +$$

$$+\sum_{\sigma=1} Z^*_\sigma(x,\tau)\partial_\tau Z_\sigma(x,\tau)-t\sum_{\mu,\sigma} b^*(x+\mu,\tau)Z^*_\sigma(x,\tau)Z_\sigma(x+\mu,\tau)b(x,\tau) -$$

$$-J\sum_{\mu,\sigma,\sigma'}^{2} \rho(x,\tau)\rho(x+\mu,\tau)Z^*_\sigma(x,\tau)Z_\sigma(x+\mu,\tau)Z^*_{\sigma'}(x+\mu,\tau)Z_{\sigma'}(x,\tau)\} \tag{1c}$$

μ is the chemical potential for the holes and $\rho(x)=1-b^*(x)b(x)$ is the number of particles. We introduce via the Hubbard Stratonovici transformation the collective $\chi_\mu(x)=\exp[iA_\mu(x)]$. Using the temporal local gauge invariance, $Z_\sigma(x,\tau)\to Z_\sigma(x,\tau)e^{i\theta(x,\tau)}$, $b(x,\tau)\to b(x,\tau)e^{i\theta(x,\tau)}$ we fix the temporal gauge field such that: $iA_o(x,\tau)=i\dot\theta(x,\tau)=-\sum_{\sigma=1}^{2} Z^*_\sigma(x,\tau)\partial_\tau Z_\sigma(x,\tau)$

$$S(\tau)=\sum_x\{b^*(x,\tau)[\partial_\tau+\mu'-iA_o(x,\tau)]b(x,\tau)+\sum_{\sigma=1}^{2} Z^*_\sigma(x,\tau)[\partial_\tau+iA_o(x,\tau)]Z_\sigma(x,\tau)$$

$$+t^2/J\sum_\mu b^*(x,\tau)b(x,\tau)b^*(x+\mu,\tau)b(x+\mu,\tau) -$$

$$-t\sum_\mu b^*(x+\mu,\tau)\chi_\mu(x,\tau)b(x,\tau) + H.C.$$

$$+\sum_{\sigma,\mu} Z^*_\sigma(x,\tau)[\rho(x,\tau)\chi_\mu(x,\tau)\rho(x+\mu,\tau)]Z_\sigma(x+\mu,\tau) + H.C.$$

$$+\sum_\mu \frac{|\chi_\mu(x,\tau)|^2}{J/2}\ \} \tag{1d}$$

In the limit of heavy holes, $t\delta<<J$ we neglect the t^2/J term (this term has to be added to the V term. See Eq. 1a). The term $\sum_{x,\mu,\sigma} Z^*_\sigma(x,\tau)[\rho(x,\tau)\chi_\mu(x,\tau)\rho(x+\mu,\tau)]Z_\sigma(x+\mu,\tau)\approx \langle\rho\rangle^2 \sum_{x,\mu,\sigma} Z^*_\sigma(x,\tau)\chi_\mu(x,\tau)$ $Z_\sigma(x+\mu,\tau)+\sum_{x,\mu,\sigma} Z^*_\sigma(x,\tau)[\rho(x,\tau)-\langle\rho\rangle]\chi_\mu(x,\tau)Z_\sigma(x+\mu,\tau) + \sum_{x,\mu,\sigma} Z^*_\sigma(x,\tau)$ $\chi_\mu(x,\tau)[\rho(x+\mu,\tau) - \langle\rho\rangle]Z_\sigma(x+\mu,\tau)$ depends on the hole concentration $\langle\rho\rangle=1-\delta$. The effect of the density fluctuations $\rho-\langle\rho\rangle$ induces next neighbour interaction. In the limit of heavy holes we find after the density integration:

$$\langle\rho\rangle^2 \sum_{x,\mu.\sigma} Z^*_\sigma(x,\tau)\chi_\mu(x,\tau)Z_\sigma(x+\mu,\tau) +$$

$$\langle[\rho(x,\tau)-\langle\rho\rangle][\rho(x+\mu,\tau)-\langle\rho\rangle]\rangle \sum_{x,\mu',\sigma,\sigma'} Z_\sigma(x,\tau)Z_\sigma(x+\mu',\tau)\langle Z^*_\sigma(x+\mu',\tau)\cdot$$

$Z_{\sigma'}(x,\tau)>$. Here μ is the nearest neighbour interaction and μ' is the next nearest neighbour interaction. The term induced by the hole fluctuation leads to frustration and to breaking of parity and time reversal symmetry. In the absence of holes $\rho=1$ the minimum is obtained for $\Sigma \bar{A}_\mu(x)=\pi$. We choose $A_\mu(x)=\bar{A}_\mu(x)+a_\mu(x)$. The Z_σ hamiltonian is diagonalized for $\bar{A}(x)=\pi/4$. The result is that one obtains two massless Dirac fermions with two point fermi surface at $(\pi/2, -\pi/2)$ and $(\pi/2, \pi/2)$. We introduce the spinor $\Psi^*_\sigma(K)=(Z^*_\sigma(K), Z^*_\sigma(K+\pi))$ and find for the Z particles the action:

$$\int d\tau \int d^2x \{ \sum_{\sigma,\alpha} \bar{\Psi}_{\sigma,\alpha} [\tau_3(\partial_\tau+iA_o)+i\vec{\tau}\cdot(\partial_x+\vec{a}_x)+M]\Psi_{\sigma,\alpha} \} .$$

$\alpha=1,2$ is the index of the two fermi surface, $\bar{\Psi}_{\sigma,\alpha}=\Psi^*_{\sigma,\alpha}\tau_3$, $\vec{\tau}=(\tau_1,\tau_2)$ τ_3, $\vec{\tau}$ represent the 2×2 Pauli matrices. The term $M\neq 0$ is induced by $\rho\neq 1$, $M\sim<(\rho-<\rho>)^2><Z^*_\sigma(x,\tau)Z_\sigma(x+\mu',\tau)>$. The integration of the Z^*_σ, Z_σ particles give rise to the following effective action:

$$S(\tau)=\sum_x b^*(x,\tau)[\partial_\tau+\mu']b(x,\tau)-t\sum_\mu b^*(x+\mu,\tau)\exp[iA_\mu(x)]b(x,\tau)$$

$$-iA_o(x,\tau)[b^*(x,\tau)b(x,\tau) - \frac{1}{\pi}\frac{M}{|M|}\varepsilon_{\mu\nu}\Delta_\mu A_\nu(x,\tau)]\} \tag{1e}$$

3. SUPERCONDUCTIVITY IN THE FRACTIONAL STATISTICAL GAS

In this chapter we study the result given in Eq. 1e. The derivative with respect to A_o shows us that the fictitious magnetic field $\nabla\times A$ is proportional to the charge density. The hamiltonian of the action given in Eq. 13 represents a hard core bose gas (in two dimensions) in the presence of statistical gauge field $A(x)$.

$$H = -t \sum_{x,\mu} b^*(x+\mu)e^{iA_\mu(x)} b(x) + H.C. + \mu'\Sigma\, b^*(x)b(x) \tag{1f}$$

$\mu'=\mu'(\delta)$ is the chemical potential for the bosons with concentration δ and $\tilde{\rho}(x)=b^*(x)b(x)=0,1$. The statistical gauge potential is given by the variation with respect to A_o in Eq. 1e

$$\tilde{\rho}(x) = \theta\vec{\nabla}x\vec{A} = \theta B(r) \tag{1g}$$

$B(r)$ is the statistical magnetic field defined on the dual lattice. θ is the strength parameter of the Chern-Simons term[10-11]. When $\theta=1/\pi$ (see Eq. 1e) one finds that the interchange of two particles multiplies the wavefunction by a phase $\exp(i(\pi/2))$. This corresponds to the semion case (half-fermion, one has to interchange twice the particles) when $\theta=1/2\pi$ the wave function is multiplied by $\exp(i\pi)$. The case corresponds to a fermion.

The coherent state representation[15], for the hard core boson is:

$$b(x) = \tilde{b}(x)\exp[i\theta_v(x)] = r(x)\exp[if(x)] \tag{2a}$$

The hard core boson can be analyzed with the use of the spin coherent representation for spin $\frac{1}{2}$. In this representation $r(x)$ is the amplitude and $f(x)$ is the phase. The amplitude[16] is given by: $r(x)=[\tilde{\rho}(x)(1-\tilde{\rho}(x)]^{1/2}$. The phase $f(x)$ can be separated into the vorticity $\theta_v(x)$ and the normal part $\Psi(x)$:

$$f(x) \equiv \Psi(x)+\theta_v(x) \tag{2b}$$

The function $\Psi(x)$ and $\theta(x)$ satisfies the conditions

$$\vec{V}^{(v)}(x) = \nabla\theta_v(x), \qquad \vec{V}^{(n)}(x) = \nabla\Psi(x) \tag{2c}$$

$$\vec{\nabla}\times\vec{V}^{(v)}(x) \neq 0, \qquad \vec{\nabla}\cdot\vec{V}^{(v)}(x) = 0, \qquad \vec{\nabla}x\vec{V}^{(n)}(x) = 0 \tag{2d}$$

$\theta_v(x)$ represents the vorticity, basically we assume that $b(x)$ vanishes at discrete points. The function $\tilde{b}(x)$ is smooth and the vanishing of $b(x)$ is absorbed into the term $\exp[i\theta_v(x)]$. The vector field $\vec{V}^{(v)}=\nabla\theta_v(x)$ can be expressed with the aid of a scalar function $\Phi^{(v)}(r)$

$$V_\mu^{(v)}(x) = \Delta_\mu\theta_v(x) = \varepsilon_{\mu\nu}\Delta_\nu\Phi^{(v)}(r) \tag{3a}$$

here we have used the definition: $\Delta_\mu f(x)=f(x+\mu)-f(x)$ and $\varepsilon_{11}=\varepsilon_{22}=0$, $-\varepsilon_{12}=\varepsilon_{21}=-1$.

The vector field $V^{(v)}(x)$ satisfies

$$\vec{\nabla}\times\vec{V}^{(v)}(x) = \varepsilon_{\mu\nu}\Delta_\mu V_\nu^{(v)}(x) = \varepsilon_{\mu\nu}\Delta_\mu\varepsilon_{\mu\nu'}\Delta_{\mu'}\Phi^{(v)}(r) =$$

$$= \Delta^2\Phi^{(v)}(r) = 2\pi q(r), \qquad q(r) = 0,\pm1, \ldots \tag{3b}$$

where $q(r)$ is the vorticity charge.

The presence of the statistical gauge fixed $A(x)$ is included by the appearance of a function $\exp(i\emptyset(x))$. The phase $\emptyset(x)$ is related to $\vec{A}(x)$ by the equation:

$$A_\mu(x) = \Delta_\mu\emptyset(x) = \emptyset(x+\mu)-\emptyset(x) \tag{3c}$$

The kinetic term can be expressed $b^*(x)\exp[iA_\mu(x)]b(x+\mu) = \tilde{b}^*(x)\exp[-i\theta_v(x)-i\emptyset(x)\exp[+i\theta_v(x+\mu)+i\emptyset(x+\mu)]\tilde{b}(x)$.

The vector field $A_\mu(x)$ can be represented with the aid of the scalar function $\Phi^{(A)}(r)$ in the same way $V_\mu(x)$ (the vortex field) has been related to the scalar function $\Phi^{(v)}(r)$. Performing a similar analysis as in equations 3a and 3b we find

$$A_\mu(x) = \Delta_\mu \emptyset(x) = \varepsilon_{\mu\nu}\Delta_\nu \Phi^{(A)}(r)$$

$$\vec{\nabla}\times\vec{A} \equiv \varepsilon_{\mu\nu}\Delta_\mu A_\nu \equiv \varepsilon_{\mu\nu}\Delta_\mu\varepsilon_{\mu\nu'}\Delta_{\mu'}\Phi^{(A)}(r)\ \Delta^2\Phi^{(A)}(r) = B(r) \tag{3d}$$

Combining the results given in equations 3a-3b and 3d we find:

$$\vec{\nabla}\times(\vec{A}+\nabla\theta_v) \equiv \vec{\nabla}\times\vec{a} = \Delta^2(\Phi^{(A)}(r)+\Phi^{(v)}(r)) =$$

$$= \Delta^2\Phi(r) = 2\pi\ [q(r) + \frac{1}{2\pi} B(r)] = 2\pi Q(r) \tag{4a}$$

where we have defined:

$$\vec{A}(x)+\nabla\theta_v(x) \equiv \vec{a}(x), \qquad \Phi(r) \equiv \Phi^{(A)}(r)+\Phi^{(v)}(r) \tag{4b}$$

with the property:

$$\vec{\nabla}\mathrm{x}\vec{a}(x) = 2\pi Q(r) = 2\pi[q(r)+(1/2\pi\Theta)\tilde{b}^*(x)\tilde{b}(x)] \tag{4c}$$

$$\vec{\nabla}\cdot\vec{a}(x) = 0 \tag{4d}$$

$Q(r)$ represents the total vorticity composed from $q(r)=0,\pm1,\ldots$ (the intrinsic vorticity) and the external vorticity $B(r)/2\pi$ induced by the electric charge $(1/\Theta)\tilde{\rho}(x)$. Equation 4c shows that the statistical field $\vec{A}(x)$ is replaced by the vector potential $\vec{a}(x)$. The hamiltonian given in equation 1a takes the form

$$H = -t\sum_{x,\mu}\tilde{b}^*(x+\mu)e^{ia_\mu(x)}\tilde{b}(x)+\mathrm{H.C.} + \mu'\tilde{b}^*(x)b(x) \tag{5a}$$

In equation 5a we perform the continuum approximation

$$\sum_{x,\mu}\tilde{b}^*(x)e^{ia_\mu(x)}\tilde{b}(x+\mu)+\mathrm{H.C.} \approx \sum_{x,\mu}\tilde{b}^*(x)b(x+\mu)+\mathrm{H.C.} -$$

$$-\sum_x \tilde{b}^+(x)\tilde{b}(x)a_\mu^2(x)+i2|\mu|\int d^2x\ \partial_\mu\tilde{b}(x)a_\mu(x) \tag{5b}$$

The function $\tilde{b}^*(x)\partial_\mu\tilde{b}(x)$ satisfies:

$$i2|\mu|\int d^2x(\tilde{b}^*(x)\partial_\mu \tilde{b}(x))a_\mu(x) \simeq i2|\mu|\overline{r^2(x)}\int d^2xV_\mu^{(n)}(x)a_\mu(x) =$$

$$=i2|\mu|\overline{r^2(x)}\int d^2x\Delta_\mu\Psi(x)a_\mu(x) = i2|\mu|\overline{r^2(x)}\ [\Psi(x)a_\mu(x)|_{-\infty}^{\infty}-\int d^2x\Psi(x)\Delta_\mu a_\mu(x)]=0 \tag{5d}$$

Equation 5a was obtained using the vanishing of $\Psi(x)\vec{a}(x)$ on the boundary and $\vec{\nabla}\cdot\vec{a} = 0$. Using equations 5b and 5d we have instead of equation 5a

$$H = H_0+H_V \tag{6a}$$

$$H_0 =-t \ \Sigma \tilde{b}^*(x+\mu)\tilde{b}(x)+H.C.+\mu' \Sigma \tilde{b}^*(x)\tilde{b}(x) + \tag{6b}$$

and

$$H_V = -t \sum_{x,\mu} \tilde{b}^+(x)\tilde{b}(x)a_\mu^2(x) = -t\int d^2xr^2(x)(\Delta_\mu\Phi(r))^2 =$$

$$\simeq \overline{tr^2(x)}\int d^2x\Phi(r)\Delta^2\Phi(r) \tag{6c}$$

Using spin coherent representation we have $\overline{r^2(x)}=\overline{\tilde{\rho}(x)(1-\tilde{\rho}(x)}\simeq\bar{\rho}(1-\bar{\rho})$ where $\bar{\rho}$ is the quasi condensate.

Equation 6c can be further simplified if we use Eq. 4a, $\Delta^2\Phi(r)=2\pi Q(r)$ with the solution $\Phi(r)=\sum_{r'} G(r,r')Q(r')$. The substitution of this solution in equation 6c gives:

$$H_V = 2\ \bar{t}\ \bar{\rho}\ \sum_{r,r'} Q(r)G(r,r')\ Q(r') \tag{7a}$$

where $\bar{t}\equiv t(1-\bar{\rho})$ is the effective hopping.

Equation 7a can be further simplified if we remove the infrared divergencies. This is obtained by imposing the charge neutrality condition[13]

$$\Sigma Q(r) = 0, \qquad \Sigma q(r) = -(1/2\pi)\Sigma B(r)$$

Instead of equation 7a we have

$$H_V = 2\pi\bar{t}\ \bar{\rho}\ [\sum_{r\neq r'} Q(r)\tilde{G}(r,r')Q(r')+(\pi/2)\sum_r Q^2(r)] \tag{7b}$$

where

$$\tilde{G}(r,r') = -Ln[K_0|r-r'|] \tag{7c}$$

$$Q(r) = q(r)+(1/2\pi\Theta)\tilde{b}^*(x)\tilde{b}(x) \quad (7d)$$

The solution of Eq. 7b changes with Θ. For $\Theta=1/2\pi$ the bosons become fermions[10]. This can be seen also from the charge equation. For $\Theta=1/2\pi$ the charge $Q(r)$ is given by: $Q(r)=q(r)+\tilde{b}^*(x)\tilde{b}(x)$.

As a result of the charge neutrality condition we find that binding will occur between the -1 intrinsic vortex and the +1 positive vortex induced by the electrical charge $\tilde{b}^*(x)\tilde{b}(x)$. Here we have the situation where the electric charge that binds with a vortex gives rise to a fermion. For $\Theta=1/\pi$ the vorticity is given by: $Q(r)=q(r)+\frac{1}{2}\tilde{b}^*(x)\tilde{b}^*(x)$. Neutrality of the vortex lattice is obtained if we choose the solutions: $Q(r)=0, \pm\frac{1}{2}, \pm 1, \ldots$ ($\frac{1}{2}$ corresponds to $q(r)=0$, $\tilde{b}^*(x)\tilde{b}(x)=1$; $-\frac{1}{2}$ corresponds to $q(r)=-1$, $\tilde{b}^*(x)\tilde{b}(x)=1$; ± 1 corresponds to $q(r)=\pm 1$, $\tilde{b}^*(x)\tilde{b}(x)=0$). For this case we can write: $Q(r)=\pm\frac{1}{2}\tilde{\rho}(x)+m(r)$, $\tilde{\rho}(x)=\tilde{b}^*(x)\tilde{b}(x)$ and $m(r)$ is an integral vorticity which satisfies $\Sigma m(r)=0$. The fractional fortices $\pm\frac{1}{2}$ can be expressed in terms of the of the electric charge $\pm\frac{1}{2}\tilde{\rho}(x)=\frac{1}{2}\tilde{\rho}_\alpha(x)$, $\alpha=+,-$, $\tilde{\rho}_+(x)=\tilde{\rho}(x)$, $\tilde{\rho}_-(x)=-\tilde{\rho}(x)$. At low temperature the $\pm\frac{1}{2}\tilde{\rho}(x)$ and $-\frac{1}{2}\tilde{\rho}(x)$ vortices are confined into dipoles. Therefore, two electric charges $\tilde{\rho}(x)$ with opposite vortices bind forming a boson of charge $2\tilde{\rho}(x)=2e$.

Since the neutrality condition $m(r)=0$ is satisfied the effect of the fluctuating vorticity $m(r)$ is evaluated by integration of these fluctuations[14]. The effect of these fluctuations is to renormalize the vortex potential $\tilde{G}(r,r')\rightarrow\tilde{G}_{eff}(r,r')$ which at low temperature is not important.

At high temperature a phase transition to a plasma phase is characterized by free vortices and therefore free electric charges[13]. The critical temperature for formation of pairs is identical to the Kosterlitz Thouless transition for $\pm\frac{1}{2}$ vortices with $T_k\approx 2\pi\bar{\rho}\,\bar{t}\approx t\delta(1-\delta)$.

In order to study the question of superconductivity one has to consider in addition to binding of vortices the kinetic part given in equation 6b.

We use the notation $\tilde{\rho}_\alpha(x)=\tilde{b}^*(x)\tilde{b}_\alpha(x)$ with the index $\alpha=+,-$ representing the $\pm$ vorticity of the electric charge. Each electric charge has $+\frac{1}{2}$ or $-\frac{1}{2}$ vorticity. We find an affective model which is similar to previous models which might have a superconducting phase

$$H = -t \sum_{x,\mu,\alpha} \tilde{b}^*(x+\mu)\tilde{b}_\alpha(x) + H.C. + (\mu' + \frac{\pi^2 \bar{\rho} \bar{t}}{4}) \sum_{x,\alpha} \tilde{b}^*_\alpha(x)\tilde{b}_\alpha(x) -$$

$$- \frac{\pi\bar{\rho}\bar{t}}{2} \sum_{x,x',\alpha} \tilde{b}^*_\alpha(x)\tilde{b}^*_{-\alpha}(x')V_{eff}(x,x')\tilde{b}_{-\alpha}(x')\tilde{b}_\alpha(x) \qquad (8)$$

The hamiltonian given in equation 8 describes a hard core boson with attractive interaction ($V_{eff}(x,x')=\tilde{G}_{eff}(r,r')-V\delta(x-x')$) for bosons with opposite spin α (vorticity). The presence of the term $V\delta(x-x')(V\to\infty)$ prevents double occupancy (the hard core boson condition for bosons with opposite vorticity). In addition we have the condition that $(\tilde{b}_\alpha)^2=(\tilde{b}^*_\alpha)^2=0$, the hard core boson condition for the same vorticity. This condition can be fulfilled if we replace the boson coherent states by Grassmanian coherent states: C_α, C^*_α (fermions). The mapping[10] is performed with the aid of a fictitious vector potential; $A^F_\mu(x)=\chi(x+\mu)-\chi(x)$, $\vec{\nabla}\times\vec{A}^{(F)}=2\pi C^*_\alpha(x)C_\alpha(x)$. The relation between the two coherent states representations is:

$$\tilde{b}^*_\alpha(x) = C^*_\alpha(x)e^{i\chi(x)}, \qquad \tilde{b}_\alpha(x) = e^{-i\chi(x)}C_\alpha(x).$$

The first condition of hard core boson with opposite vorticity is enforced by the projector:

$$P_d = \prod_x (1-n_+(x)n_-(x)), \qquad n_\pm = C^+_\pm C_\pm$$

As a result of these transformations we obtain instead of eq. 8 a usual model of fermions in a magnetic field with an attractive interaction $V_{eff}(x,x') = \tilde{G}_{eff}(r,r')$ (instead of $V_{eff}(x,x') = \tilde{G}_{eff}(r,r')-V\delta(x-x')$).

$$H = P_d\tilde{H}P_d$$

$$\tilde{H} = -t \sum_{x,\mu,\alpha=\pm} C^*_\alpha(x+\mu)e^{iA^{(F)}_\mu(x)} C_\alpha(x) + H.C. + (\mu' + \frac{\pi^2\bar{\rho}\bar{t}}{4}) \sum_{x,\alpha} C^*_\alpha(x)C_\alpha(x) -$$

$$-(\pi\bar{\rho}\bar{t}/2) \sum_{x\neq x'} C^+_+(x)C^+_-(x')\tilde{G}_{eff}(x,x')C_-(x')C_+(x)$$

If we replace $A^{(F)}$ by the mean field condition, $\vec{\nabla}\times\vec{A}^{(F)}\simeq 2\pi\langle C^*_\alpha(x)C_\alpha(x)\rangle$ and $\langle C^*_+(x)C_+(x)\rangle = \langle C^*_-(x)C_-(x)\rangle = \frac{1}{2}\langle\tilde{\rho}(x)\rangle$ we can diagonalyze the kinetic part. We obtain a complete orthonormal set of Grassmanian numbers, $\tilde{C}^*_{\ell,\alpha}$, $C_{\ell,\alpha}$ obtained from the original basis $C^*_\alpha(x), C_\alpha(x)$ by a unitary transformation, $\tilde{C}^*_{\ell,\alpha} = \sum_x \Psi_\ell(x)C^*_\alpha(x)$ where $\{\Psi_\ell(x)\}$ are eigenstates of the tight binding hamiltonian in a constant magnetic field. Due to the attractive interaction a singlet wave function for the superconducting state is suggested

$$|\emptyset_{s.c.}\rangle = Pd\,(\Pi_{\ell} \tilde{C}^{+}_{\ell,+}\tilde{C}^{+}_{\ell,-})^{N/2}|0\rangle = Pd[\sum_{x,x'}\prod_{\ell}\Psi_{\ell}(x)\Psi_{\ell}(x')$$

$$C^{+}_{+}(x)C^{+}_{-}(x')]^{N/2}|0\rangle = Pd[\sum_{x,x'} a(x,x')C^{*}_{+}(x)C^{*}_{-}(x')]^{N/2}|0\rangle$$

The form of $|\emptyset_{s.c}\rangle$ with $a(x,x')\equiv\prod_{\ell}\Psi_{\ell}(x)\Psi_{\ell}(x')$ is identical to the generalized R.V.B. states introduced by Anderson[17] for the new superconductors.

To conclude we have shown that a hard core boson with a statistical fictitious vector potential A(x) which describes the semion state $\theta=1/\pi$ leads to the following picture. The presence of the intrinsic vortices in the semion state plus the charge neutrality condition gives a binding with $\pm\frac{1}{2}$ vorticity. Below the Kosterlitz Thouless transition $T<T_k$ we find an effective model - which might have anisotropic superconductivity.

REFERENCES

1. J.G. Bednorz and K.A. Müller, Z. Phys. 64, 198 (1986).
2. P.W. Anderson, Science 235 (1987).
3. G. Kotliar, P.A. Lee and N. Read, Physica C153-542, 538 (1988).
4. N. Read and S. Sachdev, Nuclear Physics B136, 609 (1988).
5. D.R. Hofstadter, Phys. Rev. B14, 2239 (1976); G.H. Wannier, Phys. Status Solidi B88, 757 (1978).
6. Yi-Hong Chen, F. Wilczek and E. Witten, Journal of Modern Physics B, Vol. 3, No. 7, 1001 (1989).
7. D. Schmeltzer, Physica C 153-155, 1291-1292 (1988); D. Schmeltzer, Phys. Rev. B38, 893 (1988).
8. I. Affleck and Martson, Phys. Rev. B37, 3774 (1988).
9. A.N. Redlich, Phys. Rev. D29, 2366 (1984).
10. E. Fradkin, Phys. Rev. Lett. 63, 322 (1989).
11. Yi-Hong, Chen, F. Wilczak, E. Witten, B.I. Halperin, preprint, 1989.
12. N. Popov, Quantum Field Theory.
13. J.B. Kogut, Reviews of Modern Physics, Vol. 51, no. 4, October 1979, page 700.
14. S. Samuel, Phys. Rev. D18, 1916 (1978).
15. J.W. Negele and H. Orland, Quantum Many Particle Systems, Addison Wesley 1988.
16. R. Balakrishnan, R. Sridhar and R. Vasudera, Phys. Rev. B39, 174 (1989).
17. P.W. Anderson, Science 235, 1196 (1987).

ELECTRON-HOLE LIQUID MODEL FOR HIGH T_c-SUPERCONDUCTORS

A. Kallio, X. Xiong, and M. Alatalo

Department of Theoretical Physics, University of Oulu

Linnanmaa, SF-90570 Oulu 57, FINLAND

ABSTRACT

We show that the hole concentration dependence of T_c in the high-T_c superconductors of the type $La_{2-x}Sr_xCuO_4$, $YBa_2Cu_3O_{6+x}$ and $Nd_{1+x}Ba_{2-x}Cu_3O_y$ can be understood within a model electron-boson hole liquid which possesses a linear sound mode like the one in liquid-^{4}He. The electronic specific heat follows a T^3-law near $T = 0$ in addition to the linear term. The metal-insulator transition found in these compounds is proposed to be connected with the vanishing of the sound velocity at small hole concentration. The basic assumptions are confirmed by numerical model calculations. Possible connection with heavy fermion systems is discussed.

INTRODUCTION

Since the discovery of high-T_c superconductors[1], no convincing theoretical model has been presented to explain the phenomenon. In the previous paper[2] we proposed an Electron-Boson Hole Liquid (EBHL) model for high-T_c superconductivity. The basic assumption of the model was the existence of both mobile electrons and mobile hole bosons in these compounds [3–12]. Since the coherence length of the paired holes is small, $\xi_0 \sim 5-20$ Å, and hence the gap Δ is very large[13], it was assumed that one can treat them simply as composed bosons of charge $+2e$. It was further assumed that the electrons are moving mainly in the c-direction whereas the bosons are constrained to move in ab-plane[14]. From the point of view of the EBHL-model the binding mechanism is immaterial. The EBHL-model then essentially consists of a mixture of electrons and hole-bosons in analogy with the EHL in semiconductors[15]. According to our assumption electrons and holes never meet in space so the annihilation rate given by the electron-boson radial distribution function $g_{eB}(r)$ at $r = 0$ must vanish. The evidence for existence of both types of carriers in these compounds is born out by the Hall - measurements[14] and some evidence[16] show that the pairs exist even above T_c.

Recent Progress in Many-Body Theories, Volume 2
Edited by Y. Avishai
Plenum Press, New York, 1990

In the present case the bosons and fermions have opposite charges in contrast to the boson-fermion mixture model proposed recently by Friedberg and Lee[17].

It was pointed out in ref.2 that such a mixture of electrons and bosons exhibits a low-lying phonon spectrum and a high-lying plasmon spectrum which together with the usual BCS-pair breaking effect was so far neglected here in calculating the temperature dependence. These together with the non- linearity corrections to the phonon spectrum in the gap should in a more refined calculation be added to the statistical mechanics of the system.

The mobile electrons will contribute a linear term in the specific heat

$$C_e = \gamma(0)T + \beta_e(0)T^3 \tag{1}$$

with the $\gamma(0)$ proportional to electron concentration. The existence of linear term has been disputed for some time but the most recent experiments[18] seems to show that $\gamma(0) \neq 0$. Without the existence of a mobile electron component it is difficult to understand $\gamma(0)$. The cubic term in the electronic specific heat comes from the phonon spectrum and $\beta_e(0)$ is determined by the "electron sound" velocity u_e. Using the existing specific heat data[19] and a scaling argument for liquid ^{4}He it was deduced in ref.2 that u_e is larger than the crystalline sound velocity. Unfortunately there exists so far no experimental evidence for such an electron sound. Experimentally the extra sound mode should be found in Brillouin scattering. Instead of seeing three vibrational modes as in a normal crystal, one should find a fourth one. In a crystal like $La_{2-x}Sr_xCuO_4$ the dispersion $\epsilon(k) \sim \hbar u_e k$ is sensitive function of x near the end points $x = x_1$ or x_2 in Fig.5.

Within the model it is also possible to understand the nature of the metal - insulator transition: Below a certain hole concentration n_c the electrons and bosons form a "heavy fermion" bound state which is localized due to our assumption about the mobilities: the electrons move mainly in c-direction and the bosons in ab-plane. For $n_B > n_c$ one has a system of interpenetrating Fermi- and Bose fluids. At n_c the sound velocity vanishes. Such a system remains superfluid as long as the boson density is larger than the density of phonon excitations in the system. In analogy with liquid ^{4}He (or better with a mixture of ^{4}He and ^{3}He) one obtains for T_c an approximate formula

$$n_B = \frac{\hbar^2}{m_B} \frac{2\pi^2}{45} \frac{(k_B T_c)^4}{(\hbar u_e)^5} \tag{2}$$

where u_e is the electron sound velocity.

As in the ^{3}He and ^{4}He mixtures the fermion component moves[20], with the normal - component of the liquid. Hence the supercurrent is determined by the superfluid velocity

$$\vec{J}_s = 2e\vec{v}_s$$

Finally it follows from $Eq(2)$ for doped high-T_c superconductors that T_c varies with $(n_B - n_c)$ like

$$T_c = \alpha(n_B - n_c)^{3/2} \tag{3}$$

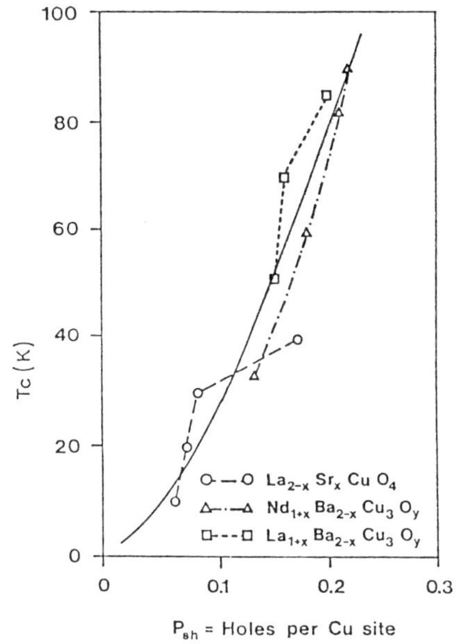

Fig.1. Transition temperature T_c as a function of p_{sh}, the mobile sheet holes per Cu site. Experimental data for $La_{2-x}Sr_xCuO_4$, $Nd_{1+x}Ba_{2-x}Cu_3O_y$ and $La_{1+x}Ba_{2-x}Cu_3O_y$ are from ref. 12. The solid line shows the theoretical result in ref.2.

This was obtained in ref.2 with the assumption that the sound velocity vanishes linearly at n_c. The exponent 3/2 is the same as the one obtained for liquid ^{4}He in porous media[21]. One has however, theoretical evidence that $u_e = \alpha(n_B - n_c)^{1/2}$ which would produce an exponent 7/8. To the extent that T_c is determined by hole concentration (or p_{sh} of ref 12) alone both exponents fit the experimental data almost equally well. In the case of $La_{2-x}Sr_xCuO_4$ it has been observed that the T_c for small hole -concentrations behaves in the predicted in ref.2 (see Fig.1) but beyond maximum at $p_{sh} \approx .22$ holes/Cu -site starts decreasing and vanishes at $p_{sh} \approx .25-.3$. Furthermore it has been shown that the coefficient of the linear term $\gamma(0)$ at the same time exhibits a very complicated behaviour. At the point of maximum T_c the electron concentration has a minimum. The purpose of the present paper is to show by model calculations that even the complicated behaviour of T_c in $La_{2-x}Sr_xCuO_4$ can be understood within the EBHL -model. The numerical model is still unrealistic in many respects: the crystalline structure is ignored and the interactions are purely Coulombic. With more realistic interactions one can start extracting parameters as like effective masses and dielectric functions etc which are still largely unknown.

MODEL CALCULATIONS

The Euler - Lagrange equations (EL) for a mixture of electrons and bosons can be written in terms of radial distribution functions $g_{\alpha\beta}(r)$ in the form

$$-\frac{\hbar^2}{2m_{\alpha\beta}}\nabla^2\sqrt{g_{\alpha\beta}} + (v_{\alpha\beta}(r) + w_{\alpha\beta}(r))\sqrt{g_{\alpha\beta}} = 0 \tag{4}$$

where $\frac{1}{m_{\alpha\beta}} = \frac{1}{m_\alpha} + \frac{1}{m_\beta}$. The 2x2 matrix of induced potentials is given by[22]

$$w(k) = -\frac{\hbar^2 k^2}{4}[M^{-1}S + SM^{-1} - 2M^{-1} + (SMS)^{-1} - (S_F M S_F)^{-1}] \quad (5)$$

and

$$S_F = \begin{cases} \frac{3}{4}\frac{k}{k_F} - \frac{1}{16}(\frac{k}{k_F})^3, & k < 2k_F \\ 1, & k \geq 2k_F \end{cases}$$

$$S = \begin{pmatrix} S_{ee} & S_{eB} \\ S_{eB} & S_{BB} \end{pmatrix}; M = \begin{pmatrix} m_e^* & 0 \\ 0 & m_B^* \end{pmatrix}. \quad (6)$$

The liquid structure factors $S_{\alpha\beta}(k)$ are the Fourier -transforms of the corresponding radial distribution functions $g_{\alpha\beta}(r)$. The main interactions $v_{\alpha\beta}(r)$ are Coulombic except at short distances in v_{eB} one has to have repulsion in order to produce $g_{eB}(0) = 0$. In the present calculation we have not considered such a repulsion. One can transform the EL -equations into k- space and prove that all $S_{\alpha\beta}(k)$ behave linearly at small k. The excitations for a mixture can be obtained by diagonalization of the matrix

$$\hbar^2 k^2 S^{-1} M^{-1} \quad (7)$$

The spectrum turns out to have a plasma branch and a phonon branch. Using purely Colomb potentials for $v_{\alpha\beta}$ we have done exploratory calculations by varying the masses and densities in order to show that it has the properties needed for EBHL -model. We define the number densities n_e, n_B and concentrations x_B and $x_e = 1 - x_B$ such that

$$\begin{aligned} n_B &= x_B \cdot n_0 \\ n_e &= (1 - x_B) \cdot n_0 \end{aligned} \quad (8)$$

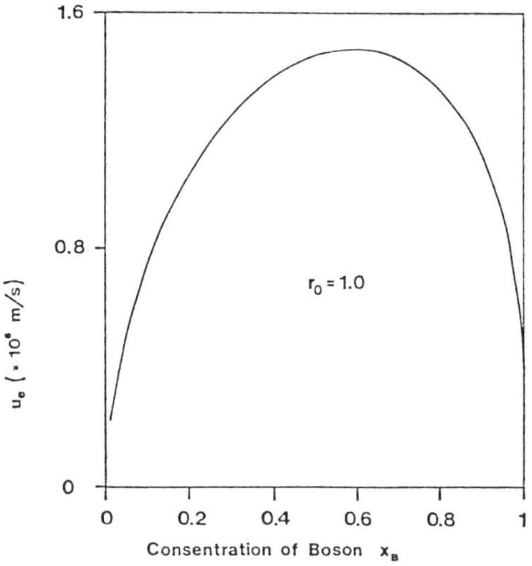

Fig.2. The velocity of sound u_e in EBHL as a function of Boson concentration x_B with the total density $r_0 = 1.0$.

where the r_o value is given by

$$\frac{1}{n_0} = 4/3\pi(r_0 a_0)^3$$

$$a_0 = \frac{\hbar^2}{m_e e^2}$$

The local charge neutrality corresponds to boson concentration $x_B = 1/3$. In Fig.2 we show the velocity of sound as a function of concentration x_B. The velocity of sound has a maximum near $x_B = 2/3$ and becomes small near both ends $x_B \approx 0$ and $x_B \cong 1$. According to $Eq(2)$ T_c is given by

$$k_B T_c = \left[\frac{45}{2\pi^2}\frac{m_B n_B}{\hbar^2}\right]^{1/4} (\hbar u_e)^{5/4} \tag{9}$$

Hence the variation of T_c with hole concentration ($=2\ n_B$) comes mainly from the variation of u_e. This scheme gives qualitatively correct behaviour for T_c: It vanishes at two points n_1 and $n_2 > n_1$ and has a maximum in between. Near the instability points the solution of the EL -equation is very difficult. For that reason the r_o -values are still unrealistic.

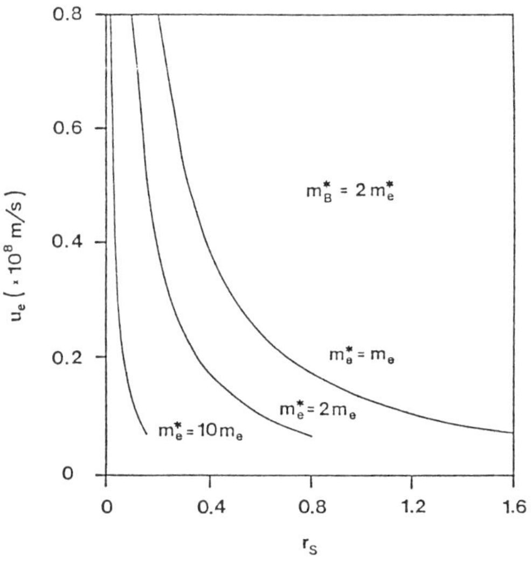

Fig.3. Density dependence of the velocity of sound in the EBHL system with the condition of local charge neutrality. The effective electron masses were shown and the Boson's mass was double of effective electron mass.

The behaviour of the velocity of sound as a function of r_o and the effective masses is shown in Fig.3 with concentration $x_B = 1/3$. The total densities n_o are still unrealistic and the speed of sound likevise very large. However with decreasing n_0 and increasing effective masses the speed of sound decreases but the $Eqs(4)$ become

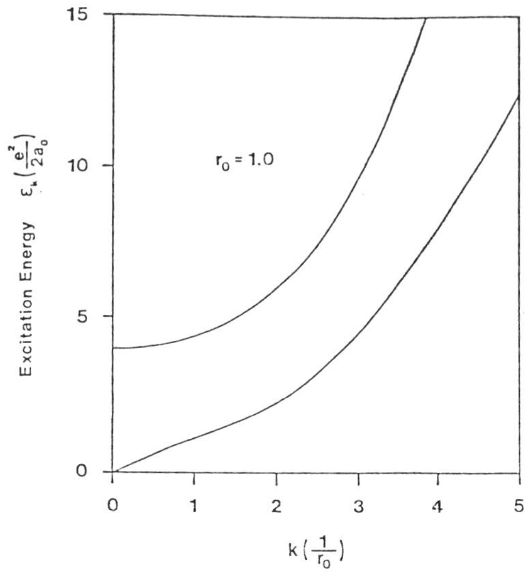

Fig.4. The excitation spectra of the EBHL model with the condition of local charge neutrality at the total density $r_0 = 1.0$.

increasingly difficult to solve. The velocities are still orders of magnitude larger than the ones estimated in ref.2. We believe that adding a repulsive component to v_{eB} would help in reaching more realistic densities n_0.

Fig.4 shows the two excitation modes already discussed. The numerical calculations reproduce the qualitative features needed for the EBHL -model. The model calculation treats the bosons as point particles which is the first thing to be corrected. Next improvement would be the formulation of the whole problem on lattice. We believe that qualitatively the excitation modes remain the same.

DOPING AND SPECIFIC HEAT

As it was mentioned before the sound velocity should vanish like $(n_B - n_c)^{1/2}$ at n_c. The numerical model shows that it vanishes at two points n_1 and n_2 while T_c has its maximum in between. Simplest formula with the desired behaviour of the velocity $u(n_B)$ would then be

$$u(n_B) = \gamma[(n_B - n_1)(n_2 - n_B)]^{1/2}. \tag{10}$$

This together with $Eq(2)$ give for T_c the behaviour

$$k_B T_c = \beta n_B^{1/4}[(n_B - n_1)(n_2 - n_B)]^{5/8}. \tag{11}$$

Assuming that in $La_{2-x}Sr_xCuO_4$ the hole concentration and hence n_B is proportional to x we have compared the experimental results of ref 23 with $Eq(11)$. As is seen from Fig.5 the agreement is very good. The high x end of the curve is seen to be abtrupt in contrast to the long tail in the theory of Aharony et al.[24]

Interestingly enough one can also make predictions about the pressure dependence of T_c using $Eq(11)$ since

$$\frac{\partial T_c}{\partial P} = \frac{\partial T_c}{\partial n_B}\frac{\partial n_B}{\partial P} = \frac{1}{\kappa}\frac{\partial T_c}{\partial n_B}, \tag{12}$$

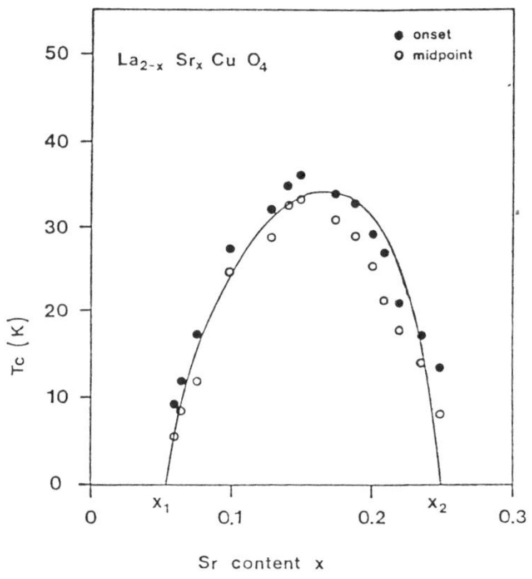

Fig.5. Transition temperature T_c as a function of Sr content x for $La_{2-x}Sr_xCuO_4$ compound. Experimental data are from ref.23. The solid line shows the theoretical result calculated from $Eq(11)$ assuming n_B to be proportional to x.

where κ is the compressibility of the lattice. For $La_{2-x}Sr_xCuO_4$ the pressure gradient should be positive near $n_B = n_1$, zero near the maximum T_c and negative past the maximum. In most other compounds only the rising part near n_1 is observed (see Fig.1) hence the gradient should be positive exept at the highest T_c where it should vanish. Systermatic measurements of the pressure gradients as a function of x are still missing.

As mentioned at the beginning the existence of the linear term in the specific heat in our model comes from the electron component. In most high-T_c superconductors $\gamma(0) > 0$, but in Bi -compounds it seems to be very small[19]. Also in $La_{2-x}Sr_xCuO_4$ it is almost zero near maximum T_c and varies in a complicated way [25] in terms of x. For small $\gamma(0)$ and hence for small electron concentration our model would predict normal BCS -behaviour because the sound mode phonons are missing and since the gap Δ is very large also T_c can be large. Another possible explanation for the complicated behaviour of $\gamma(0)$ would be the existence two topologically different, mobile bosons in the system the normal RVB -type boson in the CuO_2 - plane and another formed by holes in different CuO_2 -planes. In the superconducting state only one of them would be operative the other beeing dressed in the heavy fermion form as dicussed before. The system would have two Fermi components and one Bose component (FFB). The FBB -possibility is ruled out because the system of two Couloms bosons is known[26] to be unstable hence most propably also FBB. By $Eqs(4)$ the FFB system turns-out to have two separate sound modes and hence would naturally have two different hole concentrations n_1 and n_2 where the velocities vanish together with T_c. Such a system would have clearly much more freedom to explain the complicated behaviour of $\gamma(0)$ observed in $La_{2-x}Sr_xCuO_4$.

Finally the behaviour of specific heat near T_c would in our model be very complicated because of the corrections to the linear spectrum which is evident also in Fig.4. If the temperature point T_c (BCS) where all pairs are broken is near T_c one should observe a jump, otherwise not. The breaks observed in ref.27 could simply be connected to the non-linearities in the sound mode dispersion curve.

EBHL AND HEAVY FERMIONS

Since EBHL -model has the electron-hole symmetry the new high- T_c electron[28] superconductors would have their explanation within the same model. As mentioned in ref.2 the EBHL mechanism may also be operating in the heavy fermion (HF) systems.[29] First of all the specific heats for the heavy fermion superconductors UPt_3 and UBe_{13} have a power-law specific heat and most likely also a linear term.[30] It is possible that for the crystalline structure prevalent in HF -compounds the mobilities of electrons and holes are sufficiently "parallel" so that exiton complexes $(e^-e^-h^+)$ or $(h^+h^+e^-)$ are mobile unlike in high T_c compounds where they were assumed to be localized. The mechanism of the binding of (e^-e^-) - or (h^+h^+) -bosons could be similar in the two systems. It is well-known from the positron impurity calculations[31] that at very low electron density $r_s > 8$ a positron carries two electrons with it. In the complex $(h^+e^-e^-)$ the last electron is bound in free space by .024 Ry. The role of the crystal is to prevent the annihilation, because in a crystal the electrons and holes can reside in separate layers. Clearly the nature of constraint binding mechanism of $(h^+e^-e^-)$ on a crystal is very different in free space and also in the uniform electron gas since the separation of h^+ from the complex would require .52 Ry energy. The transition from heavy fermion exiton system to fermion-boson liquid then occurs in high-T_c superconductors through the mechanism of EBHL, as explained before.

ACKNOWLEDGMENTS

We would like to thank Dr. M. Saarela for help in the numerical calculation and Drs. P. Pietiläinen, M. Puoskari and L. Lantto for many useful discussions of the manuscript. Part of this work was supported by the Research Institute for Theoretical Physics, University of Helsinki and the Finnish Academy of Sciences.

REFERENCES

1. J.G.Bednorz and K.A.Müller, Z. Phys. B64, 189 (1986).

2. A. Kallio and X. Xiong, to be published in Phys. Rev. B, (1989).

3. A.S.Alexandrov,J.Ranninger and S.Robaszkiewicz, Phys. Rev. B**33**, 4526 (1986); B.K.Chakraverty , J. de Phys. Lett. **40**, 99 (1979).

4. P.W.Anderson, B.Baskaran, Z.Zou and T.Hsu, Phys. Rev. Lett. **58**, 2790 (1987).

5. S.K.Kivelson, D.S.Rokhsar and J.P.Sethna, Phys. Rev. B**35**, 8865 (1987).

6. J.M.Wheatley, J.C.Hsu and P.W.Anderson, Phys. Rev. B**37**, 5897 (1988).

7. M.J.Rice and Y.R.Wang, Phys. Rev. B**37**, 5893 (1988).

8. B.K.Chakraverty, D.Feinberg, Z.Hang and M.Avignon, Solid State Commun. **64**, 1147 (1987).

9. Y.Tokura, J.B.Torrance, T.C.Huang and A.I.Nazzal, Phys. Rev. B**38**,7156(1988).

10. J.M. Tranquada, S.M. Heald, A.R. Moodenbaugh and Youwen Xu, Phys. Rev. B**38**, 8893 (1988).

11. D.Y.Xing, M.Liu and C.S.Ting, Phys. Rev. B**38**, 11992 (1988).

12. M.W. Shafer, T. Penney, B.L. Olson, R.L. Greene and R.H. Koch, Phys. Rev. B**39**, 2914(1989).

13. H.Ye, W.Lu, Z.Yu, X.Shen, B.Miao, Y.Cai and Y.Qian, Phys. Rev. B**36**, 8802 (1987).

14. W.E.Pickett, Rev. Mod. Phys. **61**, 480 (1989).

15. W.F. Brinkman and T.M. Rice, Phys. Rev. B**7**, 1508 (1972); T. Chakraborty and P. Pietiläinen, Phys. Rev. Lett. **49** , 1034 (1982); T. Chakraborty, A. Kallio, L.J. Lantto and P. Pietiläinen, Phys. Rev. B**27**, 3061 (1983).

16. W.W.Warren, Jr., R.E.Walstedt, G.F.Brennert, R.J.Cava, R.Tyoko, R.F.Bell and G.Dabbagh, Phys. Rev. Lett.**62**,1193 (1989).

17. R.Friedberg and T.D.Lee, Phys. Rev. B**39**, 11482 (1989); Phys. Lett. A138, 423 (1989).

18. M.E.Reeves, S.E.Stupp, T.A.Friedmann, F.Slakey, D.M.Ginsberg and M.V.Klein, Phys. Rev. B**40**, 4573 (1989).

19. H.E. Fischer, S.K. Watson and D.G. Cahill, Comment condens. matter phys. **14**, 65 (1988); R.A. Fisher, S. Kim, S.E. Lacy, N.E. Phillips, D.E. Morris, A.G. Markelz, J. Wei and D.S. Ginley, Phys. Rev. B**38**, 11942 (1988); E.M. Forgan, C. Gibbs, C. Greaves, C.E. Gough, F. Wellhofer, S. Sutton and J.S. Abell, submitted to J.Phys. A.

20. J.G. Daunt, R.E. Probst, H.L. Johnston, L.T. Aldrich and A.O. Nier Phys. Rev. **72**, 502 (1947).

21. A. Khurana, Physics Today, Vol. **42** No. 7, 21 (1989); M.P.A. Fisher, P.B. Weichman, G. Grinstein and D.S. Fisher, Phys. Rev. B**40**, 546 (1989).

22. L. J. Lantto, Phys. Rev B**36**, 5160 (1987).

23. H. Takagi, T. Ido, S. Ishibashi, M. Uota, S. Uchida and Y. Tokura, Phys. Rev. B**40**, 2254 (1989).

24. A.Aharony, R.J.Birgeneau and M.A.Kastner, IBM J. Res. Develop. **33**, 287 (1989); A.Aharony, R.J.Birgeneau, A.Coniglio, M.A.Kastner and H.E.Stanley, Phys. Rev. Lett. **60**, 1330(1988).

25. N. Wada, H. Muro-Oka, Y. Nakamura and K. Kumagai, Physica C **157**, 453 (1989).

26. K.E. Kürten and C. E. Campbell, Phys. Rev. B**26**, 124 (1982)

27. F. Seidler, P. Böhm, H. Geus, W. Braunisch, E. Braun, W. Schnelle, Z. Drzazga, N. Wild, B. Roden, H. Schmidt and D. Wohlleben, Physica C **157**, 375 (1989).

28. H. Takagi, S. Uchida and Y.Tokura, Phys. Rev. Lett. **62**, 1197 (1989).

29. H.R. Ott and Z. Fisk, 1987, in: Handbook on the Physics and Chemistry of the Actinides, Vol. 5, eds A.J. Freeman and G.H. Lander (North-Holland, Amsterdam) p. 85.

30. See ref.29, p. 179.

31. J. Arponen and E. Pajanne, Ann. Phys. (N.Y.) **121**, 343 (1979).

STRONG ELECTRON CORRELATIONS IN CuO_2 PLANES OF HIGH TEMPERATURE SUPERCONDUCTORS

Andrzej M. Oleś[1] and Jan Zaanen[2]

[1] Institute of Physics, Jagellonian University
PL-30059 Kraków, Poland
[2] Max-Planck-Institut für Festkörperforschung
D-7000 Stuttgart 80, Federal Republic of Germany

ABSTRACT

The properties of correlated holes within a CuO_2 plane of a high temperature superconductor are investigated as a function of doping by holes. The correlated ground state of a four-band Hubbard model is found for realistic parameters by making use of ***local ansatz*** which emphasizes local correlations between holes. It is found these correlations are particularly strong within $Cu(3d_{x^2-y^2})$ orbitals. They lead to a drastic modification of the ground state, and result in only $\simeq 3\%$ population of d^8 configuration. The doped holes have $\simeq 70\%$ oxygen character, in agreement with experimental observations, as well as with the results of finite cluster diagonalization.

INTRODUCTION

Shortly after the discovery of high temperature superconducting oxides (HTSO), it became clear that the understanding of the electronic states of CuO_2 planes is a necessary step before searching for a nonphonon mechanism of pairing. The existence of quasi two-dimensional CuO_2 structures with partially filled antibonding bands is a universal feature of almost all Cu-based HTSO, as shown by the band structure calculations.[1,2] But the band structure description of these states certainly does not suffice since the interactions between holes are considerable and one has to face a many-body problem. This follows from the spectroscopic data,[3–5] as well as from the occurrence of an antiferromagnetic (AF) ground state at the hole filling close to one hole per CuO_2 unit. Our main motivation for this study comes from the fact that many mechanisms of pairing which exploit electronic degrees of freedom are strongly dependent on the charge distribution within a CuO_2 plane, when extra holes are added. Therefore, a detail understanding of the role played by different electron-electron interactions, as well as of the character of excess holes in the CuO_2 planes appears to be of fundamental importance.

In fact, the holes within the CuO_2 planes may be described by an extended Hubbard model[6,7] defined below. The Hubbard model is an outstanding many-body problem in solid state physics and so far only certain limiting cases or finite systems

Recent Progress in Many-Body Theories, Volume 2
Edited by Y. Avishai
Plenum Press, New York, 1990

described by this model could be solved exactly. A standard approach to the electron correlation problem here is the Gutzwiller wave function[8] for which ground state energy and various correlation functions may be treated analytically.[9] As a natural generalization of the Gutzwiller wave function in the present problem, we shall use a *local ansatz*[10] (LA) method which modifies the ground state found in Hartree-Fock (HF) approximation with a set of local operators which decribe correlations. In this case, however, an analytic treatment has to be replaced by certain approximations described below.

HAMILTONIAN AND CALCULATION METHOD

It is well accepted by now that a realistic Hamiltonian to describe the holes within a CuO_2 plane should include $Cu(3d_{x^2-y^2})$ and $O(2p_{x(y)})$ orbitals.[6,7] There is, however, an increasing evidence that the $Cu(3d_{3z^2-r^2})$ orbitals, labelled below as $3d_{z^2}$, also contain holes in HTSO,[11] and the local charge transfer excitations have been proposed as a possible mechanism of paring.[12] In fact, it has been measured that the population of $3d_{z^2}$ orbitals increases with the increasing value of T_c.[11] Therefore, we consider here a four-band model Hamiltonian which describes holes within $Cu(3d_{x^2-y^2})$, $Cu(3d_{z^2})$ and $O(2p_{x(y)})$ orbitals in an idealized two-dimensional CuO_2 structure,

$$H = H_0 + H_{int}, \tag{1}$$

where

$$H_0 = \varepsilon_d^0 \sum_{m\sigma} (n_{mx\sigma} + n_{mz\sigma}) + \varepsilon_p^0 \sum_{i\sigma} n_{ip\sigma} + t_{pp} \sum_{ij\sigma} (-1)^{\pi_{ij}} a_{i\sigma}^+ a_{j\sigma}$$

$$+V_x^0 \sum_{<mi>\sigma} [(-1)^{\pi_{im}} d_{mx\sigma}^+ a_{i\sigma} + h.c.] + V_z^0 \sum_{<mi>\sigma} [(-1)^{\pi_{im}} d_{mz\sigma}^+ a_{i\sigma} + h.c.], \tag{2}$$

$$H_{int} = U_d \sum_{m,\alpha=x,z} n_{m\alpha\uparrow} n_{m\alpha\downarrow} + U_p \sum_i n_{ip\uparrow} n_{ip\downarrow} + U_{xz} \sum_{m\sigma} n_{mx\sigma} n_{mz,-\sigma}$$

$$+(U_{xz} - J) \sum_{m\sigma} n_{mx\sigma} n_{mz\sigma} + U_{dp} \sum_{<mi>,\sigma\sigma'} (n_{mx\sigma} + n_{mz\sigma}) n_{pi\sigma'}, \tag{3}$$

describe the one-particle part and hole-hole interactions, respectively. The vacuum in Eq. (1) is defined as filled $Cu(3d^{10})$ and $O(2p^6)$ states. Our reference state of an *undoped* superconductor, La_2CuO_4 or $YBa_2Cu_3O_6$, contains one hole per unit cell. $d_{mx\sigma}^+$, $d_{mz\sigma}^+$, and $a_{i\sigma}^+$ are creation operators of a hole of spin σ within a $Cu(3d_{x^2-y^2})$, $Cu(3d_{z^2})$, and $O(2p)$ orbital, respectively. The density operators are $n_{mx\sigma}$, $n_{dz\sigma}$ and $n_{pi\sigma}$, the parameters ε_d^0 and ε_p^0 stand for the atomic energies of a hole in $Cu(3d)$ and $O(2p)$ orbital, respectively, and define the charge transfer energy, $\Delta = \varepsilon_p^0 - \varepsilon_d^0$. The atomic energies for a $3d_{x^2-y^2}$ and $3d_{z^2}$ hole are assumed to be equal. The hybridization elements, V_x^0 and V_z^0, describe the kinetic energy and fulfill the Slater-Koster relation, $V_z^0 = V_x^0/\sqrt{3}$.[13] The summations over $< mi >$ include nearest-neighbour Cu and O sites, and $(-1)^{\pi_{im}}$ and $(-1)^{\pi_{ij}}$ are the respective phase factors. On-site Coulomb interaction elements are defined as U_d, U_p, and U_{xz}, for the intraorbital and interorbital interactions, respectively. The Cu elements are related by an exchange element, J, in order to satisfy the rotational invariance of the interaction part in orbital space, $U_d = U_{xz} + 2J$.[14] Finally, the element U_{dp} stands for the intersite Coulomb interaction.

The HF ground state follows in the nonmagnetic case from the HF Hamiltonian,

$$H_{HF} = \varepsilon_x \sum_{m\sigma} n_{mx\sigma} + \varepsilon_z \sum_{m\sigma} n_{mz\sigma} + \varepsilon_p \sum_{i\sigma} n_{ip\sigma} + t_{pp} \sum_{ij\sigma} (-1)^{\pi_{ij}} a^{+}_{i\sigma} a_{j\sigma}$$

$$+ V_x \sum_{<mi>\sigma} [(-1)^{\pi_{im}} d^{+}_{mx\sigma} a_{i\sigma} + h.c.] + V_z \sum_{<mi>\sigma} [(-1)^{\pi_{im}} d^{+}_{mz\sigma} a_{i\sigma} + h.c.], \quad (4)$$

where

$$\varepsilon_\alpha = \varepsilon^0_d + \frac{1}{2} U_d < n_\alpha >_0 + (U_{xz} - \frac{J}{2}) < n_\beta >_0 + 4U_{dp} < n_p >_0, \quad (5a)$$

$$\varepsilon_p = \varepsilon^0_p + \frac{1}{2} U_p < n_p >_0 + 2U_{dp}(< n_x >_0 + < n_z >_0), \quad (5b)$$

$$V_\alpha = V_0 - U_{dp} < d^{+}_{m\alpha\sigma} a_{i\sigma} >, \quad (5c)$$

$\alpha = x, z$, and $\beta \neq \alpha$. All averages $< ... >_0$ in Eqs. (5a-5c) and below are calculated by using a uniform HF ground state, $|\phi_0>$, which is easily found after the diagonalization of H_{HF} in the reciprocal space. If $V_z = 0$ and $t_{pp} = 0$, the four bands reproduce the bonding, antibonding and nonbonding bands of a three-band model[15] and give a dispersionless $3d_{z^2}$ level. This qualitative picture remains valid in a general case as well, but all bands have dispersion. The bonding band is half-filled in the undoped case ($n = 1$), while doping results in the increasing hole number in this band, i.e., for $La_{2-x}Sr_xCuO_4$ or $YBa_2Cu_3O_{6+x}$ with $x > 0$.

Local interactions present in the Hamiltonian (1-3) lead to *local correlations* between holes. The correlated ground state, $|\psi_0>$, can be calculated by using a method of LA. This method was introduced by Stollhoff and Fulde[10] for molecules and recently successfully applied to the description of the ground state of undoped CuO_2 systems, by using both a model tight-binding Hamiltonian,[15] and a semiempirical method.[16] The correlated ground state, $|\psi_0>$, has in the LA the following form,

$$|\psi_0> = exp(-\sum_n \eta_n O_n) \, |\phi_0> . \quad (6)$$

The variational parameters η_n are determined by the minimization of the ground state energy,

$$E_0 = \frac{<\psi_0|H|\psi_0>}{<\psi_0|\psi_0>} = E_{HF} + E_{corr}, \quad (7)$$

where $E_{HF} =< \phi_0|H_{HF}|\phi_0 >$, and E_{corr} stands for the correlation energy. The operators O_n describe local correlations within the CuO_2 plane,[15]

$$O^{(1)}_{m\alpha} = n_{m\alpha\uparrow} n_{m\alpha\downarrow}, \qquad O^{(1)}_{i} = n_{ip\uparrow} n_{ip\downarrow}, \quad (8a)$$

$$O^{(2)}_{xz} = \sum_{\sigma\sigma'} n_{mx\sigma} n_{mz\sigma'}, \qquad O^{(2)}_{m\alpha,i} = \sum_{\sigma\sigma'} n_{m\alpha\sigma} n_{ip\sigma'}, \quad (8b)$$

$$O^{(3)}_{xz} = \vec{S}_{mx} \vec{S}_{mz}. \quad (8c)$$

They change the probability of finding: (*i*) two holes of opposite spins at the same orbital, $O^{(1)}$, (*ii*) two holes either at different orbitals of one Cu site, or at nearest neighbour Cu and O sites, $O^{(2)}$, as well as (*iii*) describe spin-spin correlations at

Cu sites, $O^{(3)}$. In addition to these operators, we introduced as well one-particle operators,

$$O_{m\alpha}^{(0)} = \sum_{\sigma} n_{m\alpha\sigma}, \qquad O_{i}^{(0)} = \sum_{\sigma} n_{ip\sigma}. \tag{8d}$$

They reoptimize the densities of $3d_{x^2-y^2}$, $3d_{z^2}$, and $2p$ holes, respectively, in the presence of correlations. All these operators include only those residual contributions which go beyond the HF approximation, as explained elsewhere.[17,18]

In order to calculate the correlation energy, we perform a local cluster expansion of Eq. (7), i.e. keep in the expansion only these terms in which the correlation operators, O_n, either refer to the orbitals within a given unit cell, or describe the couplings between the orbitals of that cell with its surrounding. Such an expansion results in

$$E_{corr} = \frac{-2\sum_n \eta_n < O_n\tilde{H}_{int} >_0 + \sum_{nn'} \eta_n\eta_{n'} < O_n(\tilde{H}_{HF} + \tilde{H}_{int})O_{n'} >_0}{1 + \sum_{nn'} \eta_n\eta_{n'} < O_nO_{n'} >_0}, \tag{9}$$

where wavy lines above the operators H_{HF} and H_{int} mean that no contractions within them are allowed when the averages are calculated. A second approximation is made to calculate the averages $< O_n\tilde{H}_{int} >_0$, $< O_n\tilde{H}O_{n'} >_0$, and $< O_nO_{n'} >_0$ which appear in Eq. (9). A satisfactory value of the correlation energy is usually obtained by employing the so-called $\vec{R} = 0$-approximation[19] in which only the leading local terms are preserved.[18] It becomes exact for dimension $d = \infty$ and is still a very good one for $d = 2$, if the system is described with one correlated orbital per site.[20] We have applied this approximation here to calculate E_{corr}; more details may be found in Refs. 15 and 18.

The values of the variational parameters η_n were found by solving selfconsistently the following system of equations,

$$\sum_{n'} (< O_n(\tilde{H}_{HF} + \tilde{H}_{int})O_{n'} >_0 - E_{corr} < O_nO_{n'} >_0)\eta_{n'} = < O_n\tilde{H}_{int} >_0 , \tag{10}$$

obtained by minimizing E_{corr}. When parameters η_n are known, one may calculate new hole densities within particular orbitals. For example, for $3d$ orbitals one has

$$< n_\alpha > = < n_\alpha >_0 + \frac{-2\sum_n \eta_n < O_n\tilde{n}_{m\alpha} >_0 + \sum_{nn'} \eta_n\eta_{n'} < O_n\tilde{n}_{m\alpha}O_{n'} >_0}{1 + \sum_{nn'} \eta_n\eta_{n'} < O_nO_{n'} >_0}, \tag{11}$$

where $\alpha = x, z$, and wavy lines above the density operator, $n_{m\alpha}$, have the same meaning as before. Analogous formula applies to $< n_p >$. The $\vec{R} = 0$ approximation does not suffice in the calculation of new densities (11) since it leads to serious deviations ($\sim 10\%$) of the total hole density in the correlated states, $|\psi_0 >$, due to the presence of one-particle operators. Therefore, we used here an improved scheme, in which also the squares of the averages $< d^{+}_{mx\sigma}a_{i\sigma} >_0$ and $< d^{+}_{mz\sigma}a_{i\sigma} >_0$ were preserved. This scheme is consistent with the performed cluster expansion and reduces the discrepancy in the total hole density to only $\sim 1\%$. The partial densities in $|\psi_0 >$ were renormalized to account for this inaccuracy.

Hole correlations within a CuO_2 plane may be quantified with the help of a parameter which expresses the reduction of charge fluctuations within an orbital of given symmetry, e.q., for $3d_{x^2-y^2}$ orbital, $< \Delta n_{mx}^2 >=< n_{mx}^2 > - < n_{mx} >^2$, with respect to the corresponding state of independent holes. The averages $< ... >$ are calculated as in Eq. (11). Such a parameter has been first introduced to quantify electron correlations within chemical bonds.[20] For example, for $3d$ orbitals one defines,

$$\Sigma_\alpha = \frac{2- < n_{m\alpha} >}{< n_{m\alpha} >} \times \frac{< \Delta n_{m\alpha}^2 >_{HF} - < \Delta n_{m\alpha}^2 >}{< \Delta n_{m\alpha}^2 >_{HF}}. \tag{12}$$

The quantity $< \Delta n_{m\alpha} >_{HF}$ is calculated for the HF state characterized by ***the same density distribution*** as in $|\psi_0>$. The factor $(2- < n_{m\alpha} >)/ < n_{m\alpha} >$ is the inverse of the maximal reduction of charge fluctuation and thus normalizes the quantity Σ_α to $0 < \Sigma_\alpha < 1$. Knowing the correlated state, $|\psi_0>$, the actual calculations of Σ_x, Σ_z, and Σ_p were performed by following the same scheme as for the correlation energy and new hole densities.

Finally, we determined also the populations of different configurations in the ground state, $P(d^n)$ and $P(p^m)$, for Cu and O atoms, respectively. These populations follow in the HF case from the binomial distribution valid for each orbital. In the correlated state, the probabilities of having one or two holes at one orbital of either a Cu or an O ion were found in a similar way to that described in Ref. 15. In addition, a correction due to the configurations $\{x\sigma, z\sigma\}$ and $\{x\sigma, z-\sigma\}$, was included in the Cu case. The global populations $P(d^n)$ were determined for $|\psi_0>$ in the restricted space, i.e. assuming that Cu sites are occupied by not more than two holes.

NUMERICAL RESULTS AND DISCUSSION

We have employed the parameters of Hamiltonian (1) as obtained from the local density approximation (LDA),[22-24] i.e., $U_d = 10.5eV$, $U_p = 4eV$, $U_{dp} = 1.2eV$, $V = 1.3eV$, and $t_{pp} = 0.65eV$. This set has been chosen after Hybertsen, Schlüter and Christensen,[22] but similar parameters have also been obtained by McMahan *et al.* in an Anderson impurity approximation,[23,24] as well as by Mila by analyzing the experimental data available from spectroscopies.[25] It is believed that such parameters are realistic, especially that they allow to understand many spectroscopic and magnetic measurements on La_2CuO_4.[22] The exchange integral J was taken to be $J = 1eV$, as suggested by the spectroscopic data.[26]

The ground state has been analyzed for different hole filling per CuO_2 unit cell, n, as presented in Fig. 1. At $n = 1$ one finds the densities of 0.49, 0.01, 0.25 in $|\phi_0>$ and 0.60, 0.02, 0.19 in $|\psi_0>$, for $3d_{x^2-y^2}$, $3d_{z^2}$, and $2p_{x(y)}$ orbitals, respectively. These numbers show strong covalency of the hole states. For this reason the band structure calculations have conventionally been interpreted as indicating a small value of Δ.[1,2] However, one should realize that the large covalency found here ***means that*** the quantity Δ_{HF}, which determines the band structure, (***and not*** Δ) is small.[15,27] Indeed, in our calculations we have $\Delta = 3.6eV$, i.e. $\Delta >> V_0$, and $\Delta_{HF} = 1.3eV$.[22] Having a positive value of $\Delta_{HF} = V_0$, one could expect that the $Cu(3d)$ orbitals absorb more than 0.5 hole. However, this simple argument is valid only under the neglect of direct interoxygen hopping, t_{pp}. Indeed, taking $t_{pp} = 0$ and keeping the values of the remaining parameters as determined from LDA,[22] we find the hole densities of 0.58, 0.02, and 0.20 for $3d_{x^2-y^2}$, $3d_{z^2}$, and $2p_{x(y)}$ orbitals in $\phi_0>$ state. The respective density distribution in the correlated state, $|\psi_0>$, 0.69, 0.03, and 0.14, is quite different from the one given above for $t_{pp} = 0.65eV$. Thus, we conclude that the parameter t_{pp} is important and contributes significantly to more covalency in the ground state. Therefore, it has a destabilizing effect on the Zhang-Rice construction

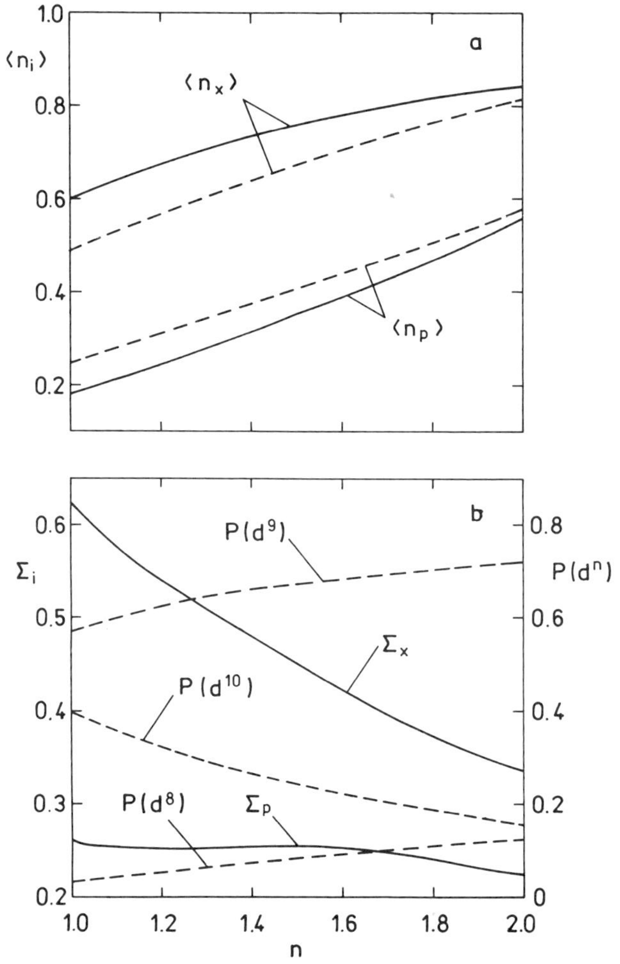

Fig. 1. (a) Hole densities $< n_i >$ for $3d_{x^2-y^2}$ and $2p_{x(y)}$ orbitals in $|\phi_0>$ and $|\psi_0>$, shown by dashed and solid lines, respectively; (b) Correlation strength parameters, Σ_x and Σ_p, for $3d_{x^2-y^2}$ and $2p_{x(y)}$ orbitals (solid line) and probabilities of different configurations at Cu ion (dashed lines), $P(d^n)$, as functions of total hole density, n.

of the $t - J$ model for HTSO.[28] Nevertheless, there are indications coming from finite system diagonalizations[29] that this construction may still be valid for the LDA parameters.

Independently of the assumed value of t_{pp}, one observes a hole transfer of $\simeq 0.12$ from $2p$ to $3d$ orbitals due to hole correlations. This is a consequence of a quite general phenomenon. Holes (or electrons) change their densities in such a way that they increase for stronger correlated orbital, if the correlations are treated explicitly. These densities are underestimated in the HF state since the only way to minimize the interaction energy there is by reducing the respective density. This phenomenon is well known in covalent bonds, where electron correlations always *reduce* the bond polarities found in selfconsistent field (HF) states. The situation is more spectacular if the average particle density per orbital is less than one. An example of a correlated system with similar filling is a π bond in O_3 molecule which consists of three orbitals filled by two holes. The hole density increases there on two external orbitals, characterized by stronger correlations (less kinetic energy), as a consequence of hole correlations.[30] In our case of CuO_2 plane, we find an increase of total $3d$ density, $< n_d >=< n_x > + < n_z >$, from 0.50 to 0.62 due to hole correlations. This change of $< n_d >$ agrees well with the respective one found for a Cu_2O_7 cluster filled by two holes, where for similar parameters, except for a somewhat smaller value of $\Delta = 1.5eV$, one finds $< n_d >= 0.52$, and 0.67, in the HF and correlated state, respectively.[31]

Surprisingly, the $3d_{z^2}$ orbitals are far less occupied than it could be expected by looking at the ratio of the respective hybridization elements, V_x^0 and V_z^0. Of course, higher values of $< n_z >$, indicated by experimental data,[11] could be obtained by assuming $\varepsilon_x^0 > \varepsilon_z^0$. It is, however, unclear why such a relation should be fulfilled for the actual octahedral coordination of Cu atoms in CuO_2 planes.

As stated above, the nature of the added holes is of decisive importance for the mechanism of superconductivity in HTSO. We find that if LDA parameters are adopted, the hole density within the $Cu(3d)$ orbitals increases only slowly with n. This increase is faster for $1 < n < 1.25$, where the doped holes have 35% $3d_{x^2-y^2}$, 5% $3d_{z^2}$, and 60% $O(2p)$ character, than for higher values of n. In the region of $1.2 < n < 1.4$, characteristic for HTSO with high values of T_c, we find only 28% and 5% of $3d_{x^2-y^2}$ and $3d_{z^2}$, character, respectively, of the doped holes, and as much as 67% of the doped holes occupy $O(2p)$ orbitals. Even more $O(2p)$ character of the doped holes is found for higher doping.

Although the LDA parameters adopted here seem to be of good quality at present, it is important to realize which of the used parameters have more influence on the hole distribution, as presented in Fig. 1a. First of all, the starting state, $|\phi_0>$, and, consequently, the correlated state, $|\psi_0>$, depend on the values of V_0 and t_{pp}. We did not vary V_0, however, since this parameter has been rather consistently determined by different calculations to be $V_0 \simeq 1.3eV$.[22–25] If the value of $t_{pp} = 0$ is taken, one finds that the doped holes are of 22% $3d_{x^2-y^2}$, 4% $3d_{z^2}$, and 74% $O(2p)$ character, respectively. These numbers reflect a more localized character of the holes on Cu sites in this case. They agree quite well with the trend observed in finite system diagonalization, where a hole added to a 2×2 CuO_2 cluster has 75% oxygen character for similar parameters ($U_d = 8eV$, $U_p = 4eV$, $U_{dp} = 0$, $\Delta = 2.6eV$, $t_{pp} = 0$.[32]

The possible inaccuracies in the values of on-site Coulomb elements have little influence on the final result. It has been already suggested that the actual density distribution is only weakly dependent on the actual value of U_p.[33] The dependence on U_d is also expected to be weak in the realistic regime of $8eV < U_d < 12eV$, where $U_d >> V$ and hole correlations are strong. Our results reproduce here the trends observed in finite cluster diagonalization.[33] The dependence on the value of U_{dp} is much stronger and one observes a crossover to the state with holes occupying mainly $O(2p)$d orbitals and flowing away from Cu sites with doping, if $U_{dp} \simeq 4eV$. This trend holds both for the LDA parameters and for smaller values of Δ, adopted by Hirsch,[34] as discussed in more detail elsewhere.[35] However, such a large value of U_{dp} is unrealistic, as shown recently by a detailed analysis of Auger spectroscopy,[36] in agreement with earlier estimates from photoemission.[4] This questions the relevance of pairing mechanisms based on the charge redistributions due to large U_{dp}, although pairing tendencies are observed also for smaller values of this parameter.[33,34,37]

The parameters quantifying hole correlations, Σ_x and Σ_p, are displayed in Fig. 1b. The correlations are strong at $3d_{x^2-y^2}$ orbital for $n = 1$, with $\Sigma_x \simeq 0.62$. They are stronger than those between π-electrons in double bonds ($\simeq 0.55$),[21] and between $3d$ electrons in transition metals ($\simeq 0.5$).[18] On the contrary, the holes at $O(2p)$ orbitals are only moderately correlated, with $\Sigma_p \simeq 0.26$ at $n = 1$. Even weaker are the correlations within the $3d_{z^2}$ orbitals, with $\Sigma_z \simeq 0.20$. The correlations within $3d_{x^2-y^2}$ orbitals, Σ_x, are found to decrease significantly under doping (see Fig. 1b), being 0.45, 0.34 for $n = 1.5$, 2, respectively. It may be argued that the physical reason for this decrease is an increased kinetic energy for larger hole fillings, but this phenomenon is not quite understood at present. On the other hand, the correlations within $O(2p)$ orbitals remain almost constant for $1 < n < 1.5$, and decrease to $\Sigma_p = 0.22$ at $n = 2$, while the ones in $3d_{z^2}$ orbitals (not shown in Fig. 1b) first increase somewhat to $\Sigma_z = 0.23$ at $n = 1.5$ and later decrease to $\Sigma_z = 0.21$ at $n = 2$.

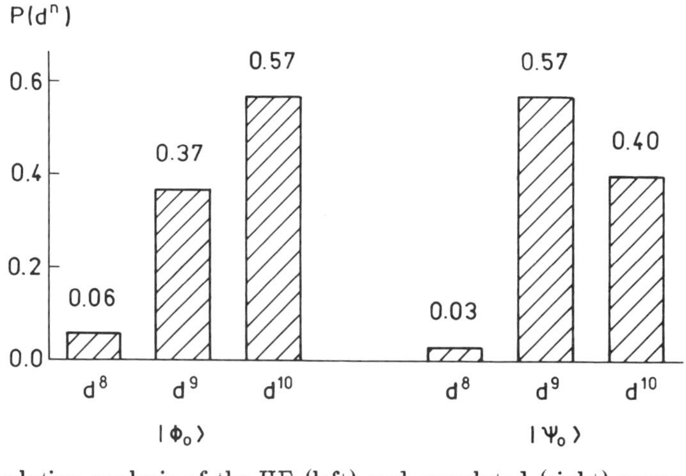

Fig. 2. Population analysis of the HF (left) and correlated (right) ground states for the LDA parameters at the filling of $n = 1$ per unit cell of CuO_2 plane.

Fig. 2 shows the result of the population analysis of $|\phi_0>$ and $|\psi_0>$, respectively, at $n = 1$. One observes a significant shift of the weight towards d^9 configuration, caused by hole correlations. The configuration d^8 has only $\simeq 3\%$ weight, in spite of increased hole density at Cu sites. We note that the weight of this configuration should not exceed 4%, as determined experimentally.[38] The changes of the respective weights with doping are shown in Fig. 1b. An increase of $P(d^9)$ corresponds to an increased probability of having a local moment at Cu sites (Cu^{2+}). Therefore, magnetic couplings should play an important role in the doped systems. At the same time, the weight of d^{10}, i.e. Cu^+, decreases. The same trend was observed by Stechel and Jennison[31] in finite cluster diagonalizations performed for a similar parameter set to the one used here.

In conclusion, we have shown that the charge distribution in the CuO_2 plane is characterized by a large covalency. At the same time, the holes within $3d_{x^2-y^2}$ are particularly strongly correlated. Thus, these new results obtained for realistic values of parameters support the earlier point of view that the HTSO are strongly correlated mixed valence systems.[15,16] In agreement with experimental observations,[39] one finds that for realistic values of $1eV < U_{dp} < 2eV$, the added holes are predominantly of oxygen character. Even more oxygen character of these holes is expected in the states which describe properly the AF correlations, either of long-range (at $n \simeq 1$), or of short-range (at $n > 1$) type. It is therefore expected that the binding mechanism due to charge redistributions induced by U_{dp} [33-34] is less likely. Nevertheless, the intersite Coulomb interaction may contribute to charge-transfer fluctuations and in this way also to the pairing.[34]

Altogether, we presented here a simple method to study the charge distribution with a tight-binding model. It reproduces the results obtained from finite system diagonalizations for realistic parameters describing CuO_2 planes in HTSO. This suggests that the latter method is quite powerful in studying model Hamiltonians, in spite of a relatively small size of the systems which could be considered so far.

ACKNOWLEDGMENTS

We thank P. Horsch, W. Stephan, G.Stollhoff, J. Dutka, P. Fulde and T.M. Rice for valuable discussions. The financial support of the Polish Research Project CPBP 01.09. is acknowledged.

REFERENCES

1. K.C. Haas, *Electronic Structure of Copper-Oxide Superconductors*, in: *Solid State Physics*, Vol. 42, H. Ehrenreich and D. Turnbull, ed., Academic Press, London, (1989).
2. W.E. Pickett, Rev. Mod. Phys. **61**:433 (1989).
3. A. Fujimori, E. Takayama-Muromachi, Y. Uchida, and B. Okai, Phys. Rev. B **35**:8814 (1987).
4. Z. Shen, J.W. Allen, J.J. Yeh, J.-S. Kang, W. Ellis, W. Spicer, I. Landau, M.B. Maple, Y.D. Dalichaouch, M.S. Torikachvili, J.Z. Sun, and T.H. Geballe, Phys. Rev. B **36**:2794 (1987).
5. H. Eskes, H. Tjeng, and G.A. Sawatzky, in: Proceedings NEC Symposium on *Mechanisms of High-T_c Superconductivity*, to be published.
6. C.M. Varma, S. Schmitt-Rink, and E. Abrahams, Solid State Commun. **62**:681 (1987).

7. V.J. Emery, Phys. Rev. Lett. **58**:2794 (1987).
8. M.C. Gutzwiller Phys. Rev. A **137**:1736 (1965).
9. W. Metzner and D. Vollhardt, Phys. Rev. Lett. **59**:121 (1987); *ibid.*, Phys. Rev. B **37**:7382 (1988).
10. G. Stollhoff and P. Fulde, Z. Phys. B **26**:257 (1977); *ibid.*, **29**:231 (1978).
11. A. Bianconi, P. Castrucci, A. Fabrizi, M. Pompa, A.M. Flank, P. Lagarde, H. Katayama-Yoshida, and G. Calestani, Physica C, to be published.
12. W. Weber, Z. Phys. B **70**:323 (1987); M. Jarrell, H.R. Krishnamurthy, and D.L. Cox, Phys. Rev. B **38**:4584 (1988); D.L. Cox, M. Jarrell, C. Jayaprakash, and H.R. Krishnamurthy, Phys. Rev. Lett. **62**:2188 (1989).
13. J.C. Slater and G.F. Koster, Phys. Rev. **94**:1498 (1954).
14. A.M. Oleś, Phys. Rev. B **28**:327 (1983).
15. A.M. Oleś, J. Zaanen, and P. Fulde, Physica B **148**:260 (1987).
16. G. Bubeck, A.M. Oleś, and M.C. Böhm, Z. Phys. B **26**:257 (1989); M.C. Böhm, G. Bubeck, and A.M. Oleś, Chem. Phys. **135**:27 (1989).
17. G. Stollhoff and P. Fulde, J. Chem. Phys. **73**:4548 (1980).
18. A.M. Oleś and G. Stollhoff, Phys. Rev. B **29**:314 (1984).
19. F. Kajzar and J. Friedel, J. Phys. (Paris) **39**:397 (1978); G. Tréglia, F. Ducastelle, and D. Spanjaard, J. Phys. (Paris) **41**:281 (1980).
20. W. Metzner and D. Vollhardt, Phys. Rev. Lett. **62**:324 (1989).
21. A.M. Oleś, F. Pfirsch, P. Fulde and M.C. Böhm, Z. Phys. B **66**:359 (1987); A.M. Oleś, P. Fulde and M.C. Böhm, Chem. Phys. **117**:385 (1987).
22. M.S. Hybertsen, M. Schlüter, and N.E. Christensen, Phys. Rev. B **39**:9028 (1989).
23. A.K. McMahan, R.M. Martin, and S. Satpathy, Phys. Rev. B **38**:6650 (1988).
24. J.F. Annett, R.M. Martin, A.K. McMahan, and S. Satpathy, Phys. Rev. B **40**:2620 (1989).
25. F. Mila, Phys. Rev. B **38**:11358 (1988).
26. D. van der Marel and G.A. Sawatzky, Phys. Rev. B **37**:10674 (1988).
27. A.M. Oleś and J. Zaanen, Int. J. Mod. Phys. B **2**:751 (1988).
28. F.C. Zhang and T.M. Rice, Phys. Rev. B **37**:3759 (1988).
29. A. Ramsak and P. Prelovsek, Phys. Rev. B **40**:2239 (1989); M.S. Hybertsen, E.B. Stechel, M. Schluter and D.R. Jennison, to be published.
30. A.M. Oleś, P. Fulde and M.C. Böhm, Z. Phys. B **76**:239 (1989).
31. E.B. Stechel and D.R. Jennison, Phys.Rev. B **38**:8873 (1988).
32. W. Stephan, private communication.
33. W.H. Stephan, W. von der Linden, and P. Horsch, Phys. Rev. B **39**:2924 (1989).
34. J.E. Hirsch, E. Loh, Jr., D.J. Scalapino, and S. Tang, Phys. Rev. B **39**:243 (1989).
35. J. Dutka and A.M. Oleś, to be published.
36. L.H. Tjeng, J. van Elp, P. Kuiper, and G.A. Sawatzky, to be published.
37. C.A. Balseiro, A.G. Rojo, E.R. Gagliano, and B. Alascio, Phys. Rev. B **38**:9315 (1989).
38. H. Eskes and G.A. Sawatzky, Phys. Rev. Lett. **62**:324 (1989).
39. J.C. Fuggle, J. Fink, and N. Nücker, Int. J. Mod. Phys. B **2**:1185 (1988); H. Rietschel, J. Fink, E. Gering, F. Gompf, N. Nücker, L. Pintschovius, B. Renker, W. Reichardt, H. Schmidt, and W. Weber, Physica C **153-155**:1067 (1988).

TWO-COMPONENT SUPERCONDUCTIVITY

Y. Bar-Yam

Materials Research Department
Weizmann Institute for Science
Rehovot 76100 Israel

There are two essential attributes of electrons for superconductivity: mobility and pairing. While this is not directly obvious, these two attributes compete against each other. High-temperature superconductivity may be obtained by combining two-components, one which provides pairing and the other mobility. In the local-strong-pairing (negative-U lattice) limit of superconductivity T_c is controlled by the hopping of electron pairs rather than by the pair binding energy. Coupling a negative-U lattice to delocalized electron states increases the hopping and thus the critical temperature. In parallel, Cooper-pair superconductivity is induced in the delocalized electrons. In the normal state both Bosonic and Fermionic states exist, and below T_c Bosonic states exist in the Fermionic gap. It is suggested that superconductivity in the new class of oxide superconductors is due to locally paired electrons on the lattice of oxygen vacancies combined with conducting metal-oxide layers. The discussion includes the superconducting properties T_c, Δ, H_c and ξ, long wave collective excitations, normal state properties including resistance and tunneling, and the isotope shift. Unusual properties are predicted including neutral Fermion excitations due to hybridization of electrons and holes, a spreading of the Femionic gap onset, a separation between the resistive transition T_c' and the evaporation of the condensate T_c, anomalies in sound and bulk modulii at T_c, linear temperature dependence of normal state resistivity, linear voltage dependence in normal state tunneling conductance, and finite zero bias conductance in superconducting state tunneling. A new signature of structural coherence which can be seen in channeling experiments and other structural probes is indicated.

I. Introduction and Motivation

Theories of superconductivity in metal-oxide superconductors have concentrated on the distinctive metal-oxide two dimensional layers in these materials. The intervening layers are treated as electrically inactive semiconductors - contributing only as an electron reservoir affected by the dopant and oxygen vacancy concentration.

The largest electron-electron effective attractions (negative correlation energies) are associated with defects in semiconductors and insulators.[1] This may be understood qualitatively since the change in occupation of a localized orbital may have a significant local effect on the structure dramatically lowering the orbital energy and thus of a second electron occupying the same orbital. Vacancy lattices in high T_c superconductors are natural locations for locally negative-correlation-energy sites.[2] Even when located in the Cu-O planes, their orbitals are localized by the semiconductor layers between the planes. For example, in $Y_1Ba_2Cu_3O_7$ one of the three Cu-O planes is incomplete, half of the oxygen atoms are

Recent Progress in Many-Body Theories, Volume 2
Edited by Y. Avishai
Plenum Press, New York, 1990

missing thus forming Cu-O chains. While it may seem natural to think about the Cu-O chains, it proves more natural to associate one localized electronic state per unit cell with the missing oxygen sites. For a negative-correlation energy, as the Fermi energy is varied these sites should have two electron transitions; changing from 2 to 0 the occupation of orbitals associated with the vacancies. The transition energy of isolated or chains of vacancies has been shown by calculations to be near the Fermi energy in these materials.[3] An ordered lattice of such defects is the extreme limit of conventional strong-coupling superconductivity - a negative-U lattice.[4]

Generally, superconductivity is discussed in two limits - weak and strong coupling. Weak coupling (Cooper pair) superconductivity is limited to low temperatures by the thermal excitation of broken pairs. For stronger coupling the corrections can be described by Eliashberg theory. However, when the electron-electron attractions are large compared to their intrinsic single particle hopping, the electrons pair and behave as a *hard core* superfluid whose Bose condensation temperature increases with the dispersion of the Bose fluid.[3] In the usual picture, the dispersion of the Boson band results from pair hopping induced by single particle hops by de-pairing and re-pairing. This limits the hopping and thus T_c because hopping is small in localized negative-U systems and pair hopping is inversely proportional to the strength of the (large) electron-electron interaction.

The extreme strong coupling superconductivity on a lattice has been studied[5] in the form:

$$H = t_b \sum_{i,\delta,\sigma} b^\dagger_{i+\delta,\sigma} b_{i,\sigma} + U \sum_i b^\dagger_{i,\uparrow} b^\dagger_{i,\downarrow} b_{i,\uparrow} b_{i,\downarrow} \tag{1}$$

Where $b^\dagger_i$ creates a particle at site i. For a large enough negative U, treating single particle excitations in second order perturbation theory leads to the effective *hard-core* Boson Hamiltonian with nearest neighbor repulsion:

$$H = -t'_b \sum_{i,\delta} (B^\dagger_{i+\delta} B_i - B^\dagger_{i+\delta} B_{i+\delta} B^\dagger_i B_i) - z\, t_b' N \tag{2}$$

$$B_i = b_{i,\uparrow} b_{i,\downarrow}; \qquad B_i^2 = 0; \qquad t'_b = t_b^2/U$$

B_i are the effective hard core Boson operators.

The critical temperature of Bose condensation (Fig. 1) is determined by t'_b and the Fermi energy or the density of particles $n_B=N/N_0$ where N_0 is the number of sites. T_c scales with the half-width of the Boson band E_0 which for this model is t'_b: $T_c=t'_b\lambda(n_B)$. The two particle hopping term t'_b arises from the single particle hopping matrix element which can be expected to be small in localized negative-U systems and is further diminished by increasing U. An additional small term may also arise from direct pair matrix elements. While the T_c of this model has been suggested to be small, in principle, there has not been developed a formal limit on the size of T_c or the size of pair hopping which can still be small on the scale of typical matrix elements and give rise to a large T_c. However, since the long range Coulomb interaction is not included in standard treatments and it controls the nature of long wavelength excitations in this model, which as q->0 is the condensate, it is not clear that such a model is actually superconducting.

In contrast to the usual mechanism for hopping, a two-component Hamiltonian combines a locally paired lattice with a free electron gas.[6] For a partially filled paired lattice, transitions to and from the free electron band cost no energy and induce pair hopping and broaden the two electron states. A correlation between single-particle electron states is induced in the free-electron states opening an excitation gap at the Fermi energy in, essentially, the usual BCS way. This has been discussed for dilute negative-U impurities in a metal for semiconductor-metal alloys.[7] The induced broadening of the paired states results in superfluid Bose condensation at relatively high temperature and, the free-electron component is a Cooper-pair superconductor with a large externally imposed gap. The highest T_c arises for half-filling of the paired or Bosonic states.

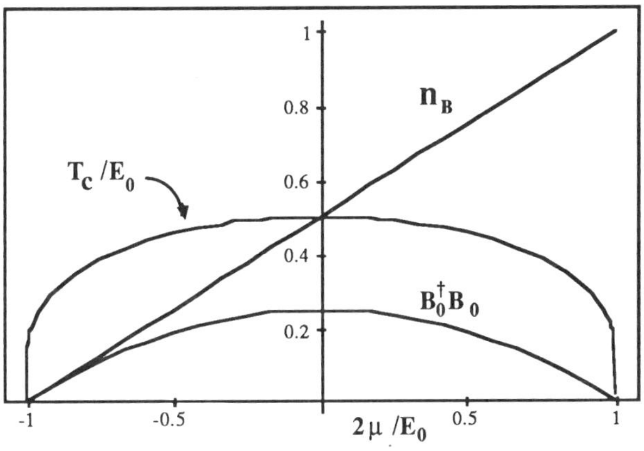

Fig. 1. Basic parameters in hard-core Boson condensation on a lattice of paired sites - negative-U lattice theory. The density of pairs on the lattice is given by n_B and is a linear function of the Fermi energy (μ). E_0, reflecting the Boson mobility, is the half-width of the Boson band. $T_c(\mu)/E_0 = \lambda(\mu)$ is shown in mean field theory and varies slowly near 1/2 filling which has the largest T_c with $\lambda(\mu)=1/2$. Below 1/2 filling the condensation can be thought of as condensation of pairs, above 1/2 filling as condensation of pair-holes. The Boson condensate at T=0: $\langle B_0^{\dagger}B_0\rangle = n_B(1-n_B)$. In conventional negative-U lattice theory there is a competing charge-density-wave order because of nearest neighbor repulsion.[5] The effect of this repulsion is not included here since the hard-core Boson model which results from the two-component theory does not intrinsically have the nearest neighbor repulsion.

II. Two-Component Theory and Mean Field Solution

This motivates consideration of a two-component theory for the electronic states in these materials: A broad metallic band of electrons (originating in the Cu-O planes) coexists with a narrow band of two particle transitions (originating in the oxygen vacancies). The lattice of the broad band is the "single particle" or "mobile" component and the lattice of two particle transitions sites is the "paired" component. Fig. 2 illustrates the two lattices and Hamiltonian terms between different orbitals.

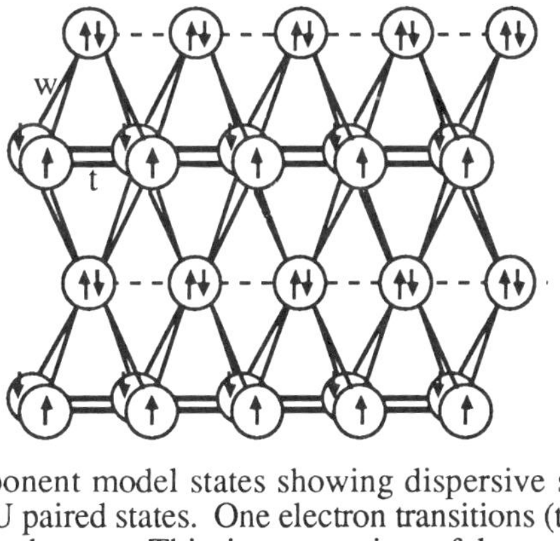

Fig. 2. Two component model states showing dispersive single particle and localized Negative-U paired states. One electron transitions (t), and two electron transitions (w) are shown. This is suggestive of layered material where negative-U sites (arising from oxygen vacancies) are essentially separate from single particle states. In $Y_1Ba_2Cu_3O_7$ two layers of single particle states interleave the paired layers in the z direction.

Diagrams for the two-component theory are shown in Fig. 3 corresponding to the Hamiltonian:

$$H = \sum_{i,\delta,\sigma} t_\delta c^\dagger_{i+\delta,\sigma} c_{i,\sigma} + E_B \sum_i b^\dagger_{i,\uparrow} b^\dagger_{i,\downarrow} b_{i,\downarrow} b_{i,\uparrow}$$

$$+ \sum_{i,\delta} w_\delta c^\dagger_{i+\delta,\uparrow} c^\dagger_{i+\delta,\downarrow} b_{i,\downarrow} b_{i,\uparrow} + \text{h.c.}$$

$$= \sum_{k,\sigma} \varepsilon_k c^\dagger_{k,\sigma} c_{k,\sigma} + E_B \sum_q B^\dagger_q B_q + \frac{1}{\sqrt{N}} \sum_{q,k} w_q c^\dagger_{k,\uparrow} c^\dagger_{-k+q,\downarrow} B_q + \text{h.c.} \tag{3}$$

$$B_q = \frac{1}{\sqrt{N}} \sum_i e^{iqR_i}\, b_{i,\uparrow} b_{i,\downarrow} \; ; \qquad w_q = \sum_\delta e^{iq\delta}\, w_\delta$$

The mobile electrons are created by the Fermion operators $c^\dagger_{i,\sigma}$. The paired states are formed out of two single particle creation operators ($B^\dagger_i = b^\dagger_{i,\uparrow} b^\dagger_{i,\downarrow}$). The energy of mobile states ε_k is related by Fourier transform to the single particle hopping t_δ. Single particle excitations in the strongly coupled (paired) lattice have been treated perturbatively since they are high above the Fermi energy which cuts through the one particle and two particle transition bands. Direct pair hopping is assumed small thus the pairs have no intrinsic hopping, only an energy E_B. Electron correlations in single particle states and direct pair hopping can be included. By choice of reference energy, $E_B/2$ is the Fermi energy of the two-particle transition. It is related to the electron-electron effective interaction on the paired lattice U_b by $E_B=2\varepsilon_b+U_b$ where ε_b would be the single particle energy. B_q are the *hard core* Boson operators. The hopping between paired and mobile bands is given by w_δ.[8]

A direct perturbation treatment of the coupling w leads to a divergent result: Hopping between different paired sites results from considering w in perturbation theory and keeping only terms which leave the electron gas unchanged from initial state to final state. The intermediate "excited states" result from the creation of free electrons at the expense of paired electrons (or the opposite). Since the two particle spectrum overlaps the single particle states, intermediate states cost no energy and perturbation theory diverges.

A self-consistent treatment leads to a nonperturbative gap in the single particle density of states at the Fermi energy. A dispersion is induced in the paired states by the hopping through the single particle states (Fig. 3d). This dispersion leads to Boson condensation of the paired component at low temperatures. $\langle B_0 \rangle$ is the macroscopic occupation of the condensed Boson state. Below the superfluid transition temperature, the q=0 term dominates the coupling Hamiltonian and the free electron part of the Hamiltonian is the same as the usual BCS mean-field-theory Hamiltonian.[9] Thus, a superconducting instability and a single particle gap arises through Cooper pairing at the Fermi energy.

The Hamiltonian can be treated using a BCS approach for T=0 and a Boguliubov transformation of the electron gas at finite temperature.

At T=0 the non-perturbative treatment using BCS mean field theory is obtained by minimizing $\langle H-\mu N \rangle$, assuming the trial wavefunction

$$\Psi = \prod_k (u_k + v_k c^\dagger_{k,\uparrow} c^\dagger_{-k,\downarrow}) \prod_i (u_i + v_i b^\dagger_{i,\uparrow} b^\dagger_{i,\downarrow}) |0\rangle \tag{4}$$

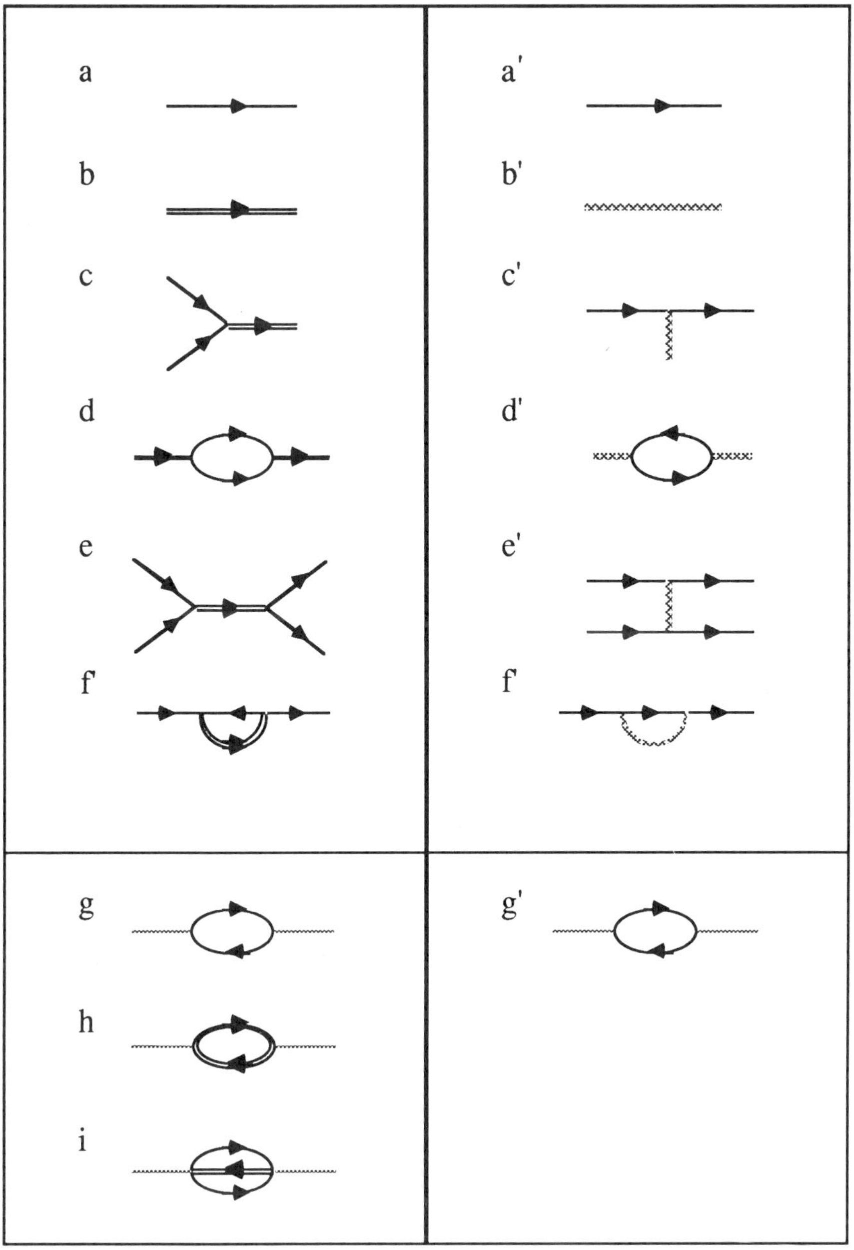

Fig. 3. Diagrams (a-i) of the two component theory in contrast to the usual diagrams (a'-g') of electrons interacting with phonons or photons. A single particle propagator is a single line with an arrow (a) and a pair by a double line with an arrow (b), the shaded line (b') is for a phonon or photon. Response diagrams are shown (g-i, g'). The essential difference between the two sets of diagrams is in the comparison of diagrams c and c'. In usual emission or absorption of Bosons electrons are conserved. In the two-component theory two electrons convert into a pair, or are created by the decay of a pair. In c' the q=0 Boson state has no meaning, while in c it is the Boson condensate. The existence of degenerate Fermionic and Bosonic states guarantees that diagrams d-f are divergent so the analysis treats non-perturbatively diagram c. The single particle self energy (f) is divergent in the normal state indicating the hybridization of electrons and holes to form neutral Fermions.

The gaps are defined as usual, but for a two-component system the equations are cross coupled. Letting the "gap" parameter Δ' be the same for all sites i, and a different Fermionic gap Δ for all k, the coupled two-component zero temperature equations become:

$$\Delta' = \Delta_i = -\frac{w}{N}\sum_k u_k v_k = -\frac{w}{2}\frac{1}{N}\sum_k \frac{\Delta}{(\Delta^2+(\varepsilon_k-\mu)^2)^{1/2}} \tag{5}$$

$$\Delta = \Delta_k = -\frac{w}{N}\sum_i u_i v_i = -\frac{w}{2}\frac{\Delta'}{(\Delta'^2+(E_B/2-\mu)^2)^{1/2}}$$

The phase between Δ' and Δ is the phase of u_iv_i compared to u_kv_k. This phase is set by the arbitrary phase of $w=w_{q=0}$ arising from the relative phases of the orbitals of the localized states and the delocalized states. Choosing w to be real and positive results in Δ' and Δ having opposite signs. Superconductivity is independent of the sign convention. In contrast to weak coupling BCS model directly coupled by phonons, Δ does not cut off at $\hbar\omega_D$.[10]

The qualitative behavior of the zero temperature mean field equations can be described for a constant free electron density of states for $(\varepsilon_k-\mu)$ in the interval $[-\alpha t,\alpha t]$; $D_c(\mu)\sim 1/2\alpha t$.[11]

$$\Delta' \approx -\Delta\, wD_c(\mu)\ln\left[\frac{2\alpha t}{|\Delta|}\right] \tag{6}$$

Solution of the coupled gap equations 5 & 6 may be obtained parametrically setting $p=|E_B/2-\mu|/\Delta'$. Fig. 4 displays the solution for $|\Delta|$, $|\Delta'|$ as a function of $|E_B/2-\mu|$ for $2\alpha t=50w$. For $\mu=E_B/2$, (half filling of the Bosonic band) Δ is always $-w/2$, where the linear dependence of Δ on w is consistent with the necessary non-perturbative treatment.

Δ' is the width of the pairing as a function of Fermi energy. As $|E_B/2-\mu|$ increases beyond Δ', both Δ and Δ' decay exponentially:

$$\Delta = -\frac{w}{2x}\Delta' = 2\alpha t\exp(-\lambda x) \tag{7}$$

$$\lambda = \frac{2}{w^2D_c(\mu)}$$

Thus, although formally the two particle transitions extend throughout the range of the single particle band, the single particle gap becomes exponentially small far from 1/2 filling of the paired band.

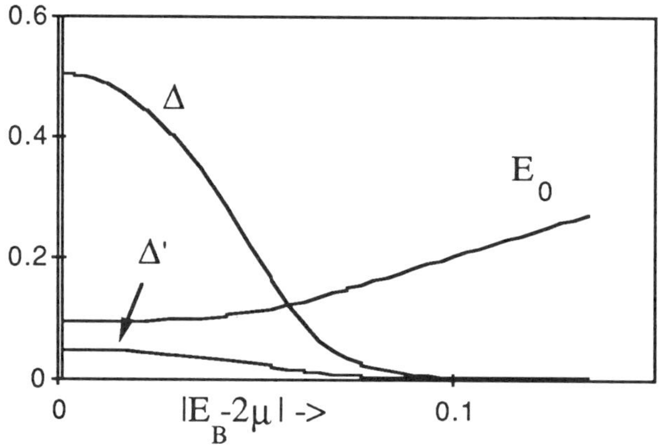

Fig. 4 Solution of the zero temperature mean field theory of the two-component theory for a smooth single particle density of states showing the two order parameters Δ and Δ' and the superfluid dispersion width E_0 (without the shift $E_q{}^0$) as a function of $|E_B-2\mu|$ (the distance of the Fermi energy from the uncoupled two particle transition energy). Units of both axis are the coupling strength w.

Treatment of elementary excitations and the transition temperature begins by constructing an approximate order parameter coupled Hamiltonian for the two components by: 1) Using a Boguliubov transformation on the free electron gas; treating $\Delta = wB_0$ and the Boguliubon density operators as number operators, and 2) Using a perturbation theory which is convergent within the single particle gap for the dispersion E_q of the paired lattice. This separates the q=0 term from the q≠0 terms of the two-component coupling. The resulting Hamiltonian and order parameters are:

$$H-\mu N = \sum_k \left((\bar{\varepsilon}_k - \mathcal{E}_k) + \mathcal{E}_k(\gamma^\dagger_{k0}\gamma_{k0}+\gamma^\dagger_{k1}\gamma_{k1})\right)$$

$$+ (\bar{E}_B - E_q^0)B_0^\dagger B_0 + \sum_{q\neq 0} (\bar{E}_B + E_q)\, B_q^\dagger B_q \qquad (8)$$

$$\mathcal{E}_k = (|\Delta(T)|^2 + (\varepsilon_k-\mu)^2)^{1/2}$$

$$|\Delta(T)|^2 = w^2 \langle B_0^\dagger B_0 \rangle$$

$$\Delta'(T) = -\langle \frac{w}{N}\sum_k c_{k,\downarrow} c_{-k,\uparrow} \rangle = -\frac{w}{2}\Delta(T)\frac{1}{N}\sum_k \frac{(1-2f(\mathcal{E}_k))}{\mathcal{E}_k}$$

$\gamma_{k,\sigma}$ are Boguliubon operators.

$$E_q \sim |w_q|^2 \frac{1}{2}\sum_k \left(1+\frac{\bar{\varepsilon}_{q-k}}{\mathcal{E}_{q-k}}\frac{\bar{\varepsilon}_k}{\mathcal{E}_k}\right)\left(\left(\frac{1}{\bar{E}_B-\mathcal{E}_k-\mathcal{E}_{q-k}}\right)+\left(\frac{1}{-\bar{E}_B-\mathcal{E}_k-\mathcal{E}_{q-k}}\right)\right)$$

$$-\left(\frac{\bar{\varepsilon}_{q-k}}{\mathcal{E}_{q-k}}-\frac{\bar{\varepsilon}_k}{\mathcal{E}_k}\right)\left(\left(\frac{1}{\bar{E}_B-\mathcal{E}_k-\mathcal{E}_{q-k}}\right)-\left(\frac{1}{-\bar{E}_B-\mathcal{E}_k-\mathcal{E}_{q-k}}\right)\right) - E_q^0 \qquad (9)$$

$$\bar{\varepsilon}_k = \varepsilon_k - \mu$$

$$\bar{E}_B = E_B - 2\mu$$

The first term in the sum over k is the ground state energy. The second term is the energy of excitation of the single particle states with the new dispersion $\mathcal{E}_k$ (Boguliubons). The Bosons now have a momentum dependent energy E_q. The q=0 term is buried in the ground state energy term. E_q^0 is adjusted so that the average of E_q is zero. This is the action of a pair back onto itself which shifts E_B and is subtracted out from the q dependence of E_q and the condensation energy. E_q is shown in Fig. 5. For $|E_B-2\mu|$ greater than 2Δ the perturbation theory expression for E_q diverges in the limit q->0. The expression for E_q is evaluated in 2nd order perturbation theory and does not include the effects of Boguliubon excitation. Expressions for the renormalization of E_B are not written.

The decoupled Hamiltonian can be solved, including the cross coupling through the q=0 term. The Fermion Hamiltonian is solved by assigning Fermion occupation to the Boguliubons. The effective Bosonic Hamiltonian may be further rewritten as a Ferromagnetic quantum x-y model in a magnetic field. In contrast with usual negative-U lattice superconductivity, the present Hamiltonian does not include a repulsive nearest neighbor term so it is not an anti-ferromagnetic Heisenberg model.[12] Crucially, the dispersion reflects longer range interactions; the range of these interactions, as controlled by hopping through single particle states, gives the coherence length of the superconductor:

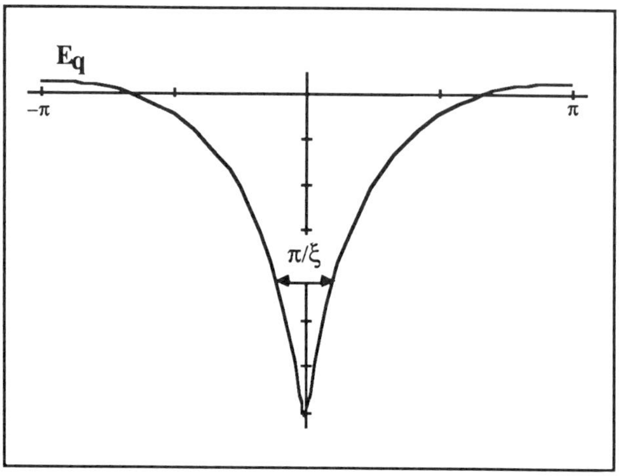

Fig. 5. Momentum dependence of the Boson energy at T=0 due to the coupling between paired and mobile states as obtained from perturbation theory for a constant density of single particle states. Without any gap in the single particle state spectrum there is a weak logarithmic divergence at q=0. An approximate (BCS) form of the cutoff was used to generate this figure. The approximate coherence length ξ (the distance of pair hopping) is indicated.

$$H_{F\text{-}xy} = \sum_q (E_B+E_q)B_q^\dagger B_q = \sum_{i,\delta} T_\delta(S_{i+\delta}\bullet S_j - S^z_{i+\delta}S^z_i) + h\sum_i S^z_i \qquad (10)$$

$$T_\delta = \frac{1}{N}\sum_q e^{-i\delta\cdot q}\, E_q$$

The value of $\Delta(T)$ is to be obtained from self-consistent solution of the hard core Boson Hamiltonian Eq. 10. The transition temperature where $\Delta(T)\text{->}0$, $T_c^0(\mu)$ of this hard core Boson Hamiltonian is proportional to E_0; $T_c^0(\mu)=j(\mu)E_0$. For half filling, $T_c^0(\mu=E_B/2)$ is comparable to E_0; specifically, in mean field $j(\mu=E_B/2)=1/2$. Note that E_0 is temperature dependent so T_c is obtained by setting $T_c = j(\mu)E_0(T_c)$.

A variational (rather than perturbative) estimate of $E_0(T)$ can be extracted from the Hamiltonian of Eq. 8 by taking the derivative of the total energy with respect to the order parameter. Keeping lowest order terms, and neglecting the dependence of E_q on Δ:

$$E_0(T) = \frac{\delta <H>}{\delta<B_0^\dagger B_0>} - E_B - E_q^0 = \frac{\delta \sum_k(-\mathcal{E}_k)(1-2f(\mathcal{E}_k))}{\delta<B_0^\dagger B_0>} - E_q^0$$

$$= w\Delta'(T)/\Delta(T) - E_q^0 \qquad (11)$$

for T=0 this may be verified in mean field theory by comparing Eq. 5 with a BCS mean field treatment of the hard core Boson Hamiltonian Eq. 10. $E_0(T=0)$ is plotted in Fig. 4. In the limit of $q\rightarrow 0$ and for $E_B=2\mu$, E_q is similar to the variational E_0 suggesting that a complete treatment would lead to a smooth Boson excitation spectrum.

The logarithmic divergence of E_0 as a function of $\Delta'(T)\text{->}0$ is cutoff by Boguliubon excitation near the transition temperature. When $\Delta'(T)<kT$ equation 11 becomes:[13]

$$E_0(T) = -\frac{w^2}{2}\int_{-\infty}^{\infty} \frac{D_c(\varepsilon)\tanh(\varepsilon/2kT)}{\varepsilon}d\varepsilon - E_q^0 \sim -w^2 D_c(\mu)\ln\left[1.13\,\frac{\alpha t}{kT}\right] - E_q^0 \qquad (12)$$

The latter expression is written for the qualitative constant single particle density of states. This expression is equivalent to the T=0 expression for E_0 with kT/1.13 replacing $\Delta'/2$ in the logarithm. Since the highest transition temperature (at $\mu=E_B/2$) is comparable to E_0, the overall increase in E_0 from T=0 to the T_c (for reasonable values) is 0-25%. The self-consistent transition temperature at which $\Delta',\Delta\to 0$ $T_c(\mu) = j(\mu)E_0(T_c(\mu))$. Thus the weak divergence of E_0 increases T_c over that obtained by considering directly the T=0 value of E_0 by a factor of 1.– 1.25. T_c can be seen to be sensitive to the single particle density, while Δ is sensitive to the paired state concentration.[14]

The thermodynamic critical field $H_c(T=0)$ can be obtained from the ground state energy (first term of Eq. 8). In contrast to BCS theory where self-consistency makes $H_c^2/8\pi$ second order in the gap and T_c, here the binding energy is first order resulting in larger critical fields. At $\mu=E_B/2$ (for small $E_q{}^0$), $H_c^2/8\pi = E_0(T=0)/4 \cong T_c/2$ per unit cell in mean field theory.

In considering the coupled Hamiltonian and the effects of dimensionality in 2+1 dimensional systems, like the layered Cu-O superconductors, it is relevant that the dispersion of the free-electron states is not the same as the dispersion of the two-particle superfluid which may be more three-dimensional (see Fig. 2). This helps superconductivity in 2+1 dimensional systems where Kosterlitz-Thouless behavior implies the importance of vortex excitations and power law decay of correlations in a strictly two-dimensional ordered phase. In the two-component theory the layering or two-dimensionality is a practical rather than fundamental aspect of the superconductivity in Cu-O based materials, it establishes the environment for the two different types of electronic states. The two different types of electronic states are essentially consistent with two different kinds of electronic environments. Extended mobile states are associated with metals. Localized negative-U states are associated with defects in semiconductors. If the regions between the Cu-O layers are characterized, they should indeed be considered as semiconductor layers. As extensively discussed in the literature, they are essentially insulating, and they can be doped. Thus, the Cu-O materials maybe characterized as layered metal-semiconductor structures.

For Fermi energies away from the paired transition energy the above treatment breaks down. A direct approach treats the virtual excitation of 2-particle states in perturbation theory. For Fermi energies above the 2-particle transitions this generates the Hamiltonian:

$$H' = -\sum_{q,k,k'} |w_q|^2 \frac{1}{\varepsilon_{q-k'}+\varepsilon_{k'}-E_B} c_{k,\downarrow} c_{q-k,\uparrow} c^{\dagger}_{q-k',\uparrow} c^{\dagger}_{k',\downarrow} \tag{13}$$

By restricting attention to near the Fermi energy the denominator may be replaced by $2\mu-E_B$. This reduces to a contribution to the on-site correlation energy of the electron gas and charge-density-wave terms contributing to the effect of the term in Footnote 8. When a particular q_0 is selected (by the Fermi surface at 1/2 Fermionic filling) long-range-interaction spin correlations are generated (however, $q_0=0$ is a usual superconducting Hamiltonian with a small interaction energy $w^2/(2\mu-E_B)$). Anti-ferromagnetism in high-T_c materials may be accounted for adequently by a positive-U in the single particle states, which would not detract from the two-component discussion of the superconductivity. However, this discussion suggests that the anti-ferromagnetism may be due to the two-component Hamiltonian itself.

III. Elementary Excitations and Transport, Isotope Shift, etc.

Conclusions from the treatment of the excitation spectrum are summarized in Fig. 6. The excitation spectrum is described primarily through the formation of a single particle excitation gap. Within this gap the two particle hard-core Boson elementary excitations are well defined. The Boson excitations generally lie within this gap for small values of $E_B-2\mu$ and low temperatures as suggested by reasonable values of E_0/Δ (Fig. 4).

The Bosonic excitations are shown in Fig. 5 at $E_B-2\mu = 0$ (Eq. 9). The long wavelength Bosonic excitations in the two-component theory are *coupled* charge density waves, where the charge oscillates between the single particle and paired states. This corresponds to a screening of the plasmons of both the negative-U lattice and the mobile states. Without long range displacements of charge, the long-range coulomb potential and the plasma frequency do not enter. Since the dispersion is controlled by the mobile states, the long wavelength Boson corresponds to a zero-sound wave in a neutral (screened) Fermi gas.[15] The zero-sound in a neutral Fermi system has a linear dispersion at $q\to 0$.

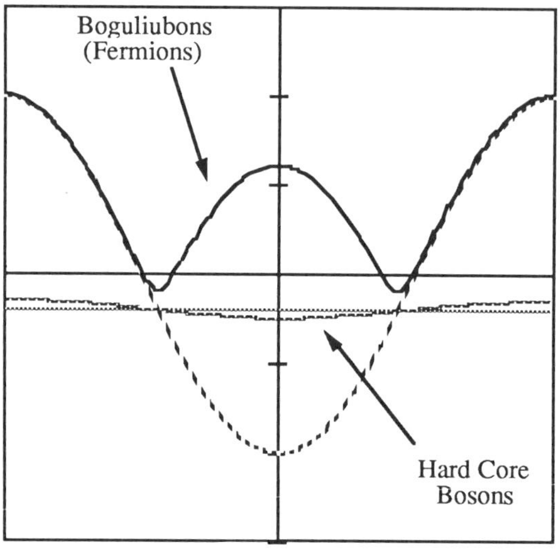

Fig. 6. Schematic elementary excitation spectrum near T=0 and $|E_B-2\mu|=0$ of the order parameter coupled two-component equations. Dashed and light shaded lines are single particle and two particle transitions before coupling. Solid and dark shaded lines are the same after coupling the two-components. As T increases the single particle gap decreases and the Boson and Fermion elementary excitations are mixed. Bosonic states continue to exist above T_c.

The other collective long wave excitation, the acoustic phonon (sound waves), couples directly to the electronic long wave Boson excitation. The primary reason for this coupling is that E_B typically varies linearly with pressure through the pressure dependence of the local structure in the negative-U center. A second reason is a change in w with pressure. Thus hydrostatic pressure should affect superconductivity, and bulk modulii and sound waves should be affected by Boson condensation.

A linear change in E_B with pressure implies that pressure affects the Bosonic occupation similar to doping. The Boson occupation, as has been discussed, affects all of the essential properties of superconductivity in a systematic way. The most observable effects are on T_c. Consider the case of a linear raising of E_B with pressure. If the Fermi energy is above Boson half-filling, pressure increases the Bosonic filling raising T_c. If the Fermi energy is at Boson half-filling very little effect should be observed, while if the Fermi energy is below Boson half-filling T_c is lowered. The effect of changes in w with pressure is a systematic shift in these properties so that the turning point from raising T_c to lowering T_c is shifted slightly from Boson-half-filling.

Direct observation of Bosonic excitations should be apparent with phonon-like absorption in Infrared and Raman spectroscopies. The distance of the average energy to the q=0 Bosonic state is $E_0 = 2kT_c$. Since Boson pairs must be excited the expected *average* energy excitation is $2E_0=4kT_c$ with the largest excitation possibly up to $8kT_c$. Phonon like absorption peaks should be observed corresponding to these Bosonic pair excitations.

The coherence and the order parameter, are determined by the hopping of pairs through the single particle states. This hopping describes the distance over which superconductivity is affected if it is suppressed locally. It is important to recognize, unlike a negative-U lattice theory, the coherence length is larger than the nearest neighbor distance because the hopping proceeds through extended single particle states, and is not exponentially suppressed on the negative-U states. The hopping probability T_δ is related by Fourier transform to the dispersion E_q of the Bosonic states (Fig. 5). There are two sources of this dispersion, the coupling w_q and the dispersion of the single particle states. The width of the dip in the dispersion is reciprocal to the coherence length ξ.

Considering now the single particle excitations. In addition to the coupling of single particle states to the q=0 condensate leading to Cooper pair formation, the single particle states are also coupled to $q \neq 0$ Bosonic excitations and the zero point motion of the Boson field. The $q \neq 0$ Bosons enable the conversion of electronic states and hole states with *non-*

zero momentum transfer. The primary effect is to hybridize electron-like Boguliubons with hole-like Boguliubons at all energies and dramatically broaden the onset of the Fermionic gap appearing as lower and upper gap features. Coupling to the zero-point motion may lead to a tail to the single particle spectrum which is exponential like an Urbach edge. When Bosonic excitations overlap the single particle excitations the coupling is more direct and perturbation theory breaks down in description of the Boson and Fermion excitations suggesting also a change in system behavior discussed below.

For the elementary excitations, finite temperature properties at low temperature follow the usual behavior of the two separate systems. The superconducting electron gas has thermodynamically excited broken pairs (Boguliubons). The paired hard-core Boson superfluid evaporates through kinetic energy excitation of Boson like excitations. Cross coupling is important since the Fermionic gap Δ is given by the paired condensate density and the dispersion of the paired superfluid E_0 is given by Δ. In this treatment, both order parameters Δ,Δ' go to zero at the same temperature. However, as the gap $\Delta(T)$ decreases the superfluid dispersion $E_0(T)$ increases.

The mixing of Boson and Fermion excitations, as indicated by the breakdown of perturbation theory, makes the analysis based on separating the two components insufficient near the transition. When Fermion and Boson coordinates are not well defined, the mixing (scattering) may preempt the transition before the order parameters are zero.[16] An estimate of the breakdown of the elementary excitation picture is at the perturbation theory overlap temperature: $|E_B-2\mu| = 2|\Delta'(T_c)|$.[17] This defines the resistive superconducting to normal transition, T_c', as a function of μ, below the thermodynamic transition T_c at which the condensate evaporates. In the perturbative treatment at $E_B=2\mu$, $T_c=T_c'$. However, the coupling of single particle states to $q\neq0$ Bosons may also lead to scattering of the condensate for $E_B=2\mu$ separating T_c and T_c'. Thus, the expected behavior of this system is an onset of resistance at the temperature $T_c'<T_c$ followed by the evaporation of the condensate and the disappearance of the mean Fermionic gap at T_c.

Above T_c the Bosonic states continue to exist, but are no longer condensed. They will affect the response of the normal state and usual Fermi liquid theory should not apply. The long range coherence disappears, local coherence imposed by the large pairing energy remains. The essential physical process is the local transition from single particle to paired states. The importance of the coupling is revealed in the divergence of the single particle self energy since the process of an electron turning into a hole plus a pair at 1/2 filling of the Boson band, still involves no change in energy. The Hamiltonian term:

$$w_{k+k'}\, c^{\dagger}_{k,\uparrow} c^{\dagger}_{k',\downarrow} B_{k+k'} \tag{14}$$

converts a hole at k' into an electron at k, where k and k' correspond to electrons and holes at the same distance from the Fermi energy. The coupling of two degenerate states, electron and hole, by the Hamiltonian implies that these states hybridize, with equal weight. The change in Boson number during this process is accounted for in the coherent ground state which does not conserve Boson number at k+k'. In mean field at 1/2 filling the Boson operators may be treated as numbers because the commutator expectation values are zero.

$$\langle [B_q, B^{\dagger}_{q'}] \rangle = \langle \delta_{q,q'} - 2\, n_B(q-q') \rangle = 0 \quad \text{(at 1/2 filling)} \tag{15}$$

where $n_B(q-q')$ is the Fourier density operator. More precisely, the Hamiltonian couples all single particle states, both electrons and holes. Wave packets which can couple to the local order (coherence) should be generated and they, in turn, induce local correlation (hopping) in the Bosonic system.

Unlike the Fermion diagram, the Bosonic diagram is convergent with the logarithmic divergence cut off by kT. The normal state is thus seen to be composed of neutral Fermion and charged Boson elementary excitations. Fermions are neutral throughout the superconducting state. The hybridization is stronger at higher temperatures because the Bosonic states are mostly responsible for hybridizing k and -k in the condensate. At T=0 the hybridization is thus likely to be more easily broken.

Since Fermions are neutral, charged Bosons carry current in the uncondensed state (even though Fermionic states originally provided the mobility). This implies that a measurement of the effective charge of carriers, such as through Aharonov-Bohm effect will detect a effective charge of 2e also in the normal state, not just in the superconducting state.

Without a complete solution of the normal state properties, transport is tentatively extracted from the Boson propagator. The imaginary part is the decay of a pair (Fig. 3c) which determines the Boson lifetime. The q averaged imaginary part of the propagator is shown in Fig. 7 as a function of energy measured from the Fermi energy. Away from the Fermi energy the decay increases linearly with energy. For an approximately constant single particle density of states:

$$\langle \mathrm{Im}\, \Sigma_B(E,q) \rangle_q = 2\pi w^2 D_c^2\, E\, \mathrm{Coth}[\beta E/2] \qquad (16)$$

E is measured with respect to 2μ. The reason for the linear dependence of the decay rate of the Bosons into two single particle states is illustrated in Fig. 8. Considering the decay of a Boson at an energy E, it can decay into two electrons of energy E/2 or one of the electrons can have less energy and the other one more. Ultimately, the largest energy difference between the two electrons is just E since the low energy electron can not have an energy lower than the Fermi energy below which the electronic states are filled. The number of possible ways to decay is then just proportional to the energy E giving the dependence of Fig. 7. The decay rate of the Boson at the Fermi energy (E=0) is proportional to kT. This is a determined by the density of available Fermion pair states at the Fermi energy which is given by the range of energies (kT) where electronic states are partially occupied.

This derivation of the Boson lifetime is essentially based on phase space considerations the result should to hold for more complete many-body treatments.

The Drude conductivity is given by

$$\sigma = (ne^2/m)\, \tau \qquad (17)$$

where n is the density, e is the charge, m is the mass and τ is the lifetime of the charged particle. The lifetime refers to the scattering of a single particle. In this case the lifetime is determined by a decay of a paired state. The temperature dependence of this expression can be obtained assuming that n and m are independent of temperature, n because the pair density is determined by μ and is pinned by the large density of paired states, m is controlled by the Boson hopping range which is (aside from logarithmic factors) independent of temperature. τ is inversely proportional to the decay rate, which at the Fermi energy is $\langle \mathrm{Im}\, \Sigma_B(0,q) \rangle_q = 4\pi w^2 D_c^2\, kT$.

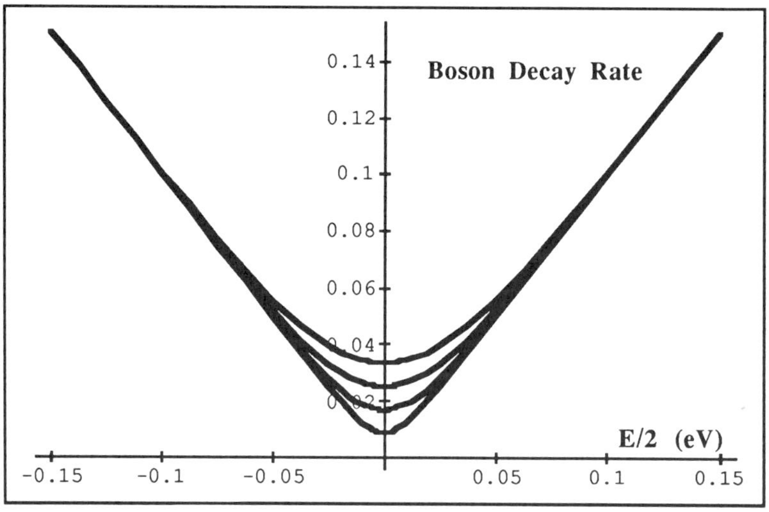

Fig. 7. The decay rate (inverse lifetime) of Bosonic pairs in the uncondensed state at an energy E away from the Fermi energy. The function plotted is $g(x)=x\, \mathrm{Coth}(x/kT) = \langle \mathrm{Im}\, \Sigma_B(E,q) \rangle_q / 4\pi w^2 D_c^2$; x=E/2. The axes are in electron volts. Four curves are plotted for different temperatures (from lower to higher) 100K, 200K, 300K, 400K. The decay rate at the Fermi energy is proportional to the temperature, this leads to a linear resistivity. Away from the Fermi energy the decay rate is proportional to |E |. This leads to an unusual behavior of the tunneling conductance. The shape of the Boson decay rate is due to phase space considerations illustrated in Fig. 8.

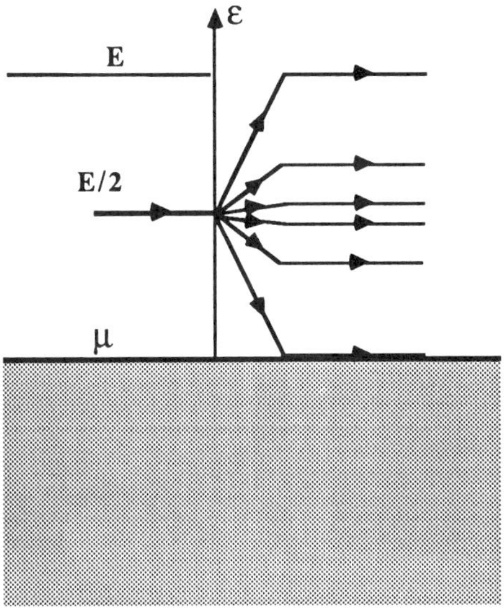

Fig. 8. Illustration of the processes by which a Bosonic pair with energy E with respect to the Fermi energy can decay into two single particle states. The pair (coming from the left) shown at the energy E/2 per electron, can decay into two electrons at E/2, or one electron can have less and the other more energy. The limit in the difference between the energy of the electrons E is established by the Fermi energy below which the electronic states are filled. The number of ways the decay is possible (the phase space) is thus proportional to E giving rise to Fig. 7.

Thus the resistance $R \sim 1/\sigma$ is proportional to kT. Substantially away from 1/2 filling, it is to be expected that a usual Fermionic dependence of the resistance should be observed.

The frequency dependent resistance $\sigma(\omega) = \sigma/(1+i\omega\tau) = (ne^2/m)\tau/(1+(\omega\tau)^2)$ can also be obtained using the frequency dependent lifetime $\tau(\omega) \sim 1/\,4\pi w^2 D_c{}^2\,\omega\,\mathrm{Coth}[\beta h\omega/2]$, which decreases as a function of ω, increasing the high frequency tail of σ(ω).

The tunneling current of electrons from a two component material into a normal metal involves charged pairs of the two-component material decaying into the single particle states of the normal metal. Thus the tunneling involves directly the decay process of Fig. 3c. A detailed discussion not given here, suggests that the tunneling behavior of the conductance is controlled by $\langle \mathrm{Im}\ \Sigma_B(V,q) \rangle_q$ including a linear G(V). A further dramatic result is suggested for tunneling in the superconducting state where pair decay into the normal metal is still possible at V=0 and T=0 (zero field conductance).

In contrast to BCS theory where the isotope shift is manifest in the formalism, there is no isotope shift apparent in the formalism of the two-component theory discussed above. In the two-component theory any isotope shift arises because of the internal workings of the Negative-U center. This is implicit in the Hamiltonian used to obtain most of the above results. The following considerations indicate that the isotope shift is sensitive to comparisons between different relaxation rates - structural and electronic, and can be small.

There are two stable electronic states of the negative-U center, unoccupied and doubly occupied. These also correspond to two different atomic positions. The transition between the electronic occupations is accompanied by a shift in atomic position, which is actually responsible for the negative-U. The quantum mechanical zero-point motion corresponds to Gaussian wavepackets in potential wells which are in first order Harmonic. The transition matrix element w includes an atomic transition which can be considered in two limits. In one limit, one considers the transition of the ground state atom wavefunction in one well into the other (Fig. 9a), w would then be reduced by the overlap of the Gaussian orbitals. In the second limit, virtual transitions into excited oscillator states are included. This corresponds to a vertical transition (Fig. 9b) followed by a relaxation using the anharmonic terms of the oscillator. The direct transition dominates when the two sites are close together, and the vertical transition when the two sites are far apart since the direct transition is exponentially suppressed by the wavefunction overlap, while the vertical transition is only suppressed inversely with the intermediate energy.[18]

If R_0 and R_2 are the lowest energy sites of the oscillators than the Gaussian overlap is:

$$\exp(-(M\kappa)^{1/2} (R_2-R_0)^2/4\hbar) \qquad (18)$$

which is quite sensitive to the mass of the Harmonic oscillator when the two locations are far apart, and not when they are close together. Thus, the isotope shift would be small for sites close together. For sites far apart when the vertical transition dominates, the Gaussian overlap does not enter and the isotope shift is also small. In any case, a large negative-U necessitates relatively large separations between the two sites indicating that vertical transitions are to be expected. This discussion has assumed that the transition rate (given by w) is very small compared to the relaxation rate of the anharmonic oscillator. The actual value of the isotope shift appears to depend on a comparison of the electronic transition rate, with the structural relaxation time of the Harmonic oscillator which depends on the size of anharmonic terms. Within the two-component theory a large electron-phonon coupling is consistent with a small isotope-shift. Conventional experiments such as IR absorption and Raman which measure the coupling between the phonons and electronic states should show large electron-phonon coupling despite the small isotope shift.

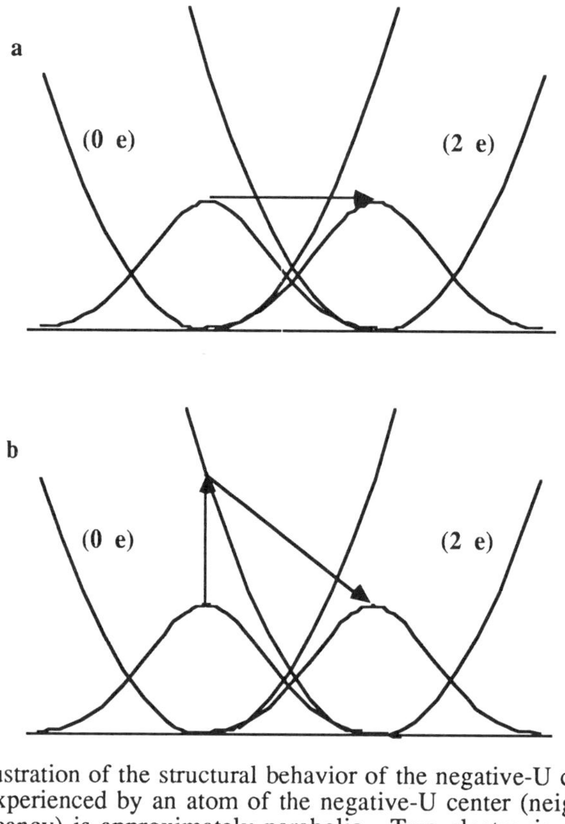

Fig. 9. Illustration of the structural behavior of the negative-U centers. The potential experienced by an atom of the negative-U center (neighboring the oxygen vacancy) is approximately parabolic. Two electronic occupations, unoccupied and doubly occupied, correspond to two different nuclear positions. The separation between them is the relaxation responsible for the negative-U. The transition between an unoccupied center and a doubly occupied center is illustrated in two limiting cases (a) direct transition between the two ground states, and (b) a vertical transition followed by a relaxation of the oscillator, assuming an anharmonic oscillator. Because of the transitions between these two states which are at the same energy, there is a quantum coherence between the two atomic locations (structural coherence). The structural coherence leads to a signature in channeling and other experiments (see text).

A dramatic signature of the two-component theory, which can take the place of the isotope shift, is "structural coherence". Structural coherence arises because of the quantum superposition of the two separated locations of the atoms in a negative-U center. The two atomic locations of the negative-U center are basic to the electronic coherence of the superconductivity. Locally, the negative-U center undergoes transitions between the two states in both the uncondensed and condensed states. The quantum state of the center is a linear combination of the occupied and unoccupied states. Thus there is a quantum coherence between these two atomic locations in the uncondensed and condensed states. However, the long range coherence of the electronic wavefunction in the condensed state leads to a long range coherence of the atomic locations as well.

The existence of two atomic locations for atoms in the negative-U center should be observable in the presence of a smearing of atomic locations in X-ray and neutron diffraction. High-accuracy experiments may be better fit by two separate atomic ovals instead of one, associated to the negative-U centers.

A direct test of structural coherence is provided by channeling experiments. In these experiments a beam of ions is directed near to a crystallographic axis. The backscattering of these ions is sensitive to the cross section for encountering an ion. If the atomic locations become correlated then the cross section should be dramatically affected, ie as the temperature is varied through T_c.

The electronic states required for the two-component theory are quite distinctive. Their signature should be found in both theoretical calculations and experiment. In theoretical electronic state calculations both a broad band and localized states (a flat band) are indeed found crossing or near the Fermi energy in Cu-O based superconductors with oxygen vacancies. In experiment, signatures of a negative-U state in single particle excitations - empty and filled states equally displaced away from the Fermi energy - may be found in photoemission and inverse photoemission.

Since the localized states in the two component theory are paired, the transitions associated with one electron states are displaced away from the Fermi energy. Two electron transitions are at the Fermi energy. This leads to dramatic changes in the amplitude of excitations which appear to be far away from the Fermi energy as the Fermi energy is shifted (Fig. 10). An important experimental tool for observations of Negative-U centers is photoinduced absorption spectra.[19,20,21] By first exciting the material, there are created single particles in the paired band which can be optically excited significantly increasing the absorption and shifting the energy of the excitation (Fig. 10).

A systematic understanding of the superconductivity in Cu-O based materials may be made using a picture based on conductive Cu-O planes, and oxygen vacancies which have three properties:

1) They introduce localized states which are Negative-U.
2) They disrupt the single particle (mobile) states.
3) If there are too many of them in a plane, localized states themselves are disrupted.

The second property arises because the vacancies are in the Cu-O planes. If they were in the semiconductor layers this might be avoided. The third property arises because, like most defects, the localized states introduced by the vacancies reside on the neighboring atoms, particularly on the 2nd neighbor oxygen atoms, which if removed disrupt the localized states.

The two-component theory suggests that a broader family of materials should be investigated as potential high-T_c materials - the family of layered metal-semiconductor materials. Among the semiconductor defects studied in detail, a large proportion have been found to be negative-U centers.[1] This suggests that many possible materials formed out of semiconductor-metal layers should be superconducting if the following conditions can be met. First that there be a large number of defects, most easily accomplished by off-stoichiometry. Second, that the defect introduce negative-U states. Third that the Fermi energy be moved to correspond with the Fermi energy of the defect two-particle transitions. In the present class of high-T_c superconductors it appears that the defects are in the metallic layers. This may have advantages in some circumstances in high matrix elements connecting the localized and extended states, however, it has the disadvantage of disrupting the superconductivity unless the materials are further staged to separate the defects from the conducting states. If proper matching between the states can be achieved, it is likely that the active defects can be in the semiconductor layers. In order to study such layered metal-semiconductor materials in general, the problems of growing monolayered metal-semiconductor superlattices must be overcome, however, there are likely to be many materials which form naturally as such superlattices which have not yet been studied in off-stoichiometric and doped regimes. These are promising materials for high-T_c studies.

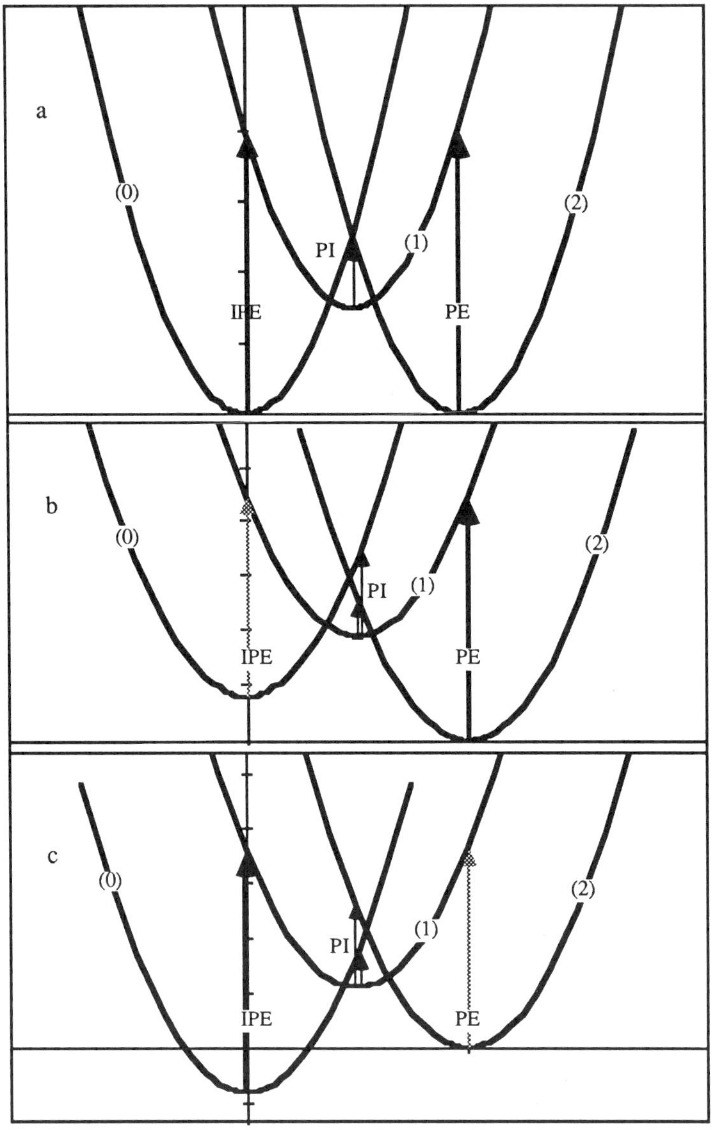

Fig. 10. Diagram representing the relationship between the negative-U center total energy curves and single particle excitations. The three curves represent the configuration coordinate energy dependence of the three possible occupations of the negative-U center. In (a) the unoccupied (0) and the doubly occupied states (2) are degenerate at their respective lowest energy configurations. This corresponds to $\mu = \mu_B$ - the two particle transitions are at the Fermi energy. A shift in the Fermi energy changes the occupation of the states to more occupied (b) or less occupied (c). Thus even though the transitions are far from the Fermi energy their intensity is dramatically affected by shifts in the Fermi energy. The energy of a defect when singly occupied is higher than the unoccupied or doubly occupied cases, this energy difference in (a) is half of the negative-U: $U_{eff} = E^2(2) + E^0(0) - 2\,E^1(1) = -\,2\,E^1(1)$ for this choice of μ and reference energy. Not shown in this figure are the effects of anharmonic terms which make the excitation energies (1/0) and (1/2) as well as the (0/1) and (2/1) inequivalent.

The strong dependence of the superconductivity on the Fermi energy, suggests, at least speculatively, that a switch may be formed with its central element a thin film of superconductor with a Fermi energy away from the optimal value for superconductivity. If it is maintained at a temperature above its own T_c and below the T_c of the optimal Fermi energy then the superconductivity of the material may be switched on by a field which shifts the Fermi level. In essence such a scheme makes use of large nonlinear effects near the transition temperature for thin films with Fermi energy dependent superconductivity.

IV. Conclusions

In summary, the two component theory differs from both independent component theories; conventional BCS superconductivity and negative-U lattice superconductivity. The expression for T_c is distinct. $2\Delta/kT_c$ is not universal. Δ mus be interpreted as a mean Fermionic gap with a broadening of the gap onset expected. Expressions are given above for these quantities and for their Fermi energy dependence. There is a greater coherence energy, critical field H_c, and related specific heat than BCS theory. The coherence length is longer than nearest neighbor negative-U lattice theories. The superconducting to normal transition is compound with a resistive transition before the complete evaporation of the condensate. The excitation spectrum is different both below and above T_c through the presence of Bosonic and Fermionic states, affecting response and transport properties. Direct excitation of Boson pairs should be possible at a well defined energy. A coupling of the condensation to sound waves and bulk modulii is expected, along with the strong pressure dependence of the superconductive transition related to the effect of doping. Charged Bosons carry the current in the normal state forming a narrow band of carriers, while the single particle Fermions become hybrid electron-hole excitations which are neutral. A linear resistance with temperature is derived from the decay of charged Boson pairs, and a linear voltage characteristic of tunneling is suggested by the Boson lifetime energy dependence. Tunneling at zero voltage and below T_c is possible through pair decay. A frequency dependence of the Drude conductivity is obtained with a shorter lifetime at larger frequencies. Three dimensional superconductors or non-antiferromagentic examples of high-T_c materials which have been found can also be understood. The strong electron-phonon coupling necessary for the negative-Us of defects are shown to be consistent with small isotope shifts but lead to a signature of structural coherence measurable in channeling experiments. The experimental effects of the strong effect of oxygen vacancies on superconductivity both enhancing and suppressing can be understood through their dual role in introducing localized paired states and decreasing the mobility of extended states. This also enables a fundamental understanding of the differences in the family of high-T_c materials. Single particle transitions are shown to be displaced from the Fermi energy even though they correspond to density of sates (two-particle) at the Fermi energy. A comparison of this phenomenology with many experiments on high-T_c materials including numerical agreement with experiments on YBaCuO is presented elsewhere.[22]

I wish to thank A. Aharony, G. Boebinger, M. L. Cohen, D. P. DiVincenzo, T. H. Geballe, Y. Gefen, W. A. Harrison, Y. Imry, M. Kardar, D.-H. Lee, P. A. Lee, S. G. Louie, E. J. Mele, N. D. Mermin, A. J. Millis, V. M. Nabutovskii, L. Pauling, J. C. Phillips, D. J. Scalapino, H.-B. Schüttler, D. Vanderbilt, for helpful comments and references. This work was supported in part by an Alon Fellowship and a grant from the Revson Foundation.

References and Notes

[1] G. A. Baraff, E. O. Kane, and M. Schluter, Phys. Rev. B 21, 5662 (1980); G. D. Watkins and J. R. Troxell, Phys. Rev. Lett. 44, 593 (1980); D. Vanderbilt and J. D. Joannopoulos, Phys. Rev. Lett. 49, 823 (1982); R. Car, P. Kelly, A. Oshiyama, S. T. Pantelides, Phys. Rev. Lett. 52, 1814 (1984); Y. Bar-Yam and J. D. Joannopoulos, Phys. Rev. Lett. 56, 2203 (1986)

[2] The direct involvement of oxygen vacancies in superconductivity has been discussed by J. C. Phillips without explicitly considering real space pairing, see Phys. Rev. Lett. 59, 1856 (1987); Phys. Rev. B 39, 7356 (1989) and unpublished.

[3] W. E. Pickett, Rev. Mod. Phys. 61, 433 (1989)

[4] P. Nozieres and S. Schmitt-Rink, J. Low Temp. Phys. 59, 195 (1985); C. M. Varma, Phys. Rev. Lett. 61, 2713 (1988); S. Robaszkiewicz, R. Micnas, and K. A. Chao, Phys. Rev. B 23, 1447 (1981) and 24, 1579 (1981).

[5] P. Nozieres and S. Schmitt-Rink, J. Low Temp. Phys. 59, 195 (1985)

[6] Metastable Bosons in a two-component formulation are discussed as early as M. R. Schafroth, Phys. Rev. 96, 1442 (1954); and more recently by J. Ranninger and S. Robaszkiewicz, Physica 125B, 468 (1985); S. P. Ionov, Izv. Akad. Nauk SSSR Ser. Fiz. 49, 310 (1985); G. M. Éliashberg, JETP Lett. Vol. 46, S81 (1987); D. M. Newns Phys. Rev. B 36, 5595 (1987); D. M. Newns, M. Rasolt, and P. C. Pattnaik, Phys. Rev. B 38, 6513 (1988); J. Ranninger, R. Micnas, and S. Robaszkiewicz, Ann. Phys. Fr. 13, 455 (1988); R. Freidberg and T. D. Lee, Phys. Lett. A 138, 423 (1989); Phys. Rev. B 40, 6745-62 (1989); L. P. Gorkov and G. M. Éliashberg, JETP Lett. Vol. 46, S84 (1987); R. Micnas, J. Ranninger, S. Robaszkiewicz, Rev. Mod. Phys. 62, 113 (1990)

[7] E. Simánek, Solid State Comm. 32, 731 (1979); C. S. Ting, D. N. Talwar, and K. L. Ngai Phys. Rev. Lett. 45, 1213 (1980); H.-B. Schüttler, M. Jarrell and D. J. Scalapino Phys. Rev. Lett. 58, 1147 (1987); Phys. Rev. B 39, 6501 (1989). In this work both components are treated equally.

[8] Other possible terms in the Hamiltonian include a coupled charge density wave promoting term $c^{\dagger}_{i,\uparrow}b^{\dagger}_{i,\downarrow}b_{i,\downarrow}c_{i,\uparrow}$. It does not change the results, near $\mu=E_B/2$, because it competes against the large t term, and the possible CDW small induced gap need not coincide with $E_B/2$.

[9] M. Tinkham, "Introduction to Superconductivity", Mcgraw-Hill, NY (1975)

[10] Structural relaxation dynamical effects have not been included.

[11] This neglects special density-of-states enhancements, particularly in quasi-two dimensional systems. see J. E. Hirsch and D. J. Scalapino, Phys. Rev. Lett. 56, 2732 (1986)

[12] Even in the special 2-D nearest neighbor case this model does not suffer from the competition between charge-density-wave and superconducting order. The charge density wave order results from the nearest neighbor repulsion. R. T. Scalettar, E. Y. Loh, J. E. Gubernatis, A. Moreo, S. R. White, D. J. Scalapino, R. L. Sugar and E. Dagotto Phys. Rev. Lett. 62, 1407 (1989)

[13] The integral is the same as that used for determining the BCS T_c but its use is significantly different.

[14] Oxygen vacancies in Cu-O based superconductors lower $D_c(\mu)$. T_c can thus be lowered as oxygen vacancies are added in the large concentration limit even though they are essential to superconductivity.

[15] "Theory of quantum liquids" D. Pines and P. Nozières (Benjamin, NY, 1966)

[16] P. A. Lee and N. Read, Phys. Rev. Lett. 58, 2691 (1987)

[17] The breakdown of the quasi-particle picture may happen earlier. In any case T_c occurs in the region of rapidly decreasing $\Delta'(T)$ so $T_c'(\mu)$ is of order $T_c(\mu)$.

[18] "Solid State Physics", W. A. Harrison, (p. 333)

[19] Z. Vardeny and J. Tauc, Phys. Rev. Lett. 54, 1844 (1985)

[20] Y. Bar-Yam, J. D. Joannopoulos and D. Adler Phys. Rev. Lett. 54, 1844 (1985)

[21] A. J. Heeger, S. Kivelson, J. R. Schrieffer, and W.-P. Su Rev. Mod. Phys. 60, 781 (1988)

[22] Y. Bar-Yam (unpublished)

COHERENT BACK SCATTERING AND OTHER OPTICAL EFFECTS IN RANDOM MEDIA

Michael J. Stephen

Physics Department
Rutgers University
Piscataway, N.J. 08855

ABSTRACT

The propagation of light in a random medium is considered. Coherent back scattering in the region of weak localization and near the mobility edge is discussed and a scaling theory is presented. Results on the fluctuations and correlations in the transmitted light are presented. The effects of inelastic scattering arising from particle motion on the spectral distribution of the scattered light is discussed. It is pointed out that there is an interesting optical analog to the RKKY interaction in metals.

1. INTRODUCTION

There has been a lot of work recently on the transport, localization and fluctuations of electrons in disordered metals.[(1)] Localization of electrons in a disordered metal is a one body problem and the statistics of the electrons is not important. What is important is that the electron behaves as a wave and localization results from the interference of the various multiply scattered waves. Thus the same phenomena which occur and methods which apply for electrons in disordered metals should be useful for other kinds of waves in disordered media e.g. phonons, spin waves, electromagnetic waves etc. The case of light scattered from a disordered medium is particularly interesting because experiments can easily be carried out and it is possible to observe the interference effects which give rise to localization. In this lecture we will discuss the theory briefly in Section 2, coherent back scattering in 3 and 4, intensity correlations and fluctuations in 5 and time dependent effects in 6 and RKKY interactions in 7.

2. THEORY

The basic equation that we begin with is the wave equation

$$[\nabla^2 + k^2(1+\epsilon(r))]\ E(r) = 0 \tag{2.1}$$

Recent Progress in Many-Body Theories, Volume 2
Edited by Y. Avishai
Plenum Press, New York, 1990

where $k = \frac{2\pi}{\lambda}$ is the wave vector of the wave, E is the wave field (either scalar or vector) and $\epsilon(r)$ is the random part of the dielectric constant. We have assumed it to be time independent so that the scattering is purely elastic (in Sect. 6 we discuss inelastic scattering) and we take ϵ to have zero mean and a correlation function of the spatial white noise form

$$\langle \epsilon(\vec{r})\ \epsilon(\vec{r}) \rangle = \Delta\ \delta(\vec{r}-\vec{r}) \tag{2.2}$$

The important assumptions that we will make are (a) the scattering from a single impurity is of the Raleigh form i.e. the scatterers are much smaller than the wavelength of the light. This is consistent with the white noise assumption (2.2). (b) We are in the region of weak localization i.e. the mean free path of the wave is much larger than the wavelength. We can then use a form of perturbation theory to calculate the field correlation functions starting from (2.1). This perturbation theory is basically an expansion in ladder graphs, the small expansion parameter being $(k\ell)^{-1}$. The ladder graphs give rise to diffusive transport of the intensity in the medium. The mean free path is determined by the density of scatterers and in the weak scattering limit is $\ell = \frac{4\pi}{\Delta k^4}$. This is of the Rayleigh form and varies with the fourth power of the wavelength. The diffusion constant is $D = \frac{c\ell}{3}$ when c is the light velocity.

The coherent part of the light beam $\langle E(r) \rangle$ (angular brackets indicate an average over impurity configurations) will only penetrate a short distance of order ℓ into the medium but provides the source for the diffusely scattered light. Properties of the diffusely scattered light are contained in the correlation function (or coherence function)

$$\Gamma(\vec{R},\vec{r}) = \langle E^*(\vec{R}+\tfrac{\vec{r}}{2})\ E(\vec{R}-\tfrac{\vec{r}}{2}) \rangle_c \tag{2.3}$$

where the subscript c indicates a cumulant. When we take the Fourier transform function on the variable $\vec{r}$ to form $\Gamma(\vec{R},\vec{q})$ this function is proportional to the average intensity of radiation at point $\vec{R}$ with wave vector $\vec{q}$ i.e. travelling in the direction $\vec{q}$. As the scattering is elastic $q \sim k$ and it is convenient to integrate over q giving the total intensity of radiation at $\vec{R}$ travelling in the direction of the unit vector $\vec{s}$ as

$$J(R,\vec{s}) = \int_0^\infty q^2 dq\ \Gamma(R,q\vec{s}) \tag{2.4}$$

The correlation function (2.3) is one of the quantities of interest.

3. COHERENT BACK SCATTERING

The diffusely scattered light from the disordered medium is the result of many multiply scattered waves. The total wave amplitude of the scattered light, E_s, is the sum of the amplitudes a_α of the different scattered waves which we label by α i.e. $E_s = \sum_\alpha a_\alpha$. The average intensity of the scattered light is then

$$\langle |E_s| \rangle = \sum_\alpha \langle |a_\alpha|^2 \rangle + \sum_{\alpha=\beta} \langle a_\alpha^* a_\beta \rangle \tag{3.1}$$

As the medium is disordered it is generally assumed that the different amplitudes a_α have phases which are completely uncorrelated in which case the second term of (3.1) averages to zero. This is what is done in the usual treatment of Rayleigh scattering from a disordered medium. Interference usually results from some symmetry and in the case where the scattering is elastic i.e. the impurities are stationary, the only global symmetry that the disordered medium possesses is that of time reversal. It is now well known that two multiply scattered waves following time reversed paths interfere constructively in the backward direction. This is true for all multiple scattering paths of two or more scatterings. This simple argument shows that the intensity reflected in the backward direction is increased by a factor 2 over the background i.e. the terms $\langle a_\alpha^* a_\beta \rangle = \langle a_\alpha^* a_\alpha \rangle$ if β is the time reverse of path α (for backscattering). The earliest reference to this effect of which I am aware is in a footnote to a paper by Watson(2) where it is attributed to Ruffine. It was also discussed in 1966 by Langer and Neal(3) for electrons. It has now been observed experimentally by a number of groups(4-8) and called coherent backscattering (cbs).

The width of the backscattered peak can be determined by a simple argument. Consider a diffusive multiple scattering path of linear dimension L. If the scattered wave and the time reversed scattered wave are emitted at a small angle θ to the backward direction they have a path difference $L \sin\theta$ and interfere constructively provided $L \sin\theta < \lambda$. The shortest diffusion path has $L \sim \ell$ and thus constructive interference only occurs if $\theta < \lambda/\ell$. The angular line shape has been given by a number of authors for scalar waves(9-11) and is

$$J(\theta) \sim 1 + \frac{1}{(1+k\ell\sin\theta)^2}$$

where the first term represents the Rayleigh scattering and the second term the cbs. The line shape is triangular for θ small.

The above argument indicates that cbs at small angles comes from long diffusion paths L. Thus any perturbation which cuts off the long diffusion paths will give rise to a rounding of the triangular peak e.g. finite size effects and absorption(10,11). Similarly any effects which break the time reversal symmetry of the scattering medium will also lead to a rounding and reduction in the cbs peak e.g. diffusion or motion of the scatterers(9), applied magnetic fields (Faraday effect) and optical activity.(12)

Up to now we have only considered scalar waves and it is found that the effects of polarization are interesting(11) and it is not difficult to show that only the polarization preserving scattering shows true cbs.

4. COHERENT BACK SCATTERING NEAR THE MOBILITY EDGE

Near the mobility edge the correlation length is ξ and within this length the multiply scattered waves are strongly correlated. On the localized side of the mobility edge, waves will only penetrate a distance of order ξ in the medium. As the scattering paths are of finite length the reflected light will not show any special feature as a function of scattering angle θ and the scattered intensity will form a smooth background.

As we pass through the mobility edge into the region of extended states multiple scattering paths of geometrical length larger than ξ become possible and give rise to cbs. We can write the reflected intensity with transverse wave vector $q = k_o \sin\theta$ where θ is the angle measured from the backward direction in the form

$$R_q \sim R_b - \left(\frac{\ell}{\xi}\right)^a f(q\xi) \tag{4.1}$$

Here R_b is a background term that does not depend strongly on q or angle. We have calculated f in $2+\epsilon$ dimensions using the results of Lee and Abrahams[12] and find the exponent $a = 3-d$. f is normalized such that $f(o) = o$ and for a semi-infinite slab we find

$$f(q\xi) \sim q\xi f^1(o) \qquad q\xi < 1 \tag{4.2}$$

so that the cbs gives a triangular peak of width $q\xi \sim 1$ i.e. $\theta \sim (k_o\xi)^{-1}$. For d=3 the height of the peak, $f^1(o)$, is of order unity. For a slab of finite thickness L, f in Eq. (4.1) is replaced by $f(q\xi, qL)$ and for small q

$$\begin{aligned} f(q\xi, qL) &\sim q^2 L\xi \qquad && L > \xi \\ &\sim q^2 L^2 && L < \xi \end{aligned} \tag{4.3}$$

so that the triangular peak is rounded.

The effects of absorption can also be important near the mobility edge. If ℓ_i is the absorption length in the medium we introduce a dimensionless variable $x = k_o^{d-1}\xi^d/2\ell_i$ and f in (4.1) is replaced by $f(q\xi, x)$. The crossover to absorption dominated behavior when the cbs peak disappears occurs when $(q\xi)^d \sim x$. As the effects of absorption scale as ξ^d this would appear to make it difficult to observe cbs near the mobility edge.

The transmitted light through a slab of thickness L would not be expected to show any interesting angular dependence. The transmission coefficient t does show interesting scaling behavior. In the region of weak localization $t \sim \ell/L$ and is analogous to the conductance of a metal. It is thus natural to generalize[13] this result (in analogy with the scaling of the conductance) to

$$t = \frac{\xi^{2-d} f_1(L/\xi)}{L} \tag{4.4}$$

Thus near the mobility edge we find

$$t \sim \frac{1}{L^{d-1}} \qquad L<\xi$$

$$\sim \frac{\xi^{2-d}}{L} \qquad L>\xi \tag{4.5}$$

The second result only applies in the region of extended states and would be replaced by an exponential dependence in the localized region.

5. INTENSITY FLUCTUATIONS AND CORRELATIONS

The speckle patterns which are observed when coherent light is scattered from a disordered surface or medium result from the large intensity fluctuations of the scattered light.[14] The gross features of a speckle pattern can be understood by again regarding the total wave amplitude of the scattered light, E_s, as the sum of many independent amplitudes a_α as in Sect. 3. Neglecting any correlations between these amplitudes the average intensity is $\langle I\rangle = \sum_\alpha \langle |a_\alpha|^2\rangle$ and the intensity fluctuations

$$\langle I^2\rangle = \sum_{\alpha\beta\gamma\delta} \langle a_\alpha^* a_\beta a_\gamma^* a_\delta\rangle = 2\langle I\rangle^2 \tag{5.1}$$

when the factor 2 comes from the two possible ways of identifying the subscripts i.e. $\alpha=\beta$, $\gamma=\delta$ or $\alpha=\delta$, $\beta=\gamma$ in (4.1). The fluctuations in the intensity are thus as large as the average intensity. This argument can be extended to show that the distribution of the intensity $P(I) = \frac{1}{\langle I\rangle} e^{-I/\langle I\rangle}$, the Raleigh or exponential distribution. There do not appear to be any interesting corrections to this law (at least in the weak scattering limit) and it is well obeyed experimentally.

We now turn to the intensity correlation function $C(R_{12}) = \langle I(R_1)I(R_2)\rangle_c$ defined as a cumulant. The simple theory of this correlation function neglecting correlations between different multiply scattered waves was worked out by Shapiro[15] who showed that

$$C_o(R_{12}) = \langle I(R_1)\rangle\langle I(R_2)\rangle \left(\frac{\sin kR_{12}}{kR_{12}}\right)^2 e^{-R_{12}/\ell} \tag{5.2}$$

when ℓ is the mean free path and the subscript zero means the uncorrelated approximation to C. This reduces to (4.1) (without the 2 because C is a cumulant) when $R_1=R_2$ and shows that the correlations decrease exponentially in a mean free path. In addition the speckle pattern has structure on the scale of the wavelength.

However, we expect that (5.2) is not correct when $R_{12}>\ell$ because the transport of the light is diffusive in nature and diffusion is a long range phenomenon. In addition the diffusing intensity has an underlying wave field so that interference can occur. These interference effects were shown by Stephen and Cwilich[16] to lead to long range correlations of the intensity decreasing like R_{12}^{-3} for $R_{12} \gg \ell$. Their result for the correlation function to order $(k\ell)^{-2}$ in perturbation theory is

$$C_r(R_{12}) = C_o(R_{12}) + \frac{1}{k^2\ell^2} \left(\frac{\ell}{R_{12}}\right)^3$$

$$C_t(R_{12}) = C_o(R_{12}) + \frac{1}{k^2\ell^2} \left(\frac{\ell}{L}\right)^3 F(R_{12}/L) \qquad (5.3)$$

when the subscript r and t indicate the correlation for the light diffusely reflected and transmitted from a slab of thickness L respectively. The function $F(R/L) = L/R-1\ldots$ for $R<L$ and $F(R/L) = e^{-R/L}$ for $R>L$. The second terms in (5.3) are only valid if $R_{12}>\ell$. Thus the intensity correlation function drops off exponentially for $R_{12}\sim\ell$ but has a power law tail for $R_{12}>\ell$. The size of this tail is reduced by the factor $(k\ell)^{-2}$. The first and second term in (4.3) join up reasonably smoothly for $R_{12}\sim\ell$. These long range correlations have not yet been observed presumably because they are quite small.

Other quantities of interest are the reflection and transmission fluctuations. For unit incident intensity on a slab of cross sectional area A and thickness L we define the reflection and transmission coefficients r and t by

$$r = \frac{1}{A}\int d^2R\; I_r(R)$$

$$t = \frac{1}{A}\int d^2R\; I_t(R) \qquad (5.4)$$

where the integrals extend over the reflecting surface and transmitting surfaces respectively and I_r and I_t are the diffusely reflected and transmitted intensities on these surfaces respectively. The average reflection coefficient $\langle r\rangle\sim 1$ for a thick slab and $\langle t\rangle\sim\ell/L$ the dependence on $1/L$ resulting because the transport is diffusive. The fluctuations in r and t can be calculated from the correlation functions (5.3) and give

$$\langle r^2\rangle_c \sim \frac{1}{k^2\ell^2}\left(\frac{\ell^2}{A}\right)$$

$$\langle t^2\rangle_c \sim \frac{1}{k^2\ell^2}\left(\frac{\ell^2}{A}\right)\left[\left(\frac{\ell}{L}\right)^2 + \left(\frac{\ell}{L}\right)\right] \qquad (5.5)$$

In the case of reflection both terms in (4.3) give contributions to $\langle r^2\rangle_c$ of the same order or magnitude and have been combined in (4.5).

For transmission fluctuations the long range correlations give rise to the second term in (4.5) which is larger than the first term by a factor $L/\ell \gg 1$. Thus an alternative experimental approach is to measure the transmission fluctuations which are larger (by L/ℓ) than predicted by simple theory.

Analogous problems to those considered here arise in the conductance and its fluctuations of electrons in disordered metals and there has been considerable experimental and theoretical interest[(17,18)] recently in this area. The relation of transmission fluctuations to conductance fluctuations has been discussed by Feng, Kane, Lee and Stone.[(19)]

6. TEMPORAL FLUCTUATIONS

In this section we discuss how the dynamical properties of the scattering medium determine the spectral properties of the scattered intensity. Often it is assumed that the scattering is weak and can be treated in Born approximation. In this case the scattering is directly related to the dynamical structure factor of the medium. Here we are interested in the case where the wave is multiply scattered in the medium which will occur if the mean free path is much less than its dimensions. Owing to the multiple scattering, the scattered intensity will not depend in an important way on the scattering angle but its spectral properties will be determined by the dynamics of the medium. The total scattered intensity is obtained by summing the contribution of all possible multiple scattering paths. The relaxation time depends on the number of scattering events i.e. on the length of the multiple scattering path. The total intensity thus contains a broad range of relaxation times.

In this case the dielectric constant in (2.1) varies slowly with the time because of the motion of the scatterers and we take its correlation function to be

$$\langle \epsilon(r,t) \rangle\ \epsilon(r^1 t^1)\rangle = C(r-r^1,\ t-t^1) \tag{6.1}$$

It is useful to consider some simple examples:

(a) The scatterers diffuse with a diffusion constant D_i. The Fourier transform of C is

$$C(q,t) = \Delta e^{-q^2 D_i |t|} \tag{6.2}$$

(b) The scatterers have a mass m and a Maxwell-Boltzmann velocity distribution

$$C(q,t) = \Delta\ e^{-q^2 t^2/2m\beta} \tag{6.3}$$

where $\beta = 1/kT$.

The condition that the variation in ϵ be slow means that we only consider times $t<\tau_\lambda$ where τ_λ is the time for a scatterer to move a wavelength. In the above two examples $\tau_\lambda = (4D_i k^2)^{-1}$ and $\tau_\lambda = (m\beta/2k^2)^{1/2}$. The frequency change on scattering is small and we can consider the propagation of an almost monochromatic wave with frequency close to $\omega = ck$. The mean free path of the wave ℓ will not depend on the motion of the scatterers and is determined by the density of scatterers, $\ell = 4\pi/\Delta k^4$.

As a simple example we consider a monochromatic point source of light at the origin in an infinite medium and examine the spectral properties of the scattered intensity at point R. In particular we consider the correlation function

$$\Gamma(R,t) = \langle E(R,t)\, E^*(R,o)\rangle_c \tag{6.4}$$

Consider a scattering path of length L which goes from the origin to R. At each scattering event we get a Doppler shift $\Delta\omega_i$ and the total Doppler shift is $\Omega_L = \sum_i \Delta\omega_i$ when i runs from 0 to L/ℓ. The contribution to (6.4) from this path is proportional to

$$\langle e^{i\Omega_L t}\rangle = (\langle e^{i\Delta\omega t}\rangle)^{L/\ell} = (f(t))^{L/\ell}$$

where f is the angular average of the correlation function (6.1)

$$f(t) = \frac{1}{4\pi\Delta}\int d\vec{s}^{\,1}\; C(k(\vec{s}-\vec{s}^{\,1}),t) \tag{6.5}$$

where $\vec{s}$ and $\vec{s}^{\,1}$ are unit vectors. In the two examples (6.2) and (6.3)

$$f^{(a)}(t) = \frac{\tau_\lambda}{|t|}\,(1-e^{-|t|/\tau_\lambda})$$

$$f^{(b)}(t) = \frac{\tau_\lambda^2}{t^2}\,(1-e^{-t^2/\tau_\lambda^2}) \tag{6.6}$$

We now assume that the light diffuses in the medium and then the probability of reaching R after L steps is proportional to $(\ell/L)^{3/2} e^{-3R^2/4\ell L}$. We now average (6.5) with respect to this distribution

$$\Gamma(t) \sim \int_o^\infty dL \left(\frac{\ell}{L}\right)^{3/2} e^{-3k^2/4\ell L}\,(f(t))^{L/\ell} \tag{6.7}$$

The integrand has a maximum at $L^2 = 3R^2/4 \log \frac{1}{f}$ which gives

$$\Gamma(t) \sim e^{-\frac{\sqrt{3}R}{\ell}(\log \frac{1}{f})^{1/2}} \tag{6.8}$$

In the two cases (6.7) we thus find

$$\Gamma_a(t) \sim e^{-\frac{R}{\ell}\left(\frac{3t}{2\tau_\lambda}\right)^{1/2}}$$

$$\Gamma_b(t) \sim e^{-\left(\frac{3}{2}\right)^{1/2}\frac{Rt}{\ell\tau_\lambda}} \tag{6.9}$$

The important features of these results are that we get a broad distribution of relaxation times which depends on the path length R of the scattered wave. For the case of diffusing scatterers the time correlation function decays as a stretched exponential and for a free gas it decays exponentially. They are very different from the case of single scattering, (6.2) and (6.3). In the multiple scattering case it is still possible to obtain information about the dynamics of the medium τ_λ. Results for different geometries have been given by Stephen[20].

A number of experimental studies of the time and frequency dependence of the field correlation function in the multiple scattering case have been carried out[21-23] generally in reasonable agreement with these results.

7. RKKY INTERACTIONS IN METALS AND THEIR OPTICAL ANALOG

The RKKY interaction in metals is the interaction behavior localized magnetic moments mediated by the conduction electrons. In a pure metal it takes the form

$$J(R) \sim \frac{(\vec{S}_1 \bullet \vec{S}_2)}{R^3} \cos\,(2k_F R + \phi) \tag{7.1}$$

where $\vec{S}_1$ and $\vec{S}_2$ are the two localized spins which are separated by a distance R, k_F is the Fermi momentum and ϕ is a phase. This interaction is believed to be responsible for spin glass behavior in dilute magnetic alloys.

The effects of scattering of the electrons on J(R) is interesting. When J(R) is averaged over the impurity configurations of the scatterers it becomes short-range and

$$\langle J(R) \rangle \sim \frac{1}{R^3}\, e^{-R/\ell} \tag{7.2}$$

where ℓ is the electron mean free path. This can be understood as follows: the different scattered electron waves have different phases ϕ in (7.1) and when we average over this random phase the interaction becomes short range.

In spin glasses the transition temperature is determined by $\langle J^2(R) \rangle$ not $\langle J(R) \rangle^2$. In this case the random phase cancels and

$$\langle J^2(R) \rangle \sim \frac{1}{R^6}\, (\vec{S}_1 \bullet \vec{S}_2)^2 \tag{7.3}$$

independent of the mean free path (provided the scattering is elastic). This result has also been verified by detailed calculations in the weak scattering limit. In the case of potential and spin-orbit scattering the mean square interaction becomes independent of the angle between $\vec{S}_1$ and $\vec{S}_2$ and

$$\langle J^2(R)\rangle \sim \frac{1}{R^6} S_1^{\ 2} S_2^{\ 2} \quad \text{(spin orbit)} \tag{7.4}$$

An interesting and important consequence of these results is that the transition temperature of the spin glass

$$(kT_c)^2 \sim \sum_j \langle J^2(R_{ij})\rangle \tag{7.5}$$

is independent of mean free path and $T_c \sim x$ the concentration of magnetic impurities. A very different conclusion is reached if we use $\langle J\rangle^2$ in (7.5). These results are a nice example of phase coherence of the scattered electron waves.

There is an optical analog to the RKKY interaction. Consider two polarizable particles in a strong monochromatic coherent laser beam. The first particle at $R_1^{\ 2}$ is polarized by the laser field and acquires a dipole moment $\vec{p}_1(t) = \vec{p}_1 \cos(\vec{k}_o\bullet\vec{R}_1 - \omega_o t)$ which radiates and sets up a field which interacts with the induced dipole on particle 2. The net effect is an induced interaction between the particles. In the case where the dipoles are small compared with the wavelength of the light this interaction is

$$V(R) = -\left[\frac{k_o^{2}\vec{p}_1\bullet\vec{p}_2}{R} + \frac{3(\vec{p}_1\bullet\vec{R})(\vec{p}_2\bullet\vec{R}) - R^2\vec{p}_1\bullet\vec{p}_2}{R^5} - \frac{k_o^2(\vec{p}_1\bullet\vec{R})(\vec{p}_2\bullet\vec{R})}{R^3}\right]$$

$$\cos(k_oR-\phi) + \frac{k_o}{R^4}\left[R^2(\vec{p}_1\bullet\vec{p}_2) - 3(\vec{p}_1\bullet\vec{R})(\vec{p}_2\bullet\vec{R})\right] \sin(k_oR - \phi)$$

where $\vec{R} = \vec{R}_2 - \vec{R}_1$ and $\phi = \vec{k}_o\bullet\vec{R}$. This interaction falls off like R^{-1} for large R and oscillation. As p_1 and p_2 are proportional to the field strength E, $V \sim E^2$. It has apparently been observed by Burns, Fournier and Golovchenko[24].

We can take over some of what we know about RKKY interactions. In the presence of elastic scattering of the photons $\langle V(R)\rangle$ would be expected to be short ranged i.e. $\sim e^{-R/\ell}$ but its mean square $\langle V^2(R)\rangle \sim \frac{k_o^4 p_1^2 p_2^2}{R^2}$ independent of the mean free path and the angle between the dipoles. It may be interesting to investigate this interaction in more detail.

REFERENCES

1. For a recent survey see "Localization, Interaction and Transport Phenomena in Impure Metals", G. Bergmann, Y. Bruynseradi and B. Kramer, Editors (Springer-Verlag, New York 1985).
2. K. M. Watson, J. Math. Phys. 10, 688 (1969).
3. J. S. Langer and T. Neal, Phys. Rev. Lett. 16, 984 (1966).
4. M. P. van Albada and A. Lagendijk, Phys. Rev. Lett. 55, 2692 (1985).

5. P. E. Wolf and G. Maret, Phys. Rev. Lett. 55, 2696 (1985).
6. S. Etemad, R. Thompson and H. J. Andrejco, Phys. Rev. Lett. 57, 575 (1986).
7. M. Kaveh, M. Rosenbluh, I. Edrei and I. Freund, Phys. Rev. Lett. 57, 2049 (1986).
8. A. Z. Genack, Phys. Rev. Lett. 58, 2043 (1987).
9. A. A. Golubentsev, Zh. Eksp. Teor. Fiz. 86, 47 (1984). (Sov. Phys. JETP 59, 26 (1984)).
10. E. Akkermans, P. E. Wolf and R. Maynard, Phys. Rev. Lett. 56, 1471 (1986).
11. M. J. Stephen and G. Cwilich, Phys. Rev. 34, 7564 (1986).
12. E. Abrahams and P. A. Lee, Phys. Rev. 33, 683 (1986).
13. P. W. Anderson, Phil. Mag. B52, 505 (1985).
14. J. W. Goodman, J. Opt. Soc. Am. 66, 1145 (1976).
15. B. Shapiro, Phys. Rev. Lett. 57, 2168 (1986).
16. M. J. Stephen and G. Cwilich, Phys. Rev. Lett. 59, 285 (1987).
17. B. L. Altschuler, Pis'ma Zh. Eksp. Teor. Fiz. 41, 530 (1985); (JETP Lett. 41, 648 (1985)).
18. P. A. Lee and A. D. Stone, Phys. Rev. Lett. 55, 1622 (1985).
19. S. Feng, C. Kane, P. A. Lee and A. D. Stone, Phys. Rev. Lett. 61, 834 (1988).
20. M. J. Stephen, Phys. Rev. B37, 1 (1988).
21. G. Maret and P. E. Wolf, Z. Phys. B65, 409 (1987).
22. D. J. Pine, D. A. Weitz, P. M. Chaikin and E. Herbolzheimer, Phys. Rev. Lett. 60, 1134 (1988).
23. I. Freund, M. Kaveh and M. Rosenbluh, Phys. Rev. Lett. 60, 1130 (1988).
24. M. M. Burns, J. M. Fournier and J. A. Golovchenko, Phys. Rev. Lett. 63, 1237 (1989).

MANY BODY TECHNIQUES FOR WAVES PROPAGATING IN RANDOM MEDIA

B. Shapiro

Department of Physics
Technion-Israel Institute of Technology
Haifa, ISRAEL

1. INTRODUCTION

A wave propagating in disordered medium undergoes multiple scattering from inhomogeneities (the wave can be acoustic, electromagnetic, or it can represent an electron wave function). The scattered waves interfere with each other and, as a result, a certain intensity pattern (a speckle pattern) is formed within the sample. This speckle pattern is highly irregular, with large intensity variations over short distances. The irregularities are not due to noise but rather reflect the specific arrangement of the inhomogeneities (impurities) in a particular sample. This is true as long as there is no inelastic scattering within the sample, i.e. the wave propagates coherently through the entire sample (mesoscopic regime).

In recent years it has become clear that such speckle patterns, as irregular as they may look, do possess a high degree of correlations. This gives rise to all kinds of unusual correlations and fluctuations in measurable quantities, like the universal conductance fluctuations in disordered metals[1,2], fluctuations in transmission through disordered slabs[3-8], or the optical memory effect[4,9,10]. The (relative) fluctuations are unusual in the sense that they can be anomalously large and do not decay as $\Omega^{-1/2}$, where Ω is the volume of the sample. For instance, for a three-dimensional metallic sample of size L, the relative fluctuation in conductance decreases with L as L^{-1} instead of the standard $L^{-3/2}$ law.

For weak disorder the problem has a small parameter $\lambda|\ell$, where λ and ℓ are, respectively, the wave length and the elastic mean free path. This small parameter enables one to develop a perturbative diagrammatic technique for calculating various averaged quantities. Indeed, most of the

Recent Progress in Many-Body Theories, Volume 2
Edited by Y. Avishai
Plenum Press, New York, 1990

theoretical results in the field - such as the universal conductance fluctuations, the optical memory effect, and many others - were first obtained by this technique. The technique turns out to be very similar to the one used in the many-particle physics, although here it is averaging over the disorder (rather than the interaction between particles) that introduces "self-energies", "ladders" and other ingredients of the technique. The technique was widely used already in the early work on wave propagation in random media (see, for instance, Ref. 11 and 12, for classical waves and electrons, respectively). Much progress on the subject has been made in recent years. New classes of diagrams were identified and summed up and new physical effects, due to interference in disordered systems, were discovered. The purpose of the paper is to discuss some of these recent developments.

Section 2 of the paper contains a brief review of the diagram technique. Sections 3 and 4 deal with two examples of anomalous fluctuations in mesoscopic systems: fluctuations in optical transmission and fluctuations in the magnetic susceptibility of disordered metals. The conclusions are summarized in Section 5.

2. THE DIAGRAMMATIC TECHNIQUE (A SHORT REVIEW)

Let us consider a scalar wave equation for a monochromatic field $\psi_\omega(\vec{r})$ of frequency ω:

$$\{\nabla^2 + k_0^2[1+\mu(\vec{r})]\}\ \psi_\omega(\vec{r}) = 0 \ . \tag{1}$$

The field $\psi_\omega(\vec{r})$ can describe the wave function of an electron or a classical wave, e.g., acoustic or (if polarization effects can be neglected) electromagnetic.[13] For classical waves $\mu(\vec{r})$ is the fluctuating part of the refraction index and $k_0=\omega/c$, where c is the speed of propagation in the average medium. For electrons $\mu(\vec{r}) = -V(\vec{r})/E$ and $k_0^2 = 2mE/\hbar^2$, where E is the particle's energy, m is its mass, and $V(\vec{r})$ is the random potential. To define the problem completely one needs, first, to specify the sources (e.g., the wave impinging on the slab) and, second, to give the statistical properties of the random function $\mu(\vec{r})$. We shall assume a white-noise Gaussian statistics, i.e.

$$\langle\mu(\vec{r})\rangle = 0, \qquad \langle\mu(\vec{r})\mu(\vec{r}\,')\rangle = u\delta(\vec{r}-\vec{r}\,') \ , \tag{2}$$

and higher order correlations factorize into products of pairs. In Eq. (2) angular brackets denote averaging over the statistical ensemble of samples, u is a constant, and both $\vec{r}$ and $\vec{r}\,'$ should be within the medium (otherwise $\mu(\vec{r})$ is identically zero).

Let us now illustrate the technique with the example of a point source, at a point $\vec{r}_0$, embedded in an infinite medium (a more realistic geometry is considered in the next section). The field at point $\vec{r}$ is then given by the Green's function of Eq. (1), which obeys the wave equation :

$$\{\nabla^2 + k_0^2[1+\mu(\vec{r})] + i\eta\}\, G_\omega(\vec{r},\vec{r}_0) = \delta(\vec{r}-\vec{r}_0) \ , \tag{3}$$

where η is a positive infinitesimal. One needs to calculate quantities like $\langle G_\omega(\vec{r},\vec{r}_0)\rangle$ (the average field at point $\vec{r}$), or $\langle|G_\omega(\vec{r},\vec{r}_0)|^2\rangle$ (the average intensity at point $\vec{r}$), or some more complicated objects involving products of G's at different points (and, possibly, different frequencies). This is done by expanding $G_\omega(\vec{r},\vec{r}_0)$ in the standard Born series and then averaging, term by term, this series or its products with other similar series. Various terms are conveniently represented by diagrams. The basic rules for generating and calculating such diagrams are described in the literature cited above. Here we give only a summary.

Diagrams for the averaged Green's function, $\langle G_\omega(\vec{r},\vec{r}_0)\rangle$, are generated from the Born series for $G_\omega(\vec{r},\vec{r}_0)$ as shown in Fig. 1.

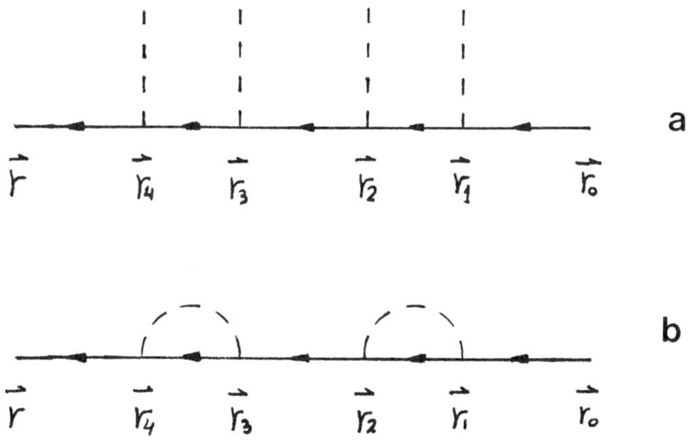

Fig. 1. (a) An example of a scattering sequence in the Born series for $G_\omega(\vec{r},\vec{r}_0)$ (propagation from $\vec{r}_0$ to $\vec{r}$ with scattering events at 4 intermediate points). (b) One of the three diagrams for $\langle G_\omega(\vec{r},\vec{r}_0)\rangle$ generated by the scattering sequence (a) upon averaging (the other two diagrams correspond to the other two possible ways of pairing the dashed lines in the sequence (a)).

A dashed line connecting two points $\vec{r}_i, \vec{r}_j$ in Fig. 1(b) corresponds to $k_0^4 \langle\mu(\vec{r}_i)\mu(\vec{r}_j)\rangle$, whereas a solid line represents a free Green's function $G_\omega^{(0)}(|\vec{r}_i-\vec{r}_j|)$. Making use of the Dyson equation one obtains, to leading order, the self-energy:

$$<G_\omega(\vec{r},\vec{r}_0)> = -\frac{1}{4\pi|\vec{r}-\vec{r}_0|}\exp[(ik_0 - \frac{1}{2\ell})\ |\vec{r}-\vec{r}_0|] \tag{4}$$

where $\ell \equiv 4\pi/uk_0^4$ is the mean free path. Eq. (4) differs from the free Green's function $G_\omega^{(0)}$ by the exponential decay factor $\exp(|\vec{r}-\vec{r}_0|/2\ell)$. An additional difference, which is of little interest and was ignored in Eq. (4), is a change in k_0).

Consider now the average intensity $<|G_\omega(\vec{r},\vec{r}_0)|^2>$, which is a particular case of a more general correlation function $<G_\omega(\vec{r},\vec{r}_0)G^*_{\omega'}(\vec{r}',\vec{r}'_0)>$. This function consists of a trivial part (product of two averaged Green's functions) plus the rest (the connected part). The most important contribution to the connected part comes from the ladder diagrams depicted in Fig. 2. The thick lines with arrows to the left correspond to the averaged (retarded) Green's functions, given in Eq. (4). The thick lines with arrows to the right correspond to the complex conjugate, i.e. advanced, Green's functions. The shaded box, $L(\vec{r}_1,\vec{r}_2,\vec{r}'_1,\vec{r}'_2;\omega,\omega')$, represents the sum of ladder diagrams, defined in Fig. 2b. It can be written as $L(\vec{r}_1,\vec{r}_2,\vec{r}'_1,\vec{r}'_2;\omega,\omega') = \delta(\vec{r}_1-\vec{r}'_1)\delta(\vec{r}_2-\vec{r}'_2)\cdot T(\vec{r}_1,\vec{r}_2;\omega,\omega')$, where the function T satisfies the integral equation

$$T(\vec{r}_1,\vec{r}_2;\omega,\omega') = \frac{4\pi}{\ell}\delta(\vec{r}_1-\vec{r}_2) + \frac{4\pi}{\ell}\int d^3r <G_\omega(\vec{r}_1,\vec{r})><G^*_{\omega'}(\vec{r}_1,\vec{r})>T(\vec{r},\vec{r}_2;\omega,\omega') \tag{5}$$

If the points $\vec{r}_1$ and $\vec{r}_2$ are not too close, i.e. $|\vec{r}_1-\vec{r}_2|>>\ell$, and the difference $|\omega-\omega'|\equiv|\Delta\omega|$ is not too large, i.e. $|\Delta\omega|<<c/\ell$, then Eq. (5) is equivalent to a diffusion equation

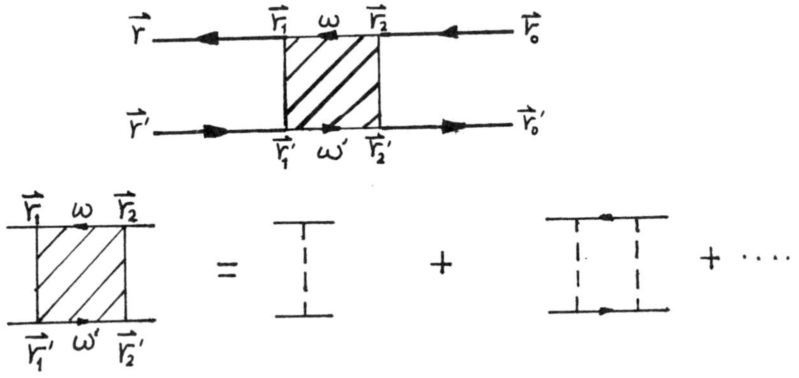

Fig. 2. The ladder diagrams

$$(\nabla^2 + i\,\frac{\Delta\omega}{D})\ T(\vec{r}_1,\vec{r}_2;\Delta\omega) = -\frac{12\pi}{\ell^3}\,\delta(\vec{r}_1-\vec{r}_2)\ , \qquad (6)$$

where $D = c\ell/3$ is the diffusion coefficient. Eq. (6) is solved by

$$T(\vec{r}_1,\vec{r}_2;\Delta\omega) = \frac{3}{\ell^3}\,\frac{1}{|\vec{r}_1-\vec{r}_2|}\,\exp[(-1+i)\sqrt{\frac{|\Delta\omega|}{2D}}\,|\vec{r}_1-\vec{r}_2|]\ . \qquad (7)$$

In particular, for $\Delta\omega=0$ the ladder T, and therefore the average intensity, is inversely proportional to the distance from the source. Note that this behavior is very different from the exponential decay for the average field [Eq. (4)].

In order to calculate fluctuations and correlations in intensity, which characterize the speckle pattern, one needs to consider more complicated objects containing products of four Green's functions. Consider, for instance, the intensity-intensity correlation function $<\Delta I_\omega(\vec{r},\vec{r}_0)\Delta I_\omega(\vec{r}\,',\vec{r}_0)> \equiv C_\omega(\vec{r},\vec{r}\,';\vec{r}_0)$ where $\Delta I_\omega \equiv |G_\omega|^2 - <|G_\omega|^2>$ is the deviation of intensity, at point $\vec{r}$, from its average value. If both $\vec{r}$ and $\vec{r}\,'$ are sufficiently far (i.e. at least a few ℓ) from the source $\vec{r}_0$, and $|\vec{r}-\vec{r}\,'|\lesssim\ell$, then this correlation function is given by the diagram in Fig. 3. Substituting the expressions for the averaged Green's function [Eq. (4)] and for the ladder [Eq. (7), with $\Delta\omega=0$], one finds[14] that this diagram gives:

$$C_\omega(\vec{r},\vec{r}\,') = <I_\omega(\vec{r})>^2\,[\frac{\sin(k_0\Delta r)}{k_0\Delta r}]^2\,\exp(-\Delta r/\ell)\ , \qquad (8)$$

where $\Delta r \equiv |\vec{r}-\vec{r}\,'|$ and $<I_\omega(\vec{r})> = 3/16\pi^2\ell|\vec{r}-\vec{r}_0|$ is the average intensity at point $\vec{r}$ (the position $\vec{r}_0$ of the source is suppressed in Eq. (8)). In particular, for $\vec{r}=\vec{r}\,'$ one obtains $<\Delta I^2> = <I>^2$ which is the well known large intensity fluctuations.[15] It was pointed out by Stephen and Cwilich[3] that Eq. (8) is valid only for $\Delta r\lesssim\ell$. For $\Delta r>>\ell$ diagrams other than that in Fig. 3 dominate the intensity-intensity correlation function C_ω. As a result, $C_\omega(\vec{r},\vec{r}\,')$ decays, under increase of $|\vec{r}-\vec{r}\,'|$, according to a power law[3] (rather than exponentially). An alternative way to study

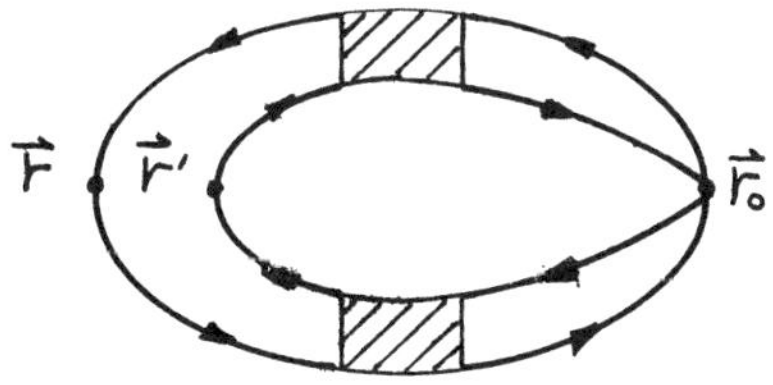

Fig. 3. This diagram describes correlations in intensity.

these long-range correlations is to use the Langevin approach as proposed by Zyusin and Spivak.[6] This will be discussed in the next Section.

3. FLUCTUATIONS IN OPTICAL TRANSMISSION

Fluctuations in optical transmission provide an instructive example of the anomalous fluctuations mentioned in the Introduction. Fluctuations in transmission were studied in Refs 3-8. In particular, Ref. 7 addressed the question of the effect of geometry (width of the beam) on transmission fluctuations. Below we define the problem and describe the results.

Consider a disordered slab of thickness L (in the z-direction) and infinite in the transverse (x,y) directions. A beam of radiation, with an arbitrary intensity profile (for instance, a uniform circular beam of diameter W) is impinging on the slab along the z-axis. The average intensity, within the slab, satisfies a diffusion equation (for $L \gg \ell$) so that the average transmission $\langle T \rangle$ is given simply by $\langle T \rangle = D/cL = \ell/3L$. Thus, $\langle T \rangle$ does not depend on the geometry (e.g. the width W) of the beam. Our purpose is to study the fluctuation $\langle \Delta T^2 \rangle$ and in particular its sensitivity to the value of the ratio L/W.

Let us first note that the expression (8) for the intensity correlation is valid for any geometry[7], as long as both $\vec{r}$ and $\vec{r}'$ are well inside the sample (a few ℓ away from the boundary) and Δr is smaller than ℓ (now the average intensity $\langle I_\omega(r) \rangle$ at point $\vec{r}$ is not given, of course, by the point source expression). It turns out that fluctuations in T are dominated not by these short-range intensity correlations but rather by the slowly decaying long-range correlations.[2,3] Diagrams responsible for the long-range correlations were identified in Refs 2,3 and, for the plane wave case (i.e. $W \gg L$), the result $\langle \Delta T^2 \rangle \simeq \ell/k_0^2 L W^2$ was obtained. Zyuzin and Spivak[6] have proposed a different approach for calculating fluctuations. In their approach (which is analogous to the Langevin approach to Brownian motion) the short-range fluctuations act as external sources in the diffusion equation. That is, large "spontaneous" fluctuations, which occur at distances smaller than ℓ , propagate by diffusion to large distances producing thus long-range correlations in various quantities. This approach was used in Ref. 7 for studying the effect of geometry on $\langle \Delta T^2 \rangle$. For instance, for a uniform circular beam with diameter W it was found that $\langle \Delta T^2 \rangle = (4/3)(k_0 \ell)^{-2} (\ell/L)^3 F(L/W)$, where $F(x)$ is a monotonically increasing function of its argument. For $x \ll 1$, $F(x) \approx (2/3)x^2$ whereas for $x \gg 1$, $F(x) \approx x/4$. This leads to

$$\langle \Delta T^2 \rangle = \begin{cases} (8/9)(k_0\ell)^{-2}\ (\ell/L)(\ell/W)^2, & \text{for } L \ll W \\ (1/3)(k_0\ell)^{-2}\ (\ell/L)^2(\ell/W), & \text{for } L \gg W \end{cases} \tag{9}$$

The case $L \ll W$ corresponds to the plane-wave limit and for this case the result coincides, up to numerical factors, with the results in Refs. 3,6. It was also supported by numerical calculations.[8]

4. FLUCTUATIONS IN THE DIAMAGNETIC SUSCEPTIBILITY

This Section is devoted to fluctuations in the diamagnetic response of disordered metals, which is another example of large anomalous fluctuations. This problem has recently been studied by a number of authors.[16-19] These authors consider a metallic sample in the presence of a weak static magnetic field. The electrons in the sample are subjected to a random potential $V(\vec{r})$ due to impurities. The electron-electron interactions and the effects related to the electron spin are neglected. For a weak magnetic field (linear response) one can derive[19] the following expression for the static magnetic susceptibility χ_{zz}:

$$\chi_{zz} = -\frac{1}{2c}\frac{1}{\Omega}\int d^3r d^3r' (x-x')^2\ Q_{yy}(\vec{r},\vec{r}') \tag{10}$$

where the integration is over the sample volume Ω and

$$Q_{yy}(\vec{r},\vec{r}') = -\frac{2}{c}\left(\frac{e\hbar}{m}\right)^2 \frac{1}{2\pi i}\int d\varepsilon f(\varepsilon)\ \left[\frac{\partial G^R_\varepsilon(\vec{r},\vec{r}')}{\partial y}\ \frac{\partial G^R_\varepsilon(\vec{r}',\vec{r})}{\partial y'} - \text{c.c.}\right] \tag{11}$$

Here G^R_ε denotes the (retarded) Green's function of the Schrödinger equation in the absence of the magnetic field. It differs from the Green's function in Eq. (3) by a factor $2m/\hbar^2$ (in addition to the changes $k_0^2 \to 2m\varepsilon/\hbar^2$, $\mu(\vec{r}) = -V(\vec{r})/\varepsilon$ already mentioned in Section 2). The function $f(\varepsilon)$ is 1 for occupied states and zero otherwise.

In the absence of disorder a straightforward calculation leads to the diamagnetic susceptibility $\chi_0 = -\mu_B^2 g_0/3$ where μ_B is the Bohr magneton and g_0 is the density of states at the Fermi level. In the presence of disorder one has to average over the ensemble of samples. For the average susceptibility $\langle\chi\rangle$ one obtains only small, of order $(k_f\ell)^{-2}$, corrections to χ_0 - consistent with the old result of Dingle[20] (k_f is the Fermi wave vector and ℓ is the electron mean free path). It is important to realize that, although having only a minor effect on the averaged susceptibility, the disorder does give rise to large sample-to-sample fluctuations. These fluctuations are manifested in the large value of the variance $\langle\Delta\chi^2\rangle$.

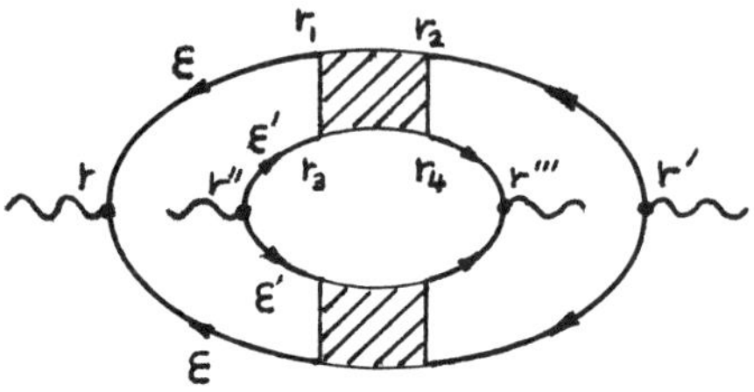

Fig. 4. A diagram for $<\Delta\chi^2>$. Arrows to the left (right) denote retarded (advanced) Green's functions.

Calculating the variance $<\Delta\chi^2>$, from Eqs. (10) and (11), requires averaging of a product of four derivatives of Green's functions. The diagrams are similar to those describing the conductance fluctuations[1], as well as to the one shown in Fig. 3. An example of a diagram contributing to $<\Delta\chi^2>$ is given in Fig. 4. The solid lines denote the averaged Green's function and a ladder box $L(\vec{r}_1,\vec{r}_2,\vec{r}_3,\vec{r}_4;\varepsilon,\varepsilon')$ is equal to $T(\vec{r}_1,\vec{r}_2;\Delta\varepsilon)\cdot\delta(\vec{r}_1-\vec{r}_3)\delta(\vec{r}_2-\vec{r}_4)$ where $\Delta\varepsilon \equiv \varepsilon-\varepsilon'$ and T is given by (compare to Eq. (7)):

$$T(\vec{r}_1,\vec{r}_2;\Delta\varepsilon) = \left(\frac{\hbar^2}{2m}\right)^2 \frac{3}{\ell^3} \frac{1}{|\vec{r}_1-\vec{r}_2|} \exp\left[(-1+i)\sqrt{\frac{|\Delta\varepsilon|}{2\hbar D}}\,|\vec{r}_1-\vec{r}_2|\,\right] \tag{12}$$

with $D = \frac{1}{3} v_F\ell$ being the diffusion coefficient (v_F is the Fermi velocity). The averaged Green's functions rapidly decay and can be integrated over, so that the contribution of the diagram in Fig. 4 reduces to[19]:

$$<\Delta\chi^2> = -2\chi_0^2 \left(\frac{8k_F^2\ell^3}{\hbar^3 v_F^3\Omega}\right)^2 \mathrm{Re}\int_{-\infty}^{0} d\varepsilon \int_{0}^{\infty} d\varepsilon' \int d^3r d^3r' (x-x')^4\, T^2(\vec{r},\vec{r}';\Delta\varepsilon) \tag{13}$$

Using Eq. (12), one obtains the Fukuyama's[17] result $<\Delta\chi^2> \simeq \chi_0^2(\ell/L)^2$, where L is the sample size (additional diagrams, i.e. others than the one in Fig. 4, also contribute to $<\Delta\chi^2>$).

It was pointed out by Serota[18] that the above estimate for $<\Delta\chi^2>$ does not take into consideration that, for an isolated sample, the diffusion equation for the ladder has an eigensolution which is constant throughout the sample. This zero-mode contributes to the ladder $T(\vec{r}_1,\vec{r}_2;\Delta\varepsilon)$ a constant $T_0(\Delta\varepsilon) = -\Omega^{-1}(\hbar^2/2m)^2(12\pi/\ell^3)(\hbar D/i\Delta\varepsilon)$ which was not included in the expression (12) for T (this expression describes a ladder in infinite space). Taking into account this constant, together with the appropriate energy cut-off $\Delta \simeq L^{-d} g_0^{-1}$ in Eq. (13), one obtains the Serota's result[18] $<\Delta\chi^2> \simeq (k_F\ell)^4\chi_0^2$. Note that $<\Delta\chi^2>$ is much larger than χ_0^2, so that fluctuations are huge and the average susceptibility does not represent a typical measurement on a specific sample. Serota's estimate for $<\Delta\chi^2>$ applies only when the sample is completely isolated. On the other hand, for a sample which is a part of the environment the much smaller Fukuyama's estimate should be valid. This will happen, if the sample can exchange electrons with reservoirs (contacts) or, for instance, if electrons (which scatter elastically within the sample) undergo an inelastic collision at the moment when they reach the boundary.

Let us emphasize that the results of this Section are restricted to weak magnetic fields (linear response). In a strong magnetic field fluctuations in diamagnetic response are considerably suppressed, as shown in Ref. 19 (for a two-dimensional electron gas in the presence of a random potential). It was assumed there that the magnetic field is so strong that the magnetic length $\lambda \equiv (\hbar c/eB)^{1/2}$ is smaller than the typical variation length L_V of the potential (and that only the first Landau level is occupied). Under these conditions the relative fluctuation δ in the magnetic moment is of order $\delta \simeq (L_V/L)(V_0 m\lambda^2/\hbar^2)$ where V_0 is the characteristic magnitude of the potential.

5. CONCLUSIONS

We have reviewed the diagrammatic technique for waves propagating in disordered systems and showed how the technique is applied for calculating fluctuations in various physical quantities. We discussed two examples of such fluctuations: fluctuations in optical transmission through disordered slabs and fluctuations in the diamagnetic susceptibility of disordered metals.

Fluctuations in transmission (in contrast to the average transmission) are sensitive to the width W of the beam. For instance, for a narrow beam (W is smaller than the sample thickness L) the square of the relative fluctuation is $\delta \equiv <\Delta T^2>/<T>^2 \simeq (k_0^2 \ell W)^{-1}$. It should be possible to observe these fluctuations experimentally.

Similar anomalous fluctuations exist also in the diamagnetic response of disordered metallic samples. Here the relative fluctuation can be even larger than unity. It would be interesting to consider the closely related problem of persistent currents in disordered metallic rings[21] for the case when the magnetic field acts not only in the annulus but also on the ring. It would also be of interest to discuss the disorder-induced fluctuations in the de-Haas-van Alphen effect and to compare their magnitude with the oscillations due to the edge states.[22]

I am indebted to my collaborators E. Akkermans and R. Pnini. I am also grateful to P.A. Lee for sending me the preprint of Serota (Ref. 18). The kind hospitality of the Centre de Recherches sur les Tres Basses Temperatures (Grenoble), where part of this work was performed, is gratefully acknowledged. The research was supported by the United States-Israel Binational Foundation (BSF) and by the Fund for Promotion of Research at the Technion.

REFERENCES

1. P.A. Lee and A.D. Stone, Phys. Rev. Lett. 55, 1622 (1985); B.L. Altshuler, Pisma Zh. Eksp. Teor. Fiz. 41, 530 (1985) [JETP Lett. 41, 648 (1985)]; B.L. Altshuler and D.E. Khmelnitskii, Pisma Zh. Eksp. Teor. Fiz. 42, 291 (1985) [JETP Lett. 42, 359 (1985)]; Y. Imry, Europhys. Lett. 1, 249 (1986).
2. J.C. Licini, D.J. Bishop, M.A. Kastner and J. Melngailis, Phys. Rev. Lett. 55, 2987 (1985); W.J. Scocpol, P.M. Mankiewich, R.E. Howard, L.D. Jackel, D.M. Tennant and A.D. Stone, Phys. Rev. Lett. 56, 2865 (1986); T.J. Thornton, M. Pepper and H. Ahmed, Phys. Rev. B36, 4514 (1987); S. Washburn and R. Webb, Adv. Phys. 35, 375 (1986).
3. M.J. Stephen and G. Cwilich, Phys. Rev. Lett. 59, 285 (1987).
4. S. Feng, C. Kane, P.A. Lee and A.D. Stone, Phys. Rev. Lett. 61, 834 (1988).
5. P.A. Mello, E. Akkermans and B. Shapiro, Phys. Rev. Lett. 61, 459 (1988).
6. A. Yu. Zyuzin and B.Z. Spivak, Zh. Eksp. Teor. Fiz. 93, 994 (1987) [Sov. Phys. - JETP 66, 560 (1987)].
7. R. Pnini and B. Shapiro, Phys. Rev. B39, 6986 (1989).
8. I. Edrei, M. Kaveh and B. Shapiro, Phys. Rev. Lett. 62, 2120 (1989).
9. I. Freund, M. Rosenbluh and S. Feng, Phys. Rev. Lett. 61, 2328 (1988).
10. R. Berkovitz and M. Kaveh, unpublished.
11. U. Frish, in Probabilistic Methods in Applied Mathematics, edited by A.T. Bharucha-Reid (Academic, New York, 1968), Vol. 1, p. 75.
12. A.A. Abrikosov, L.P. Gorkov and I.E. Dzyaloshinski, Methods of Quantum Field Theory in Statistical Physics (Prentice-Hall, 1963). Sec. 39.
13. A vector (electromagnetic) field was considered, e.g., in M.J. Stephen and G. Cwilich, Phys. Rev. B34, 7564 (1986); K. Arya, Z.B. Su and J.L. Birman, unpublished.
14. B. Shapiro, Phys. Rev. Lett. 57, 2168 (1986).
15. See, e.g., A. Ishimaru, Wave Propagation and Scattering in Random Media (Academic, New York, 1978).
16. O.D. Cheishvili, Pisma Zh. Eksp. Teor. Fiz. 48, 206 (1988) [JETP Lett. 48, 225 (1988)].
17. H. Fukuyama, J. Phys. Soc. Japan 58, 47 (1989).
18. R.A. Serota, unpublished.
19. E. Akkermans and B. Shapiro, unpublished.
20. R.B. Dingle, Proc. Roy. Soc. A211, 517 (1952).
21. H.F. Cheung, E.K. Riedel and Y. Gefen, Phys. Rev. Lett. 62, 587 (1989).
22. U. Sivan and Y. Imry, Phys. Rev. Lett. 61, 1001 (1988).

LOCALIZATION AND FOLDING OF RANDOM CHAINS

H. Orland

Service de Physique Théorique de Saclay
Laboratoire de l'Institut de Recherche Fondamentale
du Commissariat à l'Energie Atomique
F-91191 Gif-sur-Yvette Cedex

I. INTRODUCTION

Many problems in classical statistical mechanics can be described mathematically by a Schrödinger type of equation. Such is the case of one dimensional field theories, problems of interfaces in two dimensions, or problems of random chains in any dimension. In the non-disordered case the systems are usually described by a one particle time-independent Schrödinger equation. Disorder usually appears as a random time-dependence in the system. Except for very special cases, the problem cannot be studied for a given instance of the disorder, and one must resort to averages. It is indeed well known that thermodynamic properties of quenched disordered systems can be calculated (in the thermodynamic limit), by averaging the thermal average of observables over the impurity distribution[1]. For example, the quenched free energy is given by:

$$F = -T\,\overline{\text{Log}\ Z} \tag{1}$$

where the bar means an average over the disorder distribution. Similarly, if $\mathcal{A}$ is an operator, its quenched average is given by:

$$\mathcal{A}_{\mathcal{Q}} = \overline{\langle \mathcal{A} \rangle} \tag{2}$$

where the brackets denote a thermal average. A standard technique to compute quenched averages is to introduce replicas[2]. Indeed, the fundamental identity of the replica trick is:

$$\overline{\text{Log}\ Z} = \left.\frac{\mathrm{d}}{\mathrm{d}n}\,\overline{Z^n}\right|_{n=0} \tag{3}$$

Thus, in order to compute quenched averages, the procedure is the following:

i) compute $\overline{Z^n}$ for integer n.

ii) analytically continue $\overline{Z^n}$ at $n = 0$, and take its derivative at $n = 0$.

Recent Progress in Many-Body Theories, Volume 2
Edited by Y. Avishai
Plenum Press, New York, 1990

It is this averaging procedure i), as we shall see, which transforms the one-particle problem into an n-particle problem by coupling the various replicated systems. The second step ii) is to study this system in the limit when the number of particles goes to zero. Note that this is not, in any sense, a low density limit, but rather an analytic continuation in the number of particles. In this talk, we will restrict ourselves to the study of random chains in various environments.

II. CONTINUOUS DESCRIPTION OF RANDOM CHAINS

A brownian random[3] chain of N links is made by joining together N links, the position of the kth link being independent of that of the preceeding $(k-1)^{\text{th}}$.

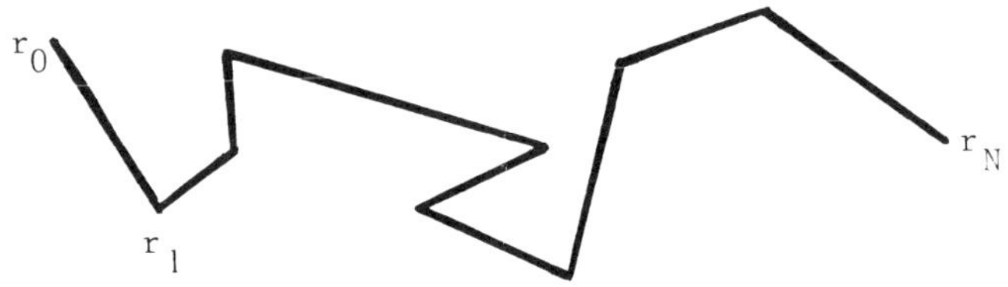

Each link is drawn from a distribution $g(\vec{r})$, so that the partition function of the chain reads:

$$Z = \int \prod_{i=1}^{N} \mathrm{d}\vec{r}_i \cdot \prod_{i=1}^{N} g(\vec{r}_i - \vec{r}_{i-1}) \tag{4}$$

Possible choices for the function $g(\vec{r})$ are:

- constant length links: $g(\vec{r}) = \delta(|\vec{r}| - a)$ where a is the length of a link.

- gaussian links: $g(\vec{r}) = \left(\frac{d}{2\pi a^2}\right)^{d/2} \mathrm{e}^{-\frac{d\vec{r}^2}{2a^2}}$ where d is the space dimension.

Many other choices of distribution are possible, however, the fundamental property of these chains is that in the large N limit (long chains), all these distribution are equivalent. Indeed, if $G\left(\vec{R}N \middle| \vec{0}0\right)$ denotes the probability to find $\vec{r}_N$ at point $\vec{R}$ (point $\vec{r}_0$ being fixed at $\vec{0}$), we have:

$$\begin{aligned} G\left(\vec{R}N \middle| \vec{0}0\right) &= \int_{\substack{\vec{r}_N = \vec{R} \\ \vec{r}_0 = \vec{0}}} \prod_{i=1}^{N-1} \mathrm{d}\vec{r}_i \cdot \prod_{i=1}^{N} g(\vec{r}_i - \vec{r}_{i-1}) \\ &= \int \frac{\mathrm{d}^d k}{(2\pi)^d} \mathrm{e}^{i\vec{k}\vec{R}} \cdot \tilde{g}^N\left(\vec{k}\right) \end{aligned} \tag{5}$$

where $\tilde{g}\left(\vec{k}\right)$ is the Fourier transform of $g(\vec{r})$. $G\left(\vec{R}N \middle| \vec{0}0\right)$ can be calculated by using the stationary phase method in (5). The expansion is manifestly around $\vec{k} = \vec{0}$, and yields:

$$\begin{aligned} G\left(\vec{R}N \middle| \vec{0}0\right) &\underset{N \longrightarrow +\infty}{\sim} \int \frac{\mathrm{d}^d k}{(2\pi)^d} \cdot \exp\left(i\vec{k}\vec{R} - \frac{a^2 N}{2d}\vec{k}^2\right) \\ &= \left(\frac{d}{2Na^2\pi}\right)^{d/2} \exp\left(-\frac{\vec{R}^2 d}{2Na^2}\right) \end{aligned} \tag{6}$$

Thus, it is a Gaussian distribution, independently of the distribution of links $g(\vec{r})$. To simplify further, we shall assume a Gaussian distribution for $g(\vec{r})$, and take the continuous limit as $N \longrightarrow \infty$.

In order to do so, we define $S = Na^2$, and take the limit $N \longrightarrow \infty$ and $a \longrightarrow 0$ so as to keep S constant. A point $\vec{r}$ along the chain is characterized by its curvilinear abcissa s, and the equivalent of (6) becomes:

$$G\left(\vec{R}S\,\middle|\,\vec{0}0\right) = \int_{\vec{r}(0)=\vec{0}}^{\vec{r}(S)=\vec{R}} \mathcal{D}\vec{r}(s)\,\exp\left(-\frac{d}{2}\int_0^S \mathrm{d}s\left(\frac{\mathrm{d}\vec{r}}{\mathrm{d}s}\right)^2\right) \tag{7a}$$

with a measure normalized in such a way that

$$G\left(\vec{R}S\,\middle|\,\vec{0}0\right) = \left(\frac{d}{2\pi S}\right)^{d/2} \exp\left(-\frac{d}{2S}\vec{r}^{\,2}\right) \tag{7b}$$

If a potential $V(\vec{r})$ is exerted on each link, of the chain, the partition function at temperature $T = \frac{1}{\beta}$ is:

$$G\left(\vec{R}S\,\middle|\,\vec{0}0\right) = \int_{\vec{r}(0)=\vec{0}}^{\vec{r}(S)=\vec{R}} \mathcal{D}\vec{r}(s)\cdot\exp\left(-\frac{d}{2}\int_0^S \mathrm{d}s\left(\frac{\mathrm{d}\vec{r}}{\mathrm{d}s}\right)^2 - \beta\int_0^S \mathrm{d}s\, V\left(\vec{r}(s)\right)\right) \tag{8a}$$

We recognize in (8) the Feynmann path integral representing the matrix element of the quantum operator:

$$G\left(\vec{R}S\,\middle|\,\vec{0}0\right) = \left\langle \vec{r}\,\middle|\mathrm{e}^{-SH}\middle|\,\vec{0}\right\rangle \tag{8b}$$

where H is given by:

$$H = -\frac{1}{2d}\frac{\partial^2}{\partial\vec{r}^2} + \beta V(\vec{r}) \tag{8c}$$

In the limit of large S, only the lowest lying states of H will contribute. In particular, if H has a gap in its spectrum, it is a situation of so-called "ground state dominance" since only the ground state contributes to (8b).

III. RANDOM CHAINS AT INTERFACES

An important problem in biophysics is that of random chains at interfaces of fluid. This is indeed related to the problem of the localisation of proteins in a cellular membrane[4].

1) One interface

Consider an interface of e.g. water and oil[5]

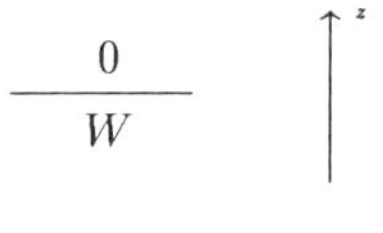

An extremely simplified model of a protein is a random chain of hydrophilic and hydrophobic molecule. The hydrophilicity (-phobicity) of a molecule, is characterized by a hydrophilic charge ξ. If $\xi > 0$, the molecule wants to sit in the water, otherwise, it prefers oil.

The potential exerted on this molecule is thus taken to be:

$$V = \xi \ \mathrm{Sgn}\ z \tag{9a}$$

The partition function of this chain is:

$$Z = \int \mathcal{D}\vec{r}(s) \exp\left(-\frac{d}{2}\int_0^S \mathrm{d}s \left(\frac{\mathrm{d}\vec{r}}{\mathrm{d}s}\right)^2 - \beta\int_0^S \mathrm{d}s\ \xi(s)\mathrm{Sgn}\ z(s)\right) \tag{9b}$$

The degrees of freedom parallel to the interface decouple, and one may consider (9b) in one spatial dimension. As it stands, (9b) can be expressed in terms of a time dependent Hamiltonian. We perform now a quenched average on $\xi(s)$, assuming a Gaussian distribution for $\xi(s)$:

$$P(\{\xi(s)\}) = \frac{1}{\mathcal{N}}\exp\left(-\frac{1}{2\bar{\xi}^2}\int_0^S \mathrm{d}s\,(\xi(s)-\xi_0)^2\right) \tag{10}$$

where $\mathcal{N}$ is a normalization constant.

After a trivial rescaling, we obtain:

$$\overline{Z^n} = \int \prod_{\alpha=1}^{n} \mathcal{D}z_\alpha(s)\mathrm{e}^{-A}\cdot\exp\left(n\beta^2\frac{\bar{\xi}^2}{2}S\right) \tag{11a}$$

where

$$\begin{aligned} A &= \frac{d}{2\beta^2\bar{\xi}^2}\sum_{\alpha=1}^{n}\int_0^S \mathrm{d}s\left(\frac{\mathrm{d}z_s}{\mathrm{d}s}\right)^2 + \beta\xi_0\int_0^S \mathrm{d}s\sum_{\alpha=1}^{n}\mathrm{Sgn}\ z_\alpha(s) \\ &-\frac{\beta^2\bar{\xi}^2}{2}\sum_{\alpha\neq\beta}\mathrm{d}s\ \mathrm{Sgn}\ z_\alpha(s)\mathrm{Sgn}\ z_\beta(s) \end{aligned} \tag{11b}$$

The width of the chain is given by:

$$w^2 = \frac{T^2}{\bar{\xi}^2}\int \mathcal{D}z_\alpha(s)\cdot z_1^2\left(\frac{S}{2}\right)\mathrm{e}^{-A}\Big|_{n=0} \tag{11c}$$

Two cases can be considered

i) $\xi_0 = 0$. We have:

$$\overline{Z^n} = \mathrm{Tr}\ \exp\left(-\beta^2\bar{\xi}^2 S H_n\right)\ \exp\left(n\beta^2\frac{\bar{\xi}^2}{2}S\right) \tag{12a}$$

where

$$H_n = -\frac{1}{2d}\sum_{\alpha=1}^{n}\frac{d^2}{\mathrm{d}z_\alpha^2} - \frac{1}{2}\sum_{\alpha\neq\beta}\mathrm{Sgn}\ z_\alpha\ \mathrm{Sgn}\ z_\beta \tag{12b}$$

Thus, the limit of large chains ($S \longrightarrow +\infty$) is equivalent to that of zero temperature. It is easy to convince oneself that the chain is localized at $T = 0$. Thus, the ground state ψ_0 of H_n

$$H_n\psi_0 = n\varepsilon_0\psi_0 \tag{13}$$

is localized, and there is ground state dominance. We thus obtain

$$\frac{F}{S} = \beta\bar{\xi}^2\left(\varepsilon_0 - \frac{1}{2}\right) \tag{14a}$$

$$w^2 = \frac{T^2}{\bar{\xi}^2}\frac{\langle\psi_0\,|z^2|\,\psi_0\rangle}{\langle\psi_0\,|\psi_0\,\rangle} \tag{14b}$$

ii) $\xi_0 \neq 0$, for instance, $\xi_0 > 0$. The previous simple scaling argument does not work. Instead, one can study the system using the Hartree approximation. The results are the following: there is a localization transition at $T_c = \frac{\bar{\xi}^2}{\xi_0}$, between a chain delocalized in the water at $T > T_c$, and a chain localized at the interface for $T < T_c$. Note that these results are obtained in the Hartree approximation, with a replica-symmetric wave-functions of the type

$$\psi(z_1,..,z_n) = \prod_{\alpha=1}^{n} \varphi(z_\alpha) \tag{15}$$

In the localized phase, the Hartree replica symmetric form is no more valid, and one must thus introduce replica symmetry breaking on correlated wave functions.

2) Two interfaces

Consider two interfaces: water-oil-water[6]

W

0

W

It is again possible to see that if $\xi_0 < 0$, the chain is localized in the oil layer (with excursions in the water layers). When $\xi_0 > 0$, there is a delocalization transition from a phase where the chain is localized around the oil to a phase where it is delocalized in the water. This transition occurs at $T_c = \frac{\bar{\xi}^2}{2\xi_0}$.

3) Lamellar system

In a lamellar system, there is a periodic arrangement of layers of the fluids. The corresponding quantum mechanical problem[6] resembles that of particles in a periodic potential. We find that the chain is delocalized throughout the fluid, with a periodic density, the period of which is that of the system.

IV. RANDOM CHAINS AROUND AN IMPURITY

Another important question is the localization of a random chain by an impurity[7,8]. Some links $\xi > 0$ are attracted by the impurity, while the others $\xi < 0$ are repelled. The impurity is modelled by a $d-$dimensional $V(\vec{r})$. The quenched average partition function reads:

$$\begin{aligned}\overline{Z^n} &= \int \prod_{\alpha=1}^{n} \mathcal{D}\vec{r}_\alpha(s) \exp\left(-\frac{d}{2}\sum_{\alpha=1}^{n}\int_0^S \mathrm{d}s \left(\frac{\mathrm{d}\vec{r}_\alpha}{\mathrm{d}s}\right)^2 + \frac{\beta^2}{2}\bar{\xi}^2 \sum_{\alpha,\beta=1}^{n}\int_0^S \mathrm{d}s \cdot V(r_\alpha(s)) \cdot V(r_\beta(s))\right) \\ &\quad \exp\left(-\beta\xi_0 \sum_{\alpha=1}^{n}\int_0^S \mathrm{d}s\ V(r_\alpha(s))\right)\end{aligned} \tag{16a}$$

Note that except in $d = 2$ (marginal dimension), the actual shape of the potential well is irrelevant, and we use $V(\vec{r}) = V_0 U(\vec{r})$ where $U(\vec{r})$ is a spherical well of unit depth, of radius a. Eq.(16a) reads:

$$\begin{aligned}\overline{Z^n} &= \int \prod_{\alpha=1}^{n} \mathcal{D}r_\alpha \cdot \exp\left(-\frac{d}{2}\sum_{\alpha=1}^{n}\int_0^S \mathrm{d}s \left(\frac{\mathrm{d}\vec{r}_\alpha}{\mathrm{d}s}\right)^2 + \left(\frac{\beta^2}{2}\bar{\xi}^2 V_0^2 - \beta\xi_0 V_0\right)\sum_{\alpha=1}^{n}\int_0^S \mathrm{d}s\ U(r_\alpha(s))\right) \\ &\quad \exp\left(\beta^2\bar{\xi}^2 V_0^2 \sum_{\alpha<\beta}\int_0^S \mathrm{d}s\ U(r_\alpha(s))\, U(r_\beta(s))\right)\end{aligned} \tag{16b}$$

The critical behaviour of this model can be calculated exactly. The results are the following:

i) Define $g = \frac{\beta^2}{2}\bar{\xi}^2 V_0^2 - \beta\xi_0 V_0$. The critical temperature of (16b) is the same as that of the annealed problem, namely when the last term in the exponent is omitted. So, for $d = 1$, $g_c = 0$, whereas in $d = 3$, g_c is finite (again the case $d = 2$ is marginal).

ii) Treating the last term in the exponent in perturbation theory, we can show by simple power counting, that the most divergent graphs are ladder diagrams

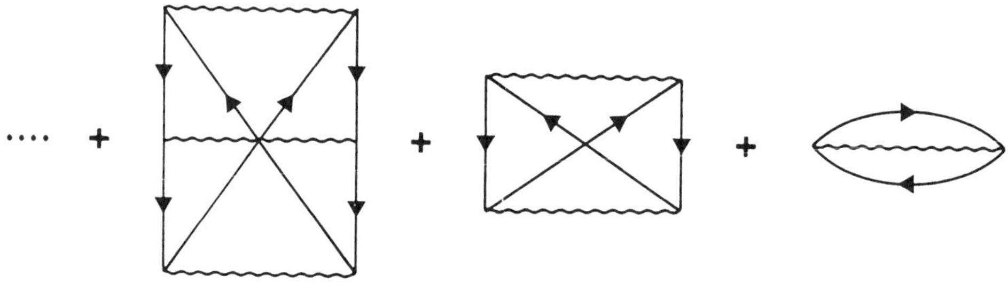

In fact, for $d > 1$, all graphs are finite, and there is thus no correction to the critical behaviour of the annealed system.

For $d = 1$, the ladder graphs have logarithmic divergences in $(g - g_c)$, and can be resummed as a geometric serie. This leads to a non analyticity in the specific heat of the type:

$$C = C_0 \left(g - g_c\right)^2 \left(1 + \frac{2\pi}{\mathrm{Log}\frac{1}{|g - g_c|}}\right) \tag{17}$$

All these results are consistent with the Harris criterion[9], which gives a negative α for $d > 1$, and $\alpha = 0$ for $d = 1$.

V. SELF-INTERACTING CHAINS

A very important question in biophysics and biochemistry is the understanding of the folding transition of proteins[4]. Proteins are biopolymers which can exist in a denaturated (unfolded coil state) or a folded state (globule). This transition can be induced by changing temperature, solvant, acidity, salinity, etc...A protein is a sequence of amino-acids, each chosen among 20 different such acids. These acids carry a charge, dipolar moment, hydryphobicity, etc... and they interact through screened Coulomb, dipolar hydrophilic, Van der Waals interactions, etc... Due to the complicated nature of these interactions and of the chemical struture of the chain, we model it as a chain of links interacting with random interactions. Furthermore, since interactions are essentially short ranged, we assume the interaction of link i at $\vec{r}_i$ with link j at $\vec{r}_j$ to be of the form:

$$v_{ij}\left(\vec{r}_i - \vec{r}_j\right) = v_{ij} \cdot \delta\left(\vec{r}_i - \vec{r}_j\right) \tag{18}$$

Two types of model can be studied:

i) The glass model. The interactions v_{ij} are taken as independent random variables, distributed, for instance, according to a Gaussian distribution :

$$P(v_{ij}) = \frac{1}{\sqrt{2\pi v^2}} \exp\left(-\frac{1}{2v^2}(v_{ij} - v_0)^2\right) \tag{19a}$$

ii) The separable model. The interaction v_{ij} are taken as a sum of M separable terms:

$$v_{ij} = \sum_{p=1}^{M} v_p \, \xi_i^p \xi_j^p \tag{19b}$$

where ξ_i^p are taken, for instance, as independent random variables with Gaussian distribution. These "charges" ξ_i^p can represent the Coulomb charge, the hydrophobic degree, etc...

In the continuous limit, the partition function of these models reads:

$$\begin{aligned} Z &= \int \mathcal{D}\vec{r}(s) \exp\left(-\frac{d}{2}\int_0^S \mathrm{d}s \left(\frac{\mathrm{d}\vec{r}}{\mathrm{d}s}\right)^2 - \frac{\beta}{2}\int_0^S \mathrm{d}s\mathrm{d}s' v(s,s')\,\delta(\vec{r}(s) - \vec{r}(s'))\right) \\ &\exp\left(-\frac{C}{6}\int_0^S \mathrm{d}s\mathrm{d}s'\mathrm{d}s''\delta(\vec{r}(s) - \vec{r}(s'))\,\delta(\vec{r}(s') - \vec{r}(s''))\right) \end{aligned} \tag{20}$$

The last term is included to avoid collapse of the chain, and ensures extensivity of the free energy (this could also have been achieved by omitting this term and scaling the random variable $v(s,s')$ properly).

We briefly present the results of both models.

1) The glass model[10a,b]

Performing the quenched average, one obtains

$$\overline{Z^n} = \int \prod_\alpha \mathcal{D}Q_{\alpha\beta}(\vec{r},\vec{r}') \exp\left(-\frac{1}{4}\sum_{\alpha<\beta}\int \mathrm{d}\vec{r}\mathrm{d}\vec{r}'\, Q_{\alpha\beta}^2(\vec{r},\vec{r}') + \operatorname{Log}\mathcal{L}(Q_{\alpha\beta}(\vec{r},\vec{r}'))\right) \tag{21a}$$

where

$$\begin{aligned} \mathcal{L}(Q_{\alpha\beta}) &= \int \prod_\alpha \mathcal{D}\vec{r}_\alpha(s) \exp\left(-\frac{1}{2}\sum_\alpha \int_0^S \mathrm{d}s \left(\frac{\mathrm{d}\vec{r}_\alpha}{\mathrm{d}s}\right)^2 - \frac{1}{2}\left(\beta v_0 - \frac{\beta^2 v^2}{a^d}\right)\right. \\ &\quad \times \sum_\alpha \int_0^S \mathrm{d}s\mathrm{d}s'\delta(\vec{r}_\alpha(s) - \vec{r}_\alpha(s')) \\ &\quad \times \exp\left(-\frac{C}{6}\int_0^S \mathrm{d}s\mathrm{d}s'\mathrm{d}s''\sum_\alpha \delta(r_\alpha(s) - r_\alpha(s'))\,\delta(r_\alpha(s') - r_\alpha(s''))\right. \\ &\quad \left. -\beta v \sum_{\alpha<\beta}\int_0^S \mathrm{d}s\, Q_{\alpha\beta}(r_\alpha(s), r_\beta(s))\right) \end{aligned} \tag{21b}$$

This integral is approximated by a mean field solution of the form:

$$Q_{\alpha\beta}(\vec{r},\vec{r}') = 2\beta v \int_0^S \mathrm{d}s\, \langle \delta(\vec{r} - \vec{r}_\alpha(s))\,\delta(\vec{r}' - \vec{r}_\beta(s))\rangle \tag{22}$$

where $1 \leq \alpha < \beta \leq n$, and $n \longrightarrow 0$.

It is easy to see that Eq.(22) implies replica symmetry breaking (RSB) similar to that of the Potts glass[11]. There is a critical temperature T_c, such that

i) for $T > T_c$, $Q_{\alpha\beta} = 0$; the overlap between two copies of the system is zero. Thus, the chain has no definite shape; it is in a coil state, analogous to the denaturated phases of proteins.

ii) for $T < T_c$, $Q_{\alpha\beta} \neq 0$. The chain has a definite folded shape, in a dense globular form.

It can be shown that, at least in dimensions $d \geq 4$ [10a] (and maybe down to $d = 2$ [10b]), this transition is of the Potts glass type[11].

2) The separable model[12a,b]

Performing the quenched average over the $\{\xi_i^p\}$, some order parameters appear naturally in the system:

$$m_{p,\alpha}(\vec{r}) = \int_0^S \mathrm{d}s \ \overline{\xi_p(s) \langle \delta(\vec{r} - r_\alpha(s)) \rangle} \tag{23}$$

These order parameters indicate the correlation between the chemical nature of a link (characterized by the $\{\xi_p(s)\}$) and its position in space $\vec{r}$. There is again a transition at a temperature T_c such that:

i) for $T > T_c$, $m_{p\alpha}(\vec{r}) = 0$; there is no correlation between the chemical sequence and the spatial structure. The chain is unfolded, with no definite shape.

ii) for $T \leq T_c$, some $m_{p\alpha}(\vec{r}) \neq 0$ condense. The chemical sequence codes for the spatial structure of the chain. It corresponds to a folded state of the chain. For instance, if there is only one type of charge (e.g. hydrophilic charge $M = 1$), the folding will translate into a spatial separation between hydrophilic and hydrophobic links.

Note that when M increases, we expect a smooth crossover from the separable form to the glass form. From simple qualitative arguments[13], it can be inferred that in real proteins, $M = 8$ (eight relevant characters). Thus, for $\alpha = \frac{M}{N}$ small (i.e. long chains), one is in the separable case, whereas for α larger (short chains), one is in the glassy situation. Experimentally, a qualitative change in the folding transition is observed around $N = 80$ [14]. Larger chains code better (the separable case) and are robust, whereas shorter chains freeze into more glassy states.

REFERENCES

1. R. Brout, Phys. Rev. **115**, 824 (1959).
2. S.F.E. Edwards and P.W. Anderson, J. Phys. F**5**, 965 (1975).
3. P.G. de Gennes, "Scaling Concepts in Polymer Physics", Cornell University Press (1979).
4. T.E. Greighton, "Proteins: Structure and Molecular properties", Ed. W.H. Freeman (New York).
5. T. Garel, D.A. Huse, S. Leibler and H. Orland, Europhys. Lett. **8**, 9 (1989).
6. T. Garel, H. Orland and E.I. Shakhnovich, submitted to J. Chem. Phys.
7. A. Yu. Grosberg and E.I. Shakhnovich, Soviet. Phys. JETP **64**, 1284 (1986).
8. T. Garel, H. Orland and E.I. Shakhnovich, in preparation.
9. A.B. Harris, J. Phys. **C7**, 1671 (1974).

10. a) T. Garel and H. Orland, Europhys. Lett. **6**, 307 (1988).
    b) E.I. Shakhnovich and A.M. Gutin, Europhys. Lett. **8**, 327 (1989).
11. D.J. Gross, I. Kanter and H. Sompolinsky, Phys. Rev. Lett. **55**, 304 (1985).
12. a)T. Garel and H. Orland, Europhys. Lett. **6**, 597 (1988).
    b) E.I. Shakhnovich and A.M. Gutin, J. Phys. France **50**, 1843 (1989).
13. J.R. Garel, T. Garel and H. Orland, J. Phys. France **50**, 3067 (1989).
14. J. Janin and S.J. Wodak, Prog. Biophys. Molec. Biol. **42**, 21 (1983).

A NEW THEORY OF COLLISIONS

B.G.Giraud
Service de Physique Théorique
CEA-IRF-CEN Saclay
BP 2, 91191 Gif/Yvette, France

Abstract: Instead of diagonalizing the many-body Hamiltonian H, we invert E-H, where E is a complex energy, eventually real. All the traditional approximations to diagonalization can be adjusted to inversion. We specially investigate mean-field methods. This lecture gives a scheme for the detailed proofs of our arguments, already published, and lists several numerically soluble cases where our new method has been sucessfully tested for the calculation of collision amplitudes.

INTRODUCTION

Consider a system of N particles, N finite, large or small, with kinetic energies t_i and two-body interactions v_{ij}. An extraordinary amount of effort has been dedicated to the diagonalization of the corresponding Hamiltonian H=T+V. Among the most successful approximations, one may quote various brands of mean-field methods and their subsequent improvements: Hartree-Fock, BCS, shell model, etc. This effort has been able to provide many reliable approximations to the spectrum of H, whether N is large or small.

Much less effort has been dedicated to the calculation of the Green's function $G=(E-H)^{-1}$, whether E is real or complex. Such an inversion problem, however, is as interesting as the diagonalization, for the propagator G occurs as a key element of the theory in many practical problems. In particular, any collision theory sooner or later introduces G in the formalism...and the numerics.

Recent Progress in Many-Body Theories, Volume 2
Edited by Y. Avishai
Plenum Press, New York, 1990

This lecture is organized as follows. In Section 1, we show how <u>any</u> diagonalization approximation which is introduced according to the Rayleigh-Ritz variational principle can be adjusted into an inversion approximation, according to a Schwinger-like variational principle. In Section 2 we recall the results, already published, of a test of this approach for a trivially soluble, two-body problem. In Section 3, we recall the success also found for this method in the case of an exactly soluble three-body problem. In Section 4, we refer to another success, for a four-body problem. In Section 5, we describe an N-body soluble model where this method can be again tested. Finally we introduce in Section 6 the numerical results of recent, extensive calculations. A brief conclusion is proposed in Section 7.

1. INVERSION VERSUS DIAGONALIZATION

The well-known Rayleigh-Ritz variational functional reads

$$\mathcal{F}=\langle\Phi'|(H-E)|\Phi\rangle,$$

where Φ',Φ are infinitely flexible trial functions and E is a Lagrange multiplier. The stationarity of $\mathcal{F}$ with respect to Φ' (resp. Φ) induces the Schroedinger equation $(H-E)\Phi=0$ (resp. $\Phi'(H-E)=0$).

Whenever Φ',Φ are not infinitely flexible, one gets an approximation. For instance, if Φ' and Φ are products of single-particle orbitals φ'_i and φ_i, one obtains the Hartree approximation: the N-body Schroedinger equation is replaced by single-particle (but coupled) diagonalizations such as

$$(t_i+U_i-\eta_i)\varphi_i=0, \tag{1.1}$$

where η_i and U_i are the well-known, self-consistent, Hartree self-energies and mean fields. (The most familiar form of U_i corresponds to the case when v is local, namely $U(r)=\int dr'v(r-r')\sum_{j\neq i}\varphi'^*_j(r')\varphi_j(r)$, with the constraints $\langle\varphi'_k|\varphi_l\rangle=\delta_{kl}$.)

Naturally, these single-particle diagonalizations mean nothing but the conditions $\delta\mathcal{F}/\delta\varphi'_i=0$. The point is, such single-particle, self-consistent diagonalizations can be performed numerically, and give a good approximation, while the full N-body diagonalization is impossible in practice.

We now investigate inversion. How can one calculate a matrix element $D=\langle\chi'|G|\chi\rangle$, where χ' and χ are arbitrary, square-integrable states? Answer: consider the functional

$$F = \langle\chi'|\Phi\rangle + \langle\Phi'|\chi\rangle - \langle\Phi'|(E-H)|\Phi\rangle,$$

the third term of which is nothing but $\mathcal{F}$. (Notice however that the parameter E is here a complex energy, with at least an infinitesimal imaginary part, rather than a norm and phase control Lagrange multiplier.)

Stationarity of F with respect to Φ' reads $\Phi=G\chi$, without any on-shell ambiguity if E is complex. Then it is easy to verify that the stationnary value of F is the number under study, namely D. An identical result occurs for the stationarity with respect to Φ, with $\chi'G=\Phi'$.

Whenever Φ',Φ are restricted, the approximaxion generated by F must be very similar to that generated by $\mathcal{F}$, since F and $\mathcal{F}$ only differ via the "source terms" $\langle\chi'|\Phi\rangle,\langle\Phi'|\chi\rangle$.

In particular, let us assume that all four functions χ',χ,Φ',Φ are products of single particle orbitals $\chi'_i,\chi_i,\varphi'_i,\varphi_i$, respectively. Then the Hartree equations, Eqs.(1.1), are hardly modified: they become

$$(\eta_i-t_i-U_i)\varphi_i=\chi_i, \tag{1.2}$$

with a non-vanishing right-hand side which comes from the functional derivative $\delta\langle\Phi'|\chi\rangle/\delta\varphi'_i$. One clearly also gets a conjugate variational equation $\chi'_i=\varphi'_i(\eta_i-t_i-U_i)$.

As will be stressed over and over again in the following Sections, variational solutions for F are almost as easy to obtain as variational solutions for $\mathcal{F}$. Approximate methods for diagonalization can be extended to the calculation of the Green's function. For more technical details, including the case of identical particles, we refer the interested reader to refs.[1)].

For the case of a calculation of a T-matrix amplitude, there is a slight complication of the theory, because the number under study reads

$\langle\chi'|V'GV|\chi\rangle$ rather than $\langle\chi'|G|\chi\rangle$. Here χ, resp. χ', describe the initial, resp. final channel, and V, resp. V', describe the corresponding channel potentials, namely the prior, resp. post interactions. These technicalities are dealt with in great detail in refs.[1]. For the sake of simplicity, they will be omitted in the present lecture.

2. A TWO-BODY SOLUBLE EXAMPLE

We refer here to the results of ref.[2]. Consider two particles, interacting via a separable potential $\langle q'|v|q\rangle = -\lambda \exp[\nu^2(q^2+q'^2)]$. Here q,q' are relative momenta, ν is a range and λ is a coupling constant. The calculation of the T-matrix for such separable potential, two-body problems is trivial.

In particular, it is easy to calculate exactly the diagonal matrix element $\mathfrak{D} = \langle\chi|v|\chi\rangle + \langle\chi|vGv|\chi\rangle$, with the following choice for χ,

$$\chi(q_1,q_2,k) = \pi^{-3}\beta^6 \exp\left[-\frac{\beta^2}{2}(q_1-k)^2\right] \exp\left[-\frac{\beta^2}{2}(q_2+k)^2\right]. \qquad (2.1)$$

Here χ describes particle 1 with momentum q_1 in a wave packet with average momentum k, and particle 2 with momentum q_2 in a wave packet with average momentum -k. The parameter β is obviously a width parameter. This wave packet representation for collisions is very convenient, for it allows calculations with square integrable states, without loss of generality or physical insight. This representation is again used systematically in the next Sections.

With this choice of χ, it is easy to find that the exact calculation of $\mathfrak{D}$ reduces to Gaussian integrations and a numerical integration for a spectral function for S-waves. Then, for the variational approximation of $\mathfrak{D}$, we choose the simplest possible ansatz: the trial functions Φ,Φ' introduced inside the functional F (or rather the modified functional which accounts for vGv instead of G alone) are also Gaussian wave packets. Namely we set

$$\Phi(q_1,q_2) = \chi(q_1,q_2,K),$$

according to Eq.(2.1). The only variational parameter is this average momentum K, and the functional becomes a function F(k,K), according to the respective parameters of χ and Φ.

It is then trivial to choose K in order to cancel the derivative $\partial F/\partial K$, calculate the variational approximation to $\mathfrak{D}$, and compare with the exact result. We have investigated many numerical ranges for all the parameters λ, ν, β, ImE, etc.

While the exact result is unique, the variational approximation is non linear and may generate several solutions. One of these is always very accurate, as shown for instance on Fig.1, which needs little comment: solution 1 is satisfactory at high energy, while solution 2 is better at low energy. We can also interpolate by a linear superposition of the variational solutions and generate a solution 3 which is always satisfactory. The interested reader will obtain all technical details in ref.[2].

In conclusion for this Section, the variational functional F is able, for a simple two-body soluble case, to provide a satisfactory G- (or T-) matrix element with just one variational parameter. It is not even necessary to use a full single-particle orbital as a variational tool, and the use of Eqs.(1.2) can be spared for more elaborate cases. Such cases are the subject of Sections 3 and 4.

3. A THREE-BODY SOLUBLE EXAMPLE

We briefly recall here the results of ref.[3]. Our model is an extension of the model of Section 2. With three particles, there are now three separable potentials $v_{ij}(q_{ij},q'_{ij})$, analogous to the potential v used in Section 2. The channel wave packet χ is the simplest possible generalization of Eq.(2.1), namely a product of three single-particle Gaussians, boosted by k_1,k_2,k_3 respectively, with the constraint $k_1+k_2+k_3=0$.

The exact matrix element $D=\langle\chi|G|\chi\rangle$ is obtained as $D=\langle\chi|\Psi\rangle$, with $\Psi=G\chi$, hence $(E-T-v_{12}-v_{13}-v_{23})\Psi=\chi$, then $\Psi=G_0\chi+\Psi_1+\Psi_2+\Psi_3$, with $G_0=(E-H_0)^{-1}$, $\Psi_1=G_0v_{23}\Psi$, $\Psi_2=G_0v_{13}\Psi$, $\Psi_3=G_0v_{12}\Psi$. As usual in manipulations leading to Faddeev-type equations, this leads to three coupled equations for Ψ_1,Ψ_2,Ψ_3. These can be easily solved numerically, for the separability of v induces separability for each component Ψ_i of Ψ. Numerical results are shown on Fig.2, for one of the many sets of parameters we have investigated. More details are available in ref.[3].

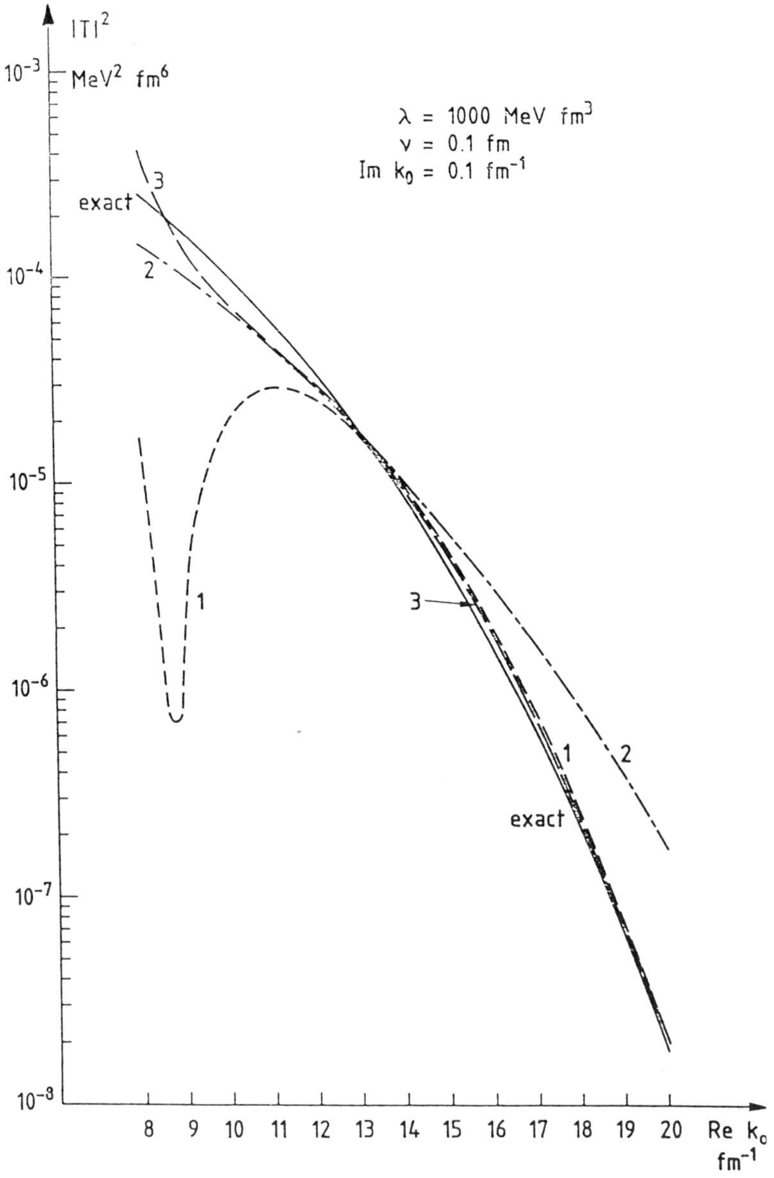

Fig. 1. Two-body case, separable potential. The full line shows the square modulus of the exact T-matrix element. The dashed line labelled "1" and the dot-dashed line "2" correspond to two different variational solutions. The long-dashed line represents a linear superposition of "1" and "2".

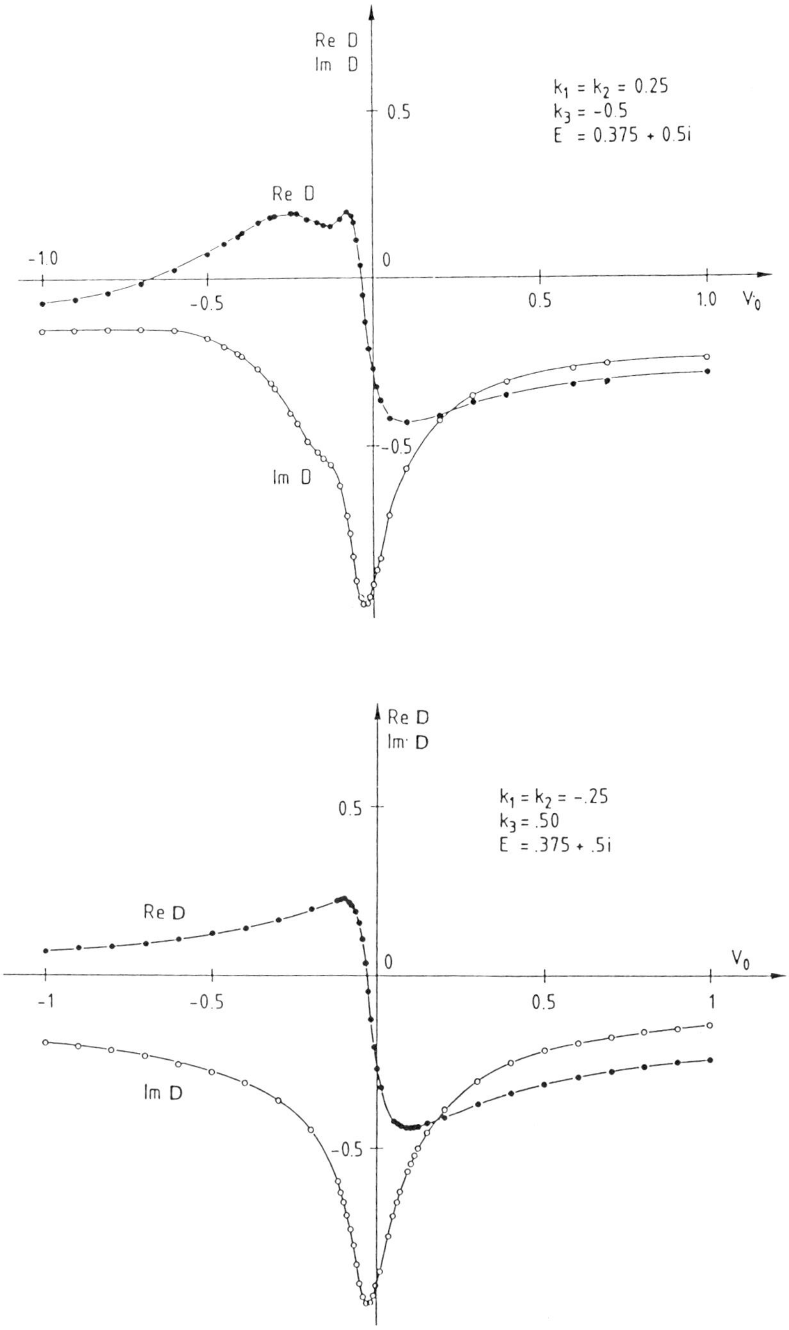

Figs. 2 and 3. Three-body case, separable potential. Good agreement between the exact amplitude (Fig.2, top) and its mean-field approximation (Fig.3, bottom), except for Re D with large, negative coupling constants.

The mean field approximation is obtained by brute force solution of Eqs.(1.2). The separability of v makes the Hartree mean field very simple. The corresponding estimate for $\langle\chi|G|\chi\rangle$ is shown on Fig.3 for the same set of parameters as for Fig.2. There is little need to comment on the good agreement between the exact and the approximate amplitudes. Notice however the wrong sign of ReD for large, negative λ.

In conclusion of this Section, there is at least one three-body problem for which our (generalized) Hartree approximation for the Green's function seems to give a satisfactory approximation. This will be extended in Section 4 to a four-body problem.

4. A FOUR-BODY SOLUBLE EXAMPLE

In ref.[4] the potential v is now taken as "super separable", namely, in momentum representation,

$$\langle q'_i q'_j | v_{ij} | q_i q_j \rangle = -\lambda \; f(q'_i) f(q'_j) f(q_i) f(q_j),$$

where f is any suitable form factor (actually a Lorentzian in our numerical application). This differs from the potential of Sections 2 and 3, where v was a function of just the relative momenta, such as $q_{ij}=(q_i-q_j)/2$.

We are again interested in the diagonal matrix element $D=\langle\chi|G|\chi\rangle$, where χ is now a product of four boosted wave packets,

$$\chi(q_1,\dots,q_4) = \Gamma(q_1-k)\;\Gamma(q_2-k)\;\Gamma(q_3+k)\;\Gamma(q_4+k).$$

With this choice for χ, many symmetries between the four particles are transparent and greatly help the solubility of the model. The single-particle wave-packets Γ are chosen as one-dimensional and Lorentzian, hence many integrations reduce to contour integrations.

The algebraic manipulations which convert the exact inversion problem $\Psi=G\chi$ into a soluble set of coupled integrodifferential equations for the various components of Ψ cannot be detailed here. They are obviously greatly simplified by the "super separability" of v. All details can be found in ref.[4]. Here we rather indicate how the mean-field equations are easy to solve:

The main simplification comes from the mean-field potential. Because of the symmetries between all four particles and the scalar nature of v, it does not depend on the label i and reads

$$\langle q'|U|q\rangle = -3\,\lambda\, f(q')\, f(q)\, \alpha^2\, n^{-1}, \text{ with } \alpha=\langle f|\varphi_1\rangle \text{ and } n=\langle\varphi_1^*|\varphi_1\rangle.$$

There is no need to solve more than one Hartree equation, for instance

$$(\eta-q^2)\ \varphi_1(q) = \Gamma(q-k) - 3\,\lambda\,\alpha^3\, n^{-1}\, f(q).$$

Hence the variational orbital φ_1 of the trial function Φ solves explicitely in terms of three parameters only, η, α and n,

$$\varphi_1(q) = (\eta-q^2)^{-1}\ [\ \Gamma(q-k) - 3\,\lambda\,\alpha^3\, n^{-1}\, f(q)\ ].$$

All practical calculations amount to the search for the self-consistency of η, α and n. The results of ref.[4] are illustrated by Fig.4, where the agreement between exact and mean-field amplitudes, for at least several ranges of parameters, is striking.

We have thus found a four-body case where a mean-field estimate of the Green's function is a good approximation.

5. THE KINETIC ENERGY MODEL

While the diagonalization of $T=\sum_{i=1}^{N} t_i$ is trivial, the calculation of a matrix element $\langle\chi|(E-T)^{-1}|\chi\rangle$ still requires an intractable multiple dimension integral when N is large. Hence our "mean-field" approach deserves investigation. We recall now the results of ref.[5].

Set χ as a product $\chi(q_1,\ldots q_N)=\Gamma(q_1-k_1)\ldots\Gamma(q_N-k_N)$, with a rotation and/or parity degeneracy between the boosts k_i. For obvious symmetry reasons, it is then sufficient to solve only one among Eqs.(1.2), for instance

$$(\eta-q^2)\ \varphi_1(q) = \Gamma(q-k_1), \text{ with } \eta=E-(N-1)\theta \text{ and } \theta=\frac{\langle\varphi_1^*|t|\varphi_1\rangle}{\langle\varphi_1^*|\varphi_1\rangle}.$$

Hence in momentum representation $\varphi_1=(\eta-t)^{-1}\Gamma_1$ is known, except for the self-consistency of η (or θ).

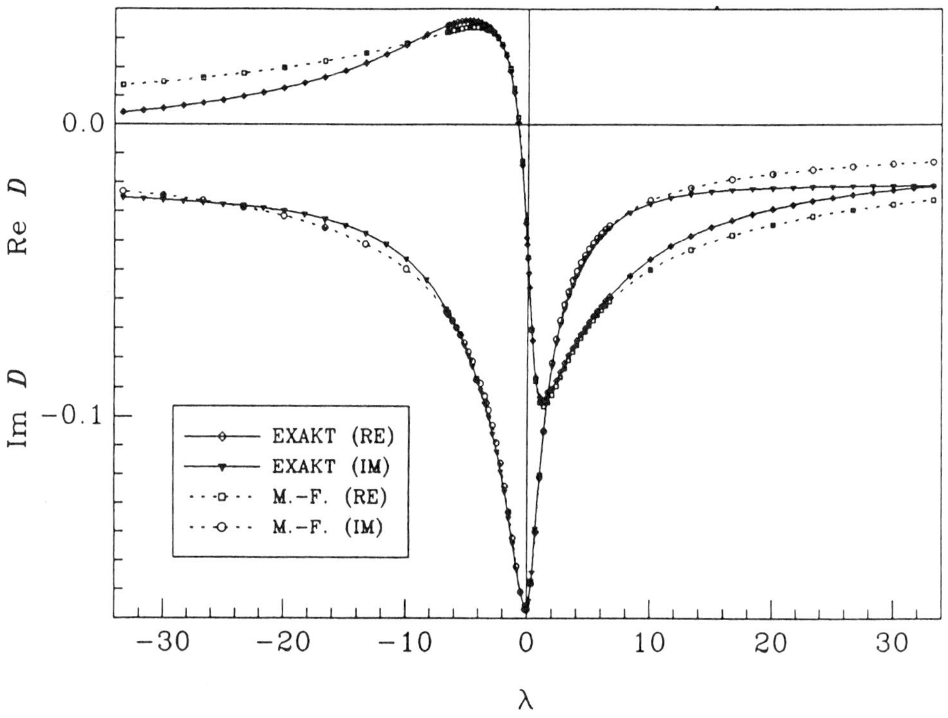

Fig. 4. Four-body case, super-separable potential. Agreement between the exact and the mean-field amplitudes, as functions of the coupling constant.

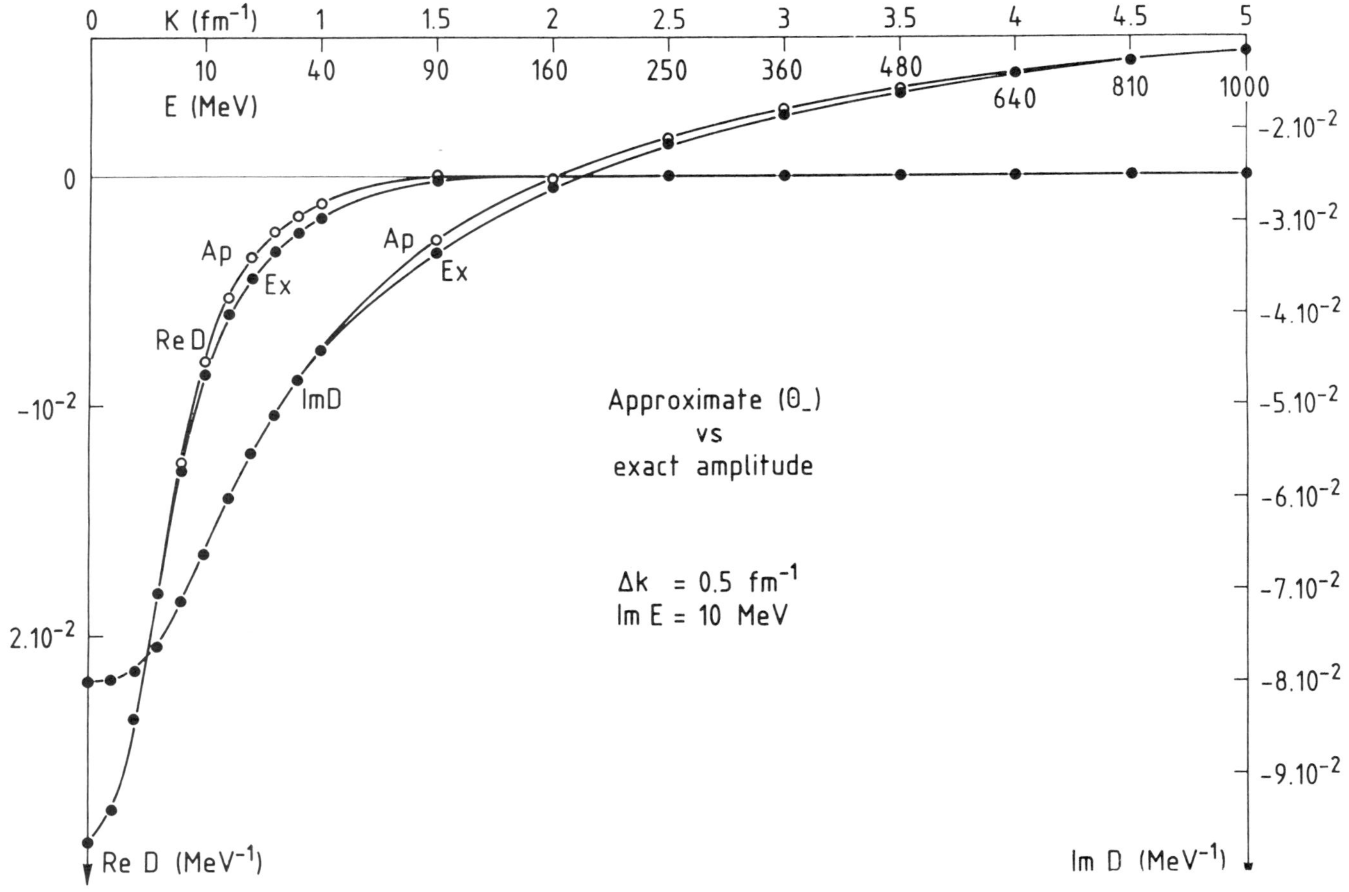

Fig. 5. Kinetic energy model. Agreement between the exact (Ex) and the mean-field (Ap) matrix elements of the free Green's function, as functions of the real part of the energy.

As shown in ref.[5], this self-consistency condition can be framed into the cancellation of a polynomial, which makes the model highly analytical. A careful analysis of the solutions shows an extremely interesting result: **when E is real, there exists a pair of solutions η,η^* with finite imaginary parts** (even though ImE=0). **One of these solutions simulates the retarded on-shell propagator**, the other simulates the advanced propagator.

Thus we get some insight into the multiplicity of solutions of the Hartree approximation to inversion. As shown by Fig.5 in one of the numerical cases investigated, this approximation is reliable in a wide range of energies.

6. THE CASE OF A SEMI-REALISTIC POTENTIAL

The confidence grown from the soluble examples described in Sections 2-5 justifies solving Eqs.(1.2) in a case where there is no exact solution for comparison. Only the consistency of the approximate results will be available for a criticism of the method.

Fig.6 is taken from ref.[6], which is a brute force, Hartree calculation of elastic proton-triton scattering. The nuclear interaction is taken as semi-realistic, namely a local, two-body Gaussian. Particle label 1 refer to projectile-like orbitals, labels 2-4 refer to target-like orbitals. There is complete symmetry between orbitals 2-4, hence labels 3 and 4 are omitted.

The Figure corresponds to final momentum at 90° (along y) away from initial momentum (along z). The plots are momentum density plots of the variational orbitals φ,φ' (which make the trial functions Φ,Φ') in the reaction plane (z y).

Not surprisingly, target orbitals φ_2,φ'_2, which carry less momentum than the projectile orbitals φ_1,φ'_1, are more centered and less deformed. It is also satisfactory that the density center of φ_1 lies "forward", while the density center of φ'_1 lies in the "tranverse" direction. **In so far as Φ,Φ' represent intermediate states of the collision, the Figure gives a picture of the reaction mechanism.**

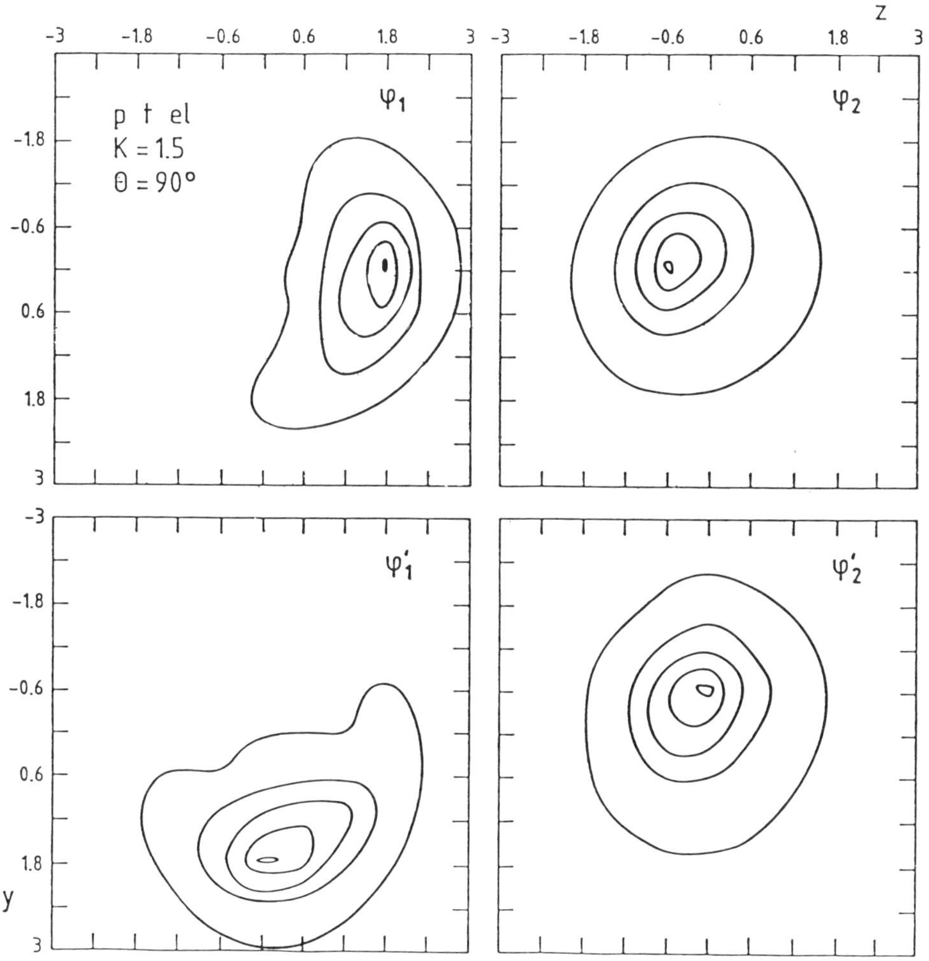

Fig.6. Momentum density patterns of the trial orbitals φ, φ' for an elastic p-t scattering at 90°. Notice how φ_1 concentrates in the forward, z-direction, while φ'_1 concentrates in the tranverse, y-direction. Notice also how both orbitals show a "horn" at 45°, which is essential for a non-vanishing overlap between these orbitals.

This Figure is only part of a whole atlas, corresponding to various energies, scattering angles, etc. At present, the numerical evidence favors a smooth behavior of the variational solution as a function of its control parameters.

To conclude this Section, we seem to have a semi-realistic case where all the results of our new, mean-field collision theory show a reasonable behaviour.

7. SUMMARY AND CONCLUSION

The guess of good trial functions has been for decades at the core of the theory of nuclear, atomic and molecular spectra. An exactly similar approach is available for the theory of propagators. One just has to slightly change the variational principle which underlies the theory.

Naturally, many auxilliary, but important questions must be investigated while solving our "extended" Hartree(-Fock, -Bogoliubov, -etc.) equations. In particular one should understand the multiplicity of solutions, their stability, the rôle of boundary conditions. This program will likely request some empiricism before all the problems are fully understood. But the same difficulty was met for a complete justification of the shell model. Our approach to propagators is not just formal, but generates numbers which can be compared with experimental data.

It is a pleasure to thank the organizers of this conference for their invitation and the opportunity of this lecture.

REFERENCES

1) B.G.Giraud, M.A.Nagarajan and I.J.Thomson, Ann.Phys.(NY) 152, 475 (1984)
B.G.Giraud, M.A.Nagarajan and C.J.Noble, Phys.Rev.A 34, 1034 (1986)

2) M.A.Nagarajan and B.G.Giraud, Phys.Rev.C 27, 232 (1983)

3) Y.Abe and B.G.Giraud, Nucl.Phys. A440, 311 (1985)

4) Joerg Lemm, Diplomarbeit (1989), Institut fuer Theoretische Physik, Universitaet Muenster, FRG

5) B.G.Giraud, Physica 19D, 112 (1986)

6) B.G.Giraud and M.A.Nagarajan, Saclay preprint SPh-T 88/89, submitted to Phys.Rev.C

CALCULATIONS OF THE GROUND STATE OF ^{16}O

Steven C. Pieper

Physics Division
Argonne National Laboratory
Argonne, IL 60439-4843

One of the central problems in nuclear physics is the description of nuclei as systems of nucleons interacting via realistic potentials. There are two main aspects of this problem: 1) specification of the Hamiltonian, and 2) calculation of the ground (or excited) states of nuclei with the given interaction. Realistic interactions must contain both two- and three-nucleon potentials and these potentials have a complicated non-central operator structure consisting, for example, of spin, isospin and tensor dependences. This structure results in formidable many-body problems in the computation of the ground states of nuclei.

At present, reliable solutions of the Faddeev equations for the A=3 nuclei with such interactions are routine.[1] Recently, Carlson[2] has made an essentially exact GFMC calculation of the ^{4}He ground state using just a two-nucleon interaction, and there are reliable variational calculations[3] for more complete potential models. Nuclear matter calculations can also be made with reasonable reliability.[4,5] However, there have been very few calculations of nuclei with A>5 using realistic interactions, and none with a modern three-nucleon interaction. In the present paper I present a new technique for variational calculations for such nuclei and apply it to the ground state of ^{16}O. The work reported was done in collaboration with V. R. Pandharipande (University of Illinois, Urbana), and R. B. Wiringa (Argonne National Laboratory).

POTENTIALS AND WAVE FUNCTION

We use the Argonne v_{14} (AV14) two-nucleon potential[6] with the Urbana VII (UVII) three-nucleon potential.[3] Variational wave functions obtained with these interactions give remarkably good agreement with the experimental data for the charge and magnetic form factors of the A=3 and A=4 nuclei.[7] The AV14 contains 13 non-central operator terms and has been fit to the np scattering and bound-state data. The UVII contains the long-range two-pion exchange three-nucleon potential and a phenomenological short-range repulsion. Its coupling constants were adjusted to give a fit to the binding energies of

Recent Progress in Many-Body Theories, Volume 2
Edited by Y. Avishai
Plenum Press, New York, 1990

A=3 and A=4 nuclei and nuclear matter in variational calculations. More recent variational (and Fadeev for ^{3}H) calculations show that these potentials somewhat overbind the few-body nuclei.[8]

We use the following variational wave function for the ground state of a closed-shell nucleus with N=Z=A/2:

$$|\Psi\rangle = \left[S \prod_{i<j}^{A} \left(1 + U_{ij}\right)\right] \left(1 + \sum_{i<j<k}^{A} U_{ijk}\right) |\Psi_J\rangle , \tag{1}$$

where

$$U_{ij} = \sum_{p=2}^{8} u_p(r_{ij})\, O_{ij}^{(p)} , \tag{2}$$

$$u_p(r_{ij}) = f_p(r_{ij})/f_c(r_{ij}) , \tag{3}$$

$$U_{ijk} = - \epsilon_1 \tilde{V}_{ijk}^{FM} - \epsilon_2 \tilde{V}_{ijk}^{R} , \tag{4}$$

and

$$|\Psi_J\rangle = \prod_{i<j}^{A} f_c(r_{ij}) |\Phi\rangle . \tag{5}$$

Here $S\Pi$ designates a symmetrized product and $|\Phi\rangle$ is a completely antisymmetric sum of products of four Slater determinants for spin-up protons, spin-down protons, spin-up neutrons, and spin-down neutrons. The wave functions in the Slater determinants are the solutions of a Woods-Saxon potential well. The sum in Eq. (2) is a sum over the first eight operators ($O^{(p)}$=1, $\tau_1 \cdot \tau_2$, $\sigma_1 \cdot \sigma_2$, $\sigma_1 \cdot \sigma_2 \tau_1 \cdot \tau_2$, S_{12}, $S_{12}\sigma_1 \cdot \sigma_2$, $L \cdot S$, and $L \cdot S \sigma_1 \cdot \sigma_2$) of the v_{14} interaction. The $f_p(r_{ij})$ and $f_c(r_{ij})$ are the solutions of coupled Euler-Lagrange equations constructed to minimize the two-body cluster energy of nucleon matter at a specific density; see Ref. 5. The $\tilde{V}_{ijk}^{FM}$ and $\tilde{V}_{ijk}^{R}$ are the two-pion and repulsive parts of UVII (see Ref. 3), except that the cut-off parameter, b, is treated as a variational parameter. The wave function Ψ contains as variational parameters: 1) the parameters of the single-particle well used to construct Φ, 2) the nuclear-matter density and the parameters used to find f_c and the f_p, and 3) the b, ϵ_1, and ϵ_2 used in U_{ijk}.

CLUSTER EXPANSION

For a given set of variational parameters, we must evaluate $\langle\Psi|H|\Psi\rangle/\langle\Psi|\Psi\rangle$ where the Hamiltonian is

$$H = \sum_{i=1}^{A} \left(-\frac{\hbar^2}{2m}\right) \nabla_i^2 + \sum_{i<j}^{A} v_{ij} + \sum_{i<j<k}^{A} V_{ijk} . \tag{6}$$

In addition we will need the expectation value of other operators such as the density. The wave function is a vector in spin-isospin space. Because of the tensor correlations, all 2^A spin components are non-zero. Charge conservation limits the number of isospin components to $A!/(Z!(A-Z)!)$. In the VMC calculations[3] of few-body nuclei this complete vector is used. Each term of the symmetrized product of pair operators contains $A(A-1)/2$ factors and the sum is sampled instead of being completely evaluated. The U_{ijk} requires some $9A(A-1)(A-2)/6$ tensor pair operators. The kinetic energy is computed by differences which requires 6A evaluations of the wave function (additional evaluations are required for the L^2 and $(L\bullet S)^2$ potentials). Table 1 summarizes these numbers and shows the total relative computer effort needed to evaluate the kinetic energy (the dominant part of the calculation) at one configuration of the particles.

Table 1. Relative effort required to compute the kinetic energy of one configuration of particles using the complete spin-isospin vector.

Nucleus	Vector Length	Pair Operations	Kinetic Energy Evaluations	Total Effort
^{3}He	24	12	18	5.2×10^3
^{4}He	96	42	24	9.7×10^4
^{6}Li	1,280	195	36	9.0×10^6
^{8}He	7,168	532	48	1.8×10^8
^{12}C	3,784,704	2,046	72	5.6×10^{11}
^{16}O	8.4×10^8	5,160	96	4.2×10^{14}
^{40}Ca	1.5×10^{23}	89,700	240	3.3×10^{30}

The few-body method outlined above clearly becomes rapidly impractical for heavier nuclei. Therefore, we use a cluster expansion in terms of the operator being evaluated and the non-central operators in Ψ to compute expectation values. For example, consider the two-body potential:

$$\frac{\langle\Psi|\sum_{i<j} v_{ij}|\Psi\rangle}{\langle\Psi|\Psi\rangle} = \frac{\sum_{i<j} n_{ij} + \sum_{i<j<k} n_{ijk} + \cdots + n_{12\cdots A}}{1 + \sum_{i<j} d_{ij} + \sum_{i<j<k} d_{ijk} + \cdots + d_{12\cdots A}}, \tag{7}$$

$$= \sum_{i<j} e_{ij} + \sum_{i<j<k} e_{ijk} + \cdots + e_{12\cdots A} . \tag{8}$$

Here

$$n_{ij} = \int d^3r_1 \ldots d^3r_A \; \Psi_J^\dagger(\vec{r}_1,\ldots,\vec{r}_A)(1 + U_{ij})^\dagger \; v_{ij}^{(2)}$$

$$(1 + U_{ij}) \; \Psi_J(\vec{r}_1,\ldots,\vec{r}_A) / \int d^3r_1 \ldots d^3r_a |\Psi_J|^2 \;, \tag{9}$$

$$n_{ijk} = \int d^3r_1 \ldots d^3r_A \; \Psi_J^\dagger (1 + U_{ijk})^\dagger \; [S(1 + U_{ij})(1 + U_{ik})(1 + U_{jk})]^\dagger$$

$$[v_{ij} + v_{ik} + v_{jk}][S(1 + U_{ij})(1 + U_{ik})(1 + U_{jk})](1 + U_{ijk})\Psi_J$$

$$/\int d^3r_1 \ldots d^3r_A |\Psi_J|^2 - n_{ij} - n_{ik} - n_{jk} \;, \tag{10}$$

and so forth, and S again indicates a symmetrized product. The equations for the d_{ij}, d_{ijk}, ... are similar to those for the n's except there are no potential operators and 1 (corresponding to zero'th order) is subtracted from d_{ij}.

We could use Eq. (7) as a cluster expansion but it converges very slowly or even diverges. Instead, we define the e_{ij}, e_{ijk}, ... of Eq. (8) by multiplying both sides by the denominator of Eq. (7) and collecting all terms involving ij, ijk, etc. We also observe from Eqs. (7-10) that the n's and d's do not depend on the specific values of i, j, k, ..., so we can write

$$n_{ij} = n_2 \;, \quad d_{ij} = d_2 \;, \tag{11}$$

$$n_{ijk} = n_3 \;, \quad d_{ijk} = d_3 \;, \tag{12}$$

etc. We then have

$$e_{ij} = e_2 = n_2/(1 + d_2) \;, \tag{13}$$

$$e_{ijk} = e_3 = \frac{n_3 - (6d_2 + 3d_3)e_2}{1 + 3d_2 + d_3} \;, \tag{14}$$

etc. As we will show, the series (8) converges well.

The complete product of central correlations and the complete single-particle part of the wave function appears in each order of our cluster expansion; thus, Eqs. (9-10) define 3A-dimensional integrals. These are evaluated in a Metropolis random walk. Ideally, the random walk would be controlled by $|\Psi_J|^2$, however generating the completely antisymmetric Ψ_J is prohibitive. Instead we compute expectation values with the antisymmetric Ψ_J on the left side and a mixed symmetry Ψ_J that contains just a single product of Slater determinants in Φ on the right side. The symmetrized product of $(1 + U_{ij})$ is also sampled.

Table 2 shows the relative computer effort required to compute the energy of one configuration at different levels of the cluster expansion. The column labelled "Pair Operations" contains all pair operations required for the wave function and kinetic and potential energy; it is smaller than the product of columns 3 and 4 of Table 1 because of simplifications realized by having U_{ijk} act on a Ψ_J that has just one component. The total of 3.2×10^7 for a four-body cluster calculation of ^{16}O can be compared to the value 4.2×10^{14} required for ^{16}O by the few-body method.

Table 2. Relative effort to compute the energy of one configuration for ^{16}O in the cluster expansion.

Cluster Order	Clusters in Expansion ^4He	^{16}O	^{40}Ca	Average Vector Length	Pair Operations	^{16}O Effort
2	6	120	780	6	9	6.5×10^3
3	4	560	9880	21	88	1.0×10^6
4	1	1820	91,390	74	236	3.1×10^7
5	-	4368	658,008	272	514	6.1×10^8

The computer program used for the calculations reported here has been extensively tested by internal checks in which the same quantity is computed in two different ways and by comparisons of calculations of ^4He with the results of our few-nucleon program that does not use a cluster expansion.

RESULTS

Table 3 shows the convergence of the cluster expansion up to the four-body cluster level using our optimal set of variational parameters. The last column shows an estimate by extrapolation from the three- and four-body cluster values of the complete 16-body cluster value. This was obtained as the average of the extrapolations $e_3/(1-x)$ and $e_3\exp(x)$ where $x=e_4/e_3$. Here e_3 and e_4 are respectively the three-body and four-body cluster values of individual components of the kinetic and potential energies. This procedure was tested by extrapolating the two-body and three-body results and comparing them to the four-body term; the extrapolation appears to have less than 25% error.

The first row of Table 3 gives the kinetic energy (with c.m. energy removed). At present we cannot include the L•S parts of U_{ij} (p=7,8), nor the $7 \leq p \leq 14$ parts of v_{ij} in the three- and four-body clusters. Therefore, the second row shows results for the truncated f_6,v_6 calculation (including the Coulomb potential). The U_{ijk} is included in the three- and four-body cluster results. We see that the expansion converges rapidly for the kinetic and two-body potential energies. The next row shows the expansion of the three-body potential, for which the first contribution is to the three-body cluster. Here the four-body cluster makes a significant contribution and the extrapolated value is significantly different from the sum of the three- and four-body cluster values. The fourth row shows the two-body cluster value for the combined effect of the p=7 and 8 parts of U_{ij} and the $7 \le p \le 14$ parts of v_{ij}. This small contribution is not extrapolated.

Table 3. Convergence of the cluster expansion for the ground-state energy of ^{16}O. The entries are in MeV/A.

Term	Cluster Expansion Term 1-body	2-body	3-body	4-body	Sum	Extrap.
T	18.2±0.2	13.9±0.2	-1.8±0.2	0.2±0.2	30.5±0.4	30.5
$v_{ij}^{p=1\text{-}6,C}$		-41.3±0.3	6.4±0.2	-0.5±0.3	-35.4±0.3	-35.4
V_{ijk}			-2.8±0.1	1.6±0.1	-1.2±0.1	-1.7
$u_{p=7,8}, v_{ij}^{p=7,14}$		-0.3±0.1			-0.3	-0.3
Total	18.2±0.2	-27.8±0.2	1.8±0.2	1.2±0.2	-6.5±0.2	-7.0

Our estimated (by extrapolation) upper bound for the ground-state energy of ^{16}O is -7.0±0.3 MeV/A where the error includes both the Monte Carlo statistical error and an estimate (0.1) of the extrapolation uncertainty. The experimental value is -7.98 MeV/A. This upper bound is the same as our upper bound for the energy of ^4He using the same Hamiltonian and a similar wave function; a better wave function gives[8] -7.47±0.06 MeV/A. Thus in this calculation we fail to demonstrate that ^{16}O is stable against decay into four alpha particles. It is not clear if this failure is due to inadequacies in the trial wave function or in the Hamiltonian, however because the wave functions have similar structure, it is reasonable to assume that part of the problem is due to the Hamiltonian.

Previous calculations of ^{16}O have also failed to obtain stability against breakup into alpha particles. Kümmel, Lührmann and Zabolitzky[9] used the coupled-cluster method to compute ^4He, ^{16}O, and ^{40}Ca with several interaction models. They retained all two- and three-cluster contributions and some of the four-cluster terms. Their binding energies for the Reid potential are -6.0 and -5.0 MeV/A for ^4He and ^{16}O. More recently, Carlson and Kalos[10] made a VMC calculation for ^{16}O using a truncated Reid potential. Their wave function is of the same form as ours except that only products of commuting U_{ij} are retained (independent-pair wave function) and there is no U_{ijk}. They have rather large sampling errors and obtain an E/A consistent with the E/A for ^4He using the same interaction.

Figure 1 shows the cluster expansion of the charge density. This was computed by folding the IJL nucleon form factors[11] (the results are very insensitive to the choice of form factor) with the point nucleon densities computed from the trial wave function (the present wave function has identical proton and neutron distributions). The dash-dot curve is obtained with $|\Psi_J|^2$, and the solid with the complete wave function. The dash-dot and dashed curves show respectively the two- and three-body cluster contributions. Because of large sampling errors, we used the cluster expansion only up to three-body clusters and extrapolated to obtain the higher contributions; this contribution is shown as the dash-dash-dot curve. For $r<1.5$ fm, the sampling errors are $\lesssim 0.05$ fm^{-3}; for larger radii they rapidly decrease. The shaded region is the experimental charge density.[12] Our rms charge radius (computed up to four-body clusters) of 2.62 fm is in reasonable agreement with the experimental average[12] of 2.73 fm.

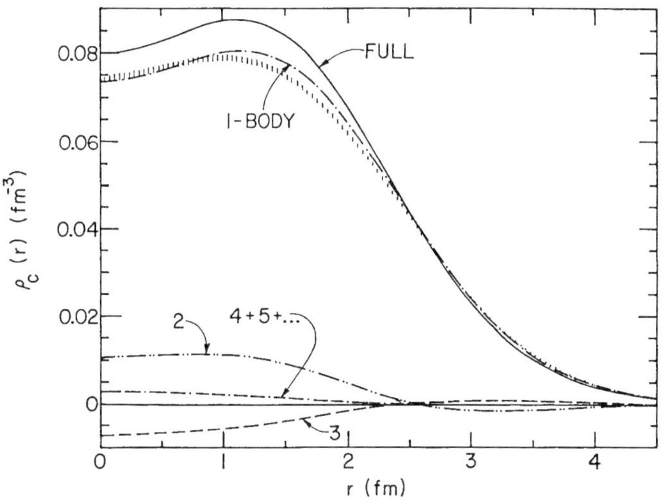

Fig. 1. Charge-density distribution of ^{16}O.

Figure 2 shows the longitudinal structure functions, $S_L(k)$, computed as described in Ref. 13. There are several interesting features of our calculation of $S_L(k)$: 1) The structure function obtained with the full wave function (solid curve) is substantially different from the mean-field result (short dashes) which was generated by making a mean-field fit to the computed $\rho(r)$ in Fig. 1. From the size of the experimental ^{12}C error bars, it appears possible to see "dynamic" correlation effects in $S_L(k)$ for $k \lesssim 1$ fm^{-1}. 2) We find that most of these "dynamic" effects are obtained with just the central correlations (Ψ_J); the converged non-central correlations have little total effect on S_L. 3) Contrary to the expectation of Refs. 13 and 14, the S_L for ^{16}O is not an interpolation of those for ^4He (dash dot) and nuclear matter (dotted) but is substantially larger for intermediate momentum transfer. We have also demonstrated this for mean-field calculations based on experimental density profiles for ^4He, ^{16}O and ^{40}Ca; both the ^{16}O and ^{40}Ca S_L are not interpolations of the results for ^4He and Fermi gas nuclear matter.

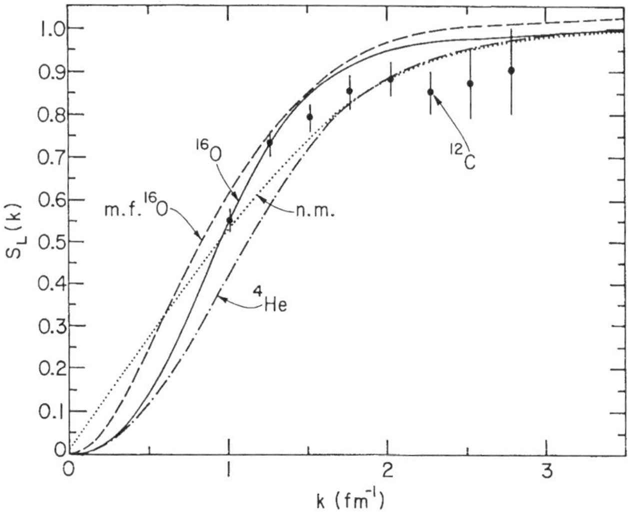

Fig. 2. Longitudinal structure functions. The data[15] are for ^{12}C with a theoretical correction.[14]

CONCLUSION

A cluster expansion in terms of the non-central terms of the wave function is a practical method for computing the ground-state energies of nuclei. The convergence is good, with the four-body clusters really being needed only for the expectation value of the three-body potential. The CPU time need to achieve acceptable statistical errors is reasonable (but not small--the calculation in Table I contains 4,500 samples and took 16.3 CPU hours on a Cray-2S running at an average speed of 125 MFLOPS).

Our upper bound for the ^{16}O ground-state energy fails to show that ^{16}O is stable against decay into four alpha particles. We can assume that further improvements to the form of the wave function will somewhat lower the bound, but it appears that some adjustment of the Urbana VII three-nucleon potential will be necessary to produce accurate ground-state energies for few-body and light nuclei. Work on both of these points is in progress. The longitudinal structure function obtained in this calculation shows significant effects due to dynamical correlations.

ACKNOWLEDGEMENTS

I thank V. R. Pandharipande for his hospitality during a sabbatical at the University of Illinois, and Professors S. Fantoni and A. Fabrocini for their hospitality during an extended visit to the University of Pisa. The calculations reported here were made possible by grants of time on the Cray computers at the National Center for Supercomputing Applications, University of Illinois at Urbana and on the Energy Research Cray computers at NMFECC. This work is supported by the National Science Foundation Grant No. DMR-83-1698 and the U. S. Department of Energy, Nuclear Physics Division, under contract W-31-109-ENG-38.

REFERENCES

1. C. R. Chen et al., Phys. Rev. C **33**, 1740 (1986); T. Sasakawa and S. Ishikawa, Few-Body Sys. **1**, 3 (1986).
2. J. Carlson, Phys. Rev. C **36**, 2026 (1987); C **38**, 1879 (1988).
3. R. Schiavilla, V. R. Pandharipande, and R. B. Wiringa, Nucl. Phys. **A449**, 219 (1986).
4. B. D. Day and R. B. Wiringa, Phys. Rev. C **32**, 1057 (1985).
5. R. B. Wiringa, V. Fiks, and A. Fabrocini, Phys. Rev. C **38**, 1010 (1988).
6. R. B. Wiringa, R. A. Smith, and T. L. Ainsworth, Phys. Rev. C **29** (1984).
7. R. Schiavilla, V. R. Pandharipande, and D. O. Riska, Phys. Rev. C (in press).
8. R. B. Wiringa, to be published.
9. H. Kümmel, K, H. Lührmann, and J. G. Zabolitzky, Phys. Reports **36C**, 1 (1978).
10. J. Carlson and M. H. Kalos, Phys. Rev. C **32**, 2105 (1985).
11. F. Iachello, A. Jackson, and A. Landé, Phys. Lett. **B43**, 1919 (1973).
12. H. DeVries, C. W. DeJager, and C. DeVries, Atomic Data and Nucl. Data Tables **36**, 495 (1987).
13. R. Schiavilla et al., Nucl. Phys. **A473**, 267 (1987).
14. R. Schiavilla, A. Fabrocini, and V. R. Pandharipande, Nucl. Phys. **A473**, 290 (1987).
15. P. Barreau et al., Nucl. Phys. **A402**, 515 (1983).

A NEW STATE OF NUCLEAR MATTER†

W. H. Dickhoff, C. C. Gearhart, and B. E. Vonderfecht

Department of Physics, Washington University
St. Louis, Missouri 63130, USA

A. Polls
Departament d'Estructura i Constituents de la Matèria,
Universitat de Barcelona, E-8028 Barcelona, Spain

A. Ramos
TRIUMF, 4004 Wesbrook Mall
Vancouver BC, Canada V6T2A3

Abstract: Recent developments, mainly related to exclusive electron scattering experiments, clearly establish that the nucleus is not a mean field system. The investigation of nuclear systems with special emphasis on dynamical aspects is therefore a natural consequence. A convenient description of non mean field dynamics is provided by a self-consistent formulation of the Green's function (SCGF) method. A study of nuclear matter at the level of ladder diagrams is made with special emphasis on a symmetric treatment of particles and holes (implied by the method) which is important for the treatment of long-range phenomena. A strong pairing instability in the $^3S_1$–$^3D_1$ channel is observed in a large density range. A self-consistent treatment of these correlations together with a proper treatment of short-range correlations is discussed. It is suggested that the actual ground state of nuclear matter around the empirical saturation density corresponds to a superfluid. It is also pointed out that similar coherence in particle-particle and hole-hole propagation is observed in finite nuclei. The importance of these investigations for a non-relativistic saturation mechanism is pointed out.

I. INTRODUCTION

In recent years the perspective on the naive single-particle (sp) shell model has changed. Experimental evidence has been gathered that nuclei are systems that cannot be described in terms of a mean-field picture. Partial occupation of previously thought fully occupied orbitals is now well established.[1,2] Even more dramatic is the clear

†This research was supported by the Condensed Matter Theory Program of the Division of Materials Research of the U.S. National Science Foundation under Grant No. DMR-8519077 which also provided computer time at the Pittsburgh Supercomputer Center. Additional support was provided by NATO under Grant No. RG 85/0684 and by CAYCIT Grant No. PB85-0072-C02-00 (Spain).

evidence of quasi-particle features associated with strongly correlated Fermi systems. Results from the (e,e′p) reaction clearly demonstrate that orbitals corresponding to one major shell below the Fermi energy, are already strongly fragmented and orbitals close to the Fermi energy have strength concentrated in one state but this strength typically carries only 50% of the sum rule.[3–5]

Theoretical investigation of the nucleon spectral function in finite nuclei indicates that one can explain all these features qualitatively, by solving the Dyson equation for the finite system with the inclusion of an energy dependent self-energy.[6–9] This self-energy is calculated in second order using a microscopic G matrix derived from nuclear matter. By limiting the energy of the intermediate states to within 50 MeV from the Fermi surface one focuses on the influence of the low-lying excitations, on the fragmentation of the sp strength. The distribution of strength from these calculations has the same features as found experimentally. The difference with the experimental distribution is two-fold. Firstly the experimental distribution is reduced with respect to the theoretical results by roughly a global factor of 2/3 and secondly there is a stronger fragmentation at very low energy. This last feature will naturally be improved when long-range correlations at the level of the Random Phase Approximation (RPA), are included in the self-energy since this leads to a stronger coupling to the collective states at low energy which has the required effect. The first feature should be related to the influence of short-range and tensor correlations[10] which deplete a sizable fraction of the strength and move it to high energy in the A+1 system.

The fragmentation of sp strength also has important consequences for the understanding of the nuclear response. When the mean field picture would suffice this response can be calculated using the standard RPA. Since this starting point is obviously incorrect, one has to include the dynamical fragmentation of strength into the calculation. A convenient extension of the RPA method based on this picture is dicussed in Ref. 11. Again including low energy excitations, good global agreement for all types of nuclear response functions is obtained up to 100 MeV excitation energy. Also here, the missing ingredient for complete quantitative agreement with experiment is a global overestimate of the strength in the theoretical calculation. This is consistent with the observations pointed out for the spectral functions and also points towards the influence of short-range correlations which can move, in this case, particle-hole strength to very high energies.

The influence of short-range and tensor correlations can be investigated in nuclear matter.[12] It is expected that such results have a bearing on the situation in finite nuclei when nuclear matter is considered around the empirical saturation density. The importance of a dynamical treatment is clearly indicated by the above discussion. A convenient method for dealing with dynamical correlations is provided by the Green's function method. A self-consistent formulation of this method is discussed in sect. 2. It is applied to the nuclear matter problem by incorporating the ladder diagrams in a symmetric treatment of particles and holes which allows the calculation of particle and hole spectral functions for nuclear matter. Calculations for the central part of the Reid $^3S_1$–$^3D_1$ interaction have focused on the role of short-range correlations which leads to a 10–15% overall depletion of strength below k_F to very high energy which is consistent with the observations for finite nuclei.[13]

A symmetric treatment of particles and holes is also important for the description of long-range correlations. Indeed it is found that realistic nuclear interactions exhibit a strong pairing instability in the solution of the ladder equation.[14] It has also been shown that the self-consistent formulation of the theory between the interaction approximated by ladder diagrams and the sp propagator including such self-energy insertions, contains under simplifying assumptions the non-linear BCS gap equation.[15] The dynamical features as well as the importance of pairing phenomena deserve further investigation also in relation to the nuclear saturation problem. It has for example been demonstrated that a symmetric treatment of particles and holes can lead to repulsive contributions to the energy which increase with density.[16] Results for a realistic interaction in connection with pairing correlations are discussed in sect. 3.

II. LADDERS AND SELF-CONSISTENT GREEN'S FUNCTIONS

An outline of the Self-Consistent Green's Function approach will be given first. The physical content of the method can be visualized as follows. A particle propagating through the medium experiences interactions with the other particles which result in modification of the properties of that particle. These interactions between particles immersed in the medium are naturally different compared to the interaction between free particles. Consequently, one deals with a non-linear problem in which self-consistency plays a crucial role. The physical concept outlined here can be captured conveniently using propagator (Green's function) language. The influence of the medium on the particle is expressed in terms of the self-energy. The propagator is then obtained by solving the Dyson equation. This equation will generate the distribution of single-particle strength which is proportional to the imaginary part of the propagator.

In this form of the theory it is then necessary to make a choice concerning the physical correlations that should be included in the interaction between particles in the medium. Such physical correlations can then be expressed in terms of generic diagrams and the numerical problem to be solved is then completely defined. In the case of nucleons a dominant correlation is the short-range repulsion the nucleons experience. These short-range correlations can be included in the interaction by incorporating the ladder diagrams as a minimum.

First the exact version of the theory is summarized in Figs. 1 and 2. In Fig. 1(a) the Dyson equation is shown.

$$g(\mathbf{k},\omega) = g^{(0)}(\mathbf{k},\omega) + g^{(0)}(\mathbf{k},\omega)\ \Sigma(\mathbf{k},\omega)\ g(\mathbf{k},\omega) \quad , \tag{1}$$

where $g^{(0)}$ is the unperturbed sp propagator and Σ the self-energy. All medium effects on the particle are contained in the self-energy and in turn this self-energy must reflect the interaction of the particle with the other particles. This relation is determined by the coupling of the sp propagator to the two-particle propagator which is represented graphically in Fig. 1(b). The vertex function Γ includes all possible interactions between the sp propagators and its relation to the two-particle propagator is shown in Fig. 2.[17] Clearly this formulation incorporates the physical notions discussed above, relating the medium effects on the particle to its interaction which is also medium modified, with the other particles.

Any approximation to the vertex function Γ or equivalently to g^{II} will nevertheless incorporate these same physical features albeit at a certain level of approximation. A minimum approximation to Γ incorporates the ladder diagrams in the nuclear case. This is related to the familiar ideas of Brueckner.[18] The corresponding Γ^L is the medium equivalent of the free T-matrix. The ladder equation is shown graphically in Fig. 3(a). This equation can be written as

$$\Gamma^L = V + Vgg\Gamma^L \equiv V + \Delta\Gamma^L = V + Vg_L^{II}V \quad , \tag{2}$$

where g is written as in Eq. (1) to avoid a proliferation of superscripts. The two-particle propagator g_L^{II} is the ladder approximation to the exact two-particle propagator in Fig. 2. For later purposes $\Delta\Gamma^L$ is also introduced in Eq. (2). With this minimum choice for the interaction in the medium the problem to be solved is completely specified. In Fig. 3(b) the resulting approximation to the Dyson equation is displayed in terms of the self-energy which has to be used in Eq. (1). This problem is non-linear since the self-energy in the Dyson equation depends on the solution g. This is the same non-linearity which is encountered in the Hartree-Fock problem.

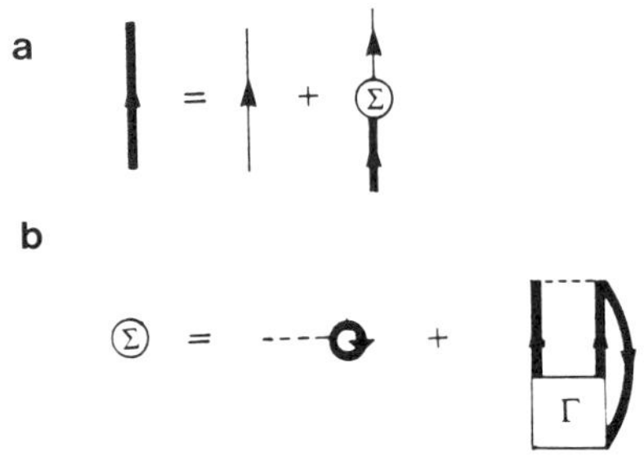

Figure 1. Diagrammatic representation of the Dyson equation (a) for the sp propagator. The structure of the proper self-energy Σ is given in (b) where the coupling to the vertex function Γ is shown. The V contribution contains both direct and exchange.

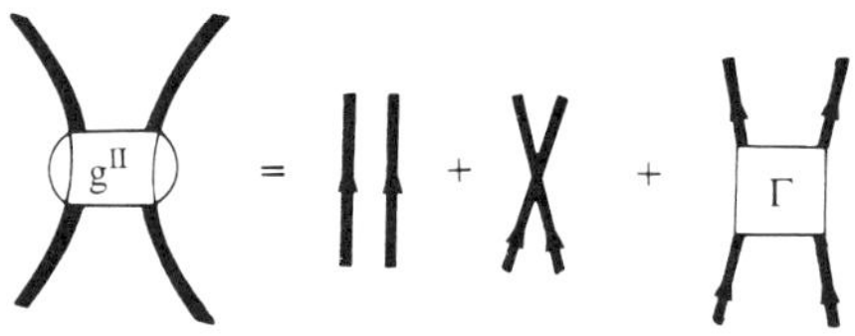

Figure 2. Diagrammatic representation of the two-particle propagator g^{II}. The first contributions contain the product of fully dressed sp propagators which have no interaction with each other. The remaining term includes all possible interactions between the fully dressed propagators and defines the reducible vertex function Γ.

The contents of the scheme outlined in Fig. 3 are very different from the conventional treatment of the short-range correlations in terms of the hole-line expansion[19] or the coupled cluster method.[20] In these approaches one makes a clear distinction between particles and holes which are treated quite differently. As a consequence a substantial gap in the sp spectrum can result at the Fermi momentum since the hole spectrum is lowered due to on-shell self-energy corrections whereas the particle spectrum is untouched. In neutron matter the situation is even worse, since the hole self-energy contributions are repulsive, leading to a sp spectrum with lower energies for particle states just above k_F than for hole states just below k_F. The large gap in nuclear matter leads to a suppression of correlations that are sensitive to particle-hole energy differences which include all types of long-range correlations. Using a propagator description, the ladder equation allows the propagation of both particles and holes in pairs. The presence of holes can lead to the appearance of pairing instabilities which can only be treated by a symmetric treatment of particles and holes.

In Ref. 13 the scheme of Fig. 3 is applied to the central part of the $^3S_1$–$^3D_1$ component of the Reid interaction.[21] This avoids the appearance of pairing instabilities since this reduced interaction is not attractive enough. Nevertheless, a detailed study of the influence of short-range correlations on sp properties can be carried out.[13] The solution of the ladder equation does not have complex eigenvalues and leads to a continuous sp spectrum. The self-consistency problem of Fig.3 was solved by only taking the on-shell self-energy contribution into account in the self-consistency procedure

$$\varepsilon(k) = \frac{k^2}{2m} + \mathrm{Re}\ \Sigma^C(k,\varepsilon(k)) \quad . \tag{3}$$

Here the self-energy is labeled with a C for continuum contribution. This leads to a simplified treatment of the ladder equation but allows a full study of the nucleon self-energy and the related sp spectral functions. An important feature of these spectral functions is that they show where the sp strength is distributed. The influence of short-range correlations on this distribution is to shift 10–15% of the strength to very high energy as a featureless distribution.[13] A comparison of the strength distribution below ε_F with a Correlated Basis Function calculation[22] using the full Urbana v_{14} interaction,[23] reveals that similar results are obtained. The not too large differences between the calculations can be attributed to the difference in the short-range part of the interaction as well as the use of only central correlations in Ref. 13. An important result of such calculations is that these spectral functions show quasi-particle behavior[24] in the presence of strong short-range correlations. This result together with the results for spectral functions obtained for finite nuclei demonstrate that this is the basis for the relevance of the nuclear shell model.[6,7]

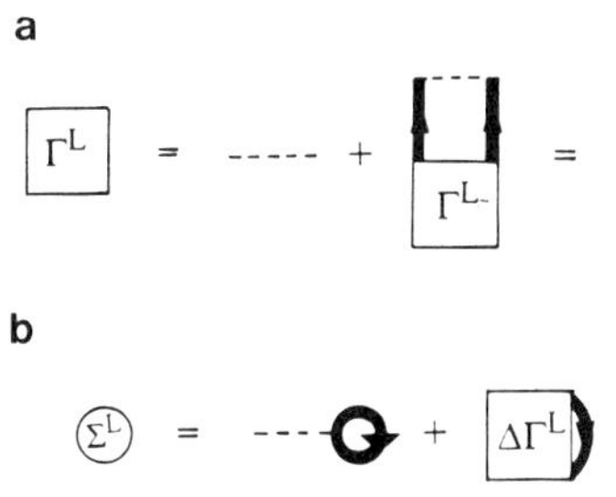

Figure 3. Ladder approximation to Γ (a) and the corresponding self-energy (b).

The spectral functions for the infinite system also contain the information to obtain the momentum distribution of nuclear matter. Treating only the continuum part of the interaction in the medium as in Ref. 13 one preserves the discontinuity in the momentum distribution at k_F. A paired system displays a continuous momentum distribution. Nevertheless the bulk effect of depletion below k_F and partial occupation above will be due to the strong short-range and tensor correlations which move sp strength to high energy.

A treatment of the full Reid interaction has to rely on a different approach to the solution of the self-consistency problem posed by Fig. 3. In Ref. 15 it was shown that a BCS treatment of pairing is in fact included in the self-consistent scheme of Fig. 3 and one can combine the ladder equation together with the Dyson equation into the non-linear gap equation. One must therefore incorporate some features of pairing into the solution procedure. Obviously, this involves the presence of a gap in the sp spectrum which has to be self-consistently generated. Instead of solving the non-linear gap equation, one can separately solve for the bound state contributions of the ladder equation. With these contributions to the self-energy added to the usual continuum terms, the Dyson equation must then be solved. The solution of this equation can be tailored into the form of the BCS propagator depending on the degree of sophistication in the treatment of the self-energy. When bound state solutions to the ladder equation are found, however, a gap in the sp spectrum is guaranteed.[14] This procedure must then be repeated until the spectrum is self-consistent. In this form the treatment of pairing can be incorporated directly in the description of strongly correlated systems.

As in the case of the treatment of the normal system an average self-consistency for the sp spectrum will be sought. This spectrum will have a discontinuity and an ansatz for the solution of the propagator in Fig. 3 of the form

$$\begin{aligned} \varepsilon(k) &= \frac{k^2}{2m} + \mathrm{Re}\,\Sigma^C(k,\varepsilon(k)) + U_+^B(k) \qquad k>k_F \\ &= \frac{k^2}{2m} + \mathrm{Re}\,\Sigma^C(k,\varepsilon(k)) + U_-^B(k) \qquad k<k_F \end{aligned} \tag{4}$$

is used. The functions U_+^B and U_-^B will be obtained from the bound pair state contributions to the self-energy (see below).

As a starting point of the iteration scheme one simply chooses a gap in the sp spectrum that will prevent the occurrence of complex eigenvalues. With such an ansatz the continuum of hole-hole excitations ends at $2\varepsilon_F^-$ which is lower than the beginning of the continuum of particle-particle excitations denoted by $2\varepsilon_F^+$. The energy of the bound pair states can then be found by solving the ladder equation for the two-particle propagator (see Fig. 2)

$$g_L^{II} = g_{(0)}^{II} + g_{(0)}^{II}\, V\, g_L^{II} \quad . \tag{5}$$

Assuming the presence of bound states in the Lehmann representation for g_L^{II} one obtains

$$\begin{aligned} &g_L^{II}(\mathbf{k},\mathbf{k}';\mathbf{K},\Omega,S) \\ &= \frac{z^*(\mathbf{k};\mathbf{K},E_0^{A+2}(\mathbf{K},S),S)\,z(\mathbf{k}';\mathbf{K},E_0^{A+2}(\mathbf{K},S),S)}{\Omega - (E_0^{A+2}(\mathbf{K},S) - E_0) + i\eta} + \int_{2\varepsilon_F^+}^{\infty} d\Omega' \frac{z^*(\mathbf{k};\mathbf{K},\Omega',S)\,z(\mathbf{k}';\mathbf{K},\Omega',S)}{\Omega - \Omega' + i\eta} \\ &- \frac{z^*(\mathbf{k}';\mathbf{K},E_0^{A-2}(\mathbf{K},S),S)\,z(\mathbf{k};\mathbf{K},E_0^{A-2}(\mathbf{K},S),S)}{\Omega - (E_0 - E_0^{A-2}(\mathbf{K},S)) - i\eta} - \int_{-\infty}^{2\varepsilon_F^-} d\Omega' \frac{z^*(\mathbf{k}';\mathbf{K},\Omega',S)\,z(\mathbf{k};\mathbf{K},\Omega',S)}{\Omega - \Omega' - i\eta} \end{aligned} \tag{6}$$

Only one energy variable is necessary since the ladder approximation is considered here, which requires merely two external times. In Eq. (6) the momenta $\mathbf{k},\mathbf{k}'$ denote the relative momenta $\frac{1}{2}(\mathbf{k}_1-\mathbf{k}_2)$ and $\frac{1}{2}(\mathbf{k}_3-\mathbf{k}_4)$. The center of mass (cm) momentum $\mathbf{K}$ is conserved and corresponds to $\mathbf{k}_1+\mathbf{k}_2 = \mathbf{k}_3+\mathbf{k}_4$. The pair transition amplitudes are denoted by z and the total spin S also represents total isospin. The bound states in the A±2 systems have energies which depend on the total spin S (and isospin) and the cm momentum $\mathbf{K}$. The subscript 0 therefore denotes the lowest states in the corresponding systems which have total momentum $\mathbf{K}$, only for $|\mathbf{K}| = 0$ they correspond to the ground states.

One can then investigate if the bound state contributions to Eq. (6) indeed solve Eq. (5). At this point it is helpful to remember the result for the unperturbed propagator $g_{(0)}^{II}$. The word unperturbed here of course refers to the fact that the sp propagators contributing to $g_{(0)}^{II}$ do not interact with each other although dressing of the individual propagators is included according to Eq. (4).

$$\begin{aligned} &g_{(0)}^{II}(\mathbf{k},\mathbf{k}';\mathbf{K},\Omega) = (2\pi)^3(\delta(\mathbf{k}-\mathbf{k}') - \delta(\mathbf{k}+\mathbf{k}')) \\ &\left\{ \frac{\theta(|\frac{1}{2}\mathbf{K}+\mathbf{k}| - k_F)\,\theta(|\frac{1}{2}\mathbf{K}-\mathbf{k}| - k_F)}{\Omega - \varepsilon(\frac{1}{2}\mathbf{K}+\mathbf{k}) - \varepsilon(\frac{1}{2}\mathbf{K}-\mathbf{k}) + i\eta} - \frac{\theta(k_F - |\frac{1}{2}\mathbf{K}+\mathbf{k}|)\,\theta(k_F - |\frac{1}{2}\mathbf{K}-\mathbf{k}|)}{\Omega - \varepsilon(\frac{1}{2}\mathbf{K}+\mathbf{k}) - \varepsilon(\frac{1}{2}\mathbf{K}-\mathbf{k}) - i\eta} \right\} \end{aligned} \tag{7}$$

Here spin and isospin indices are implied. The imaginary part of this propagator extends to $2\varepsilon_F^-$, is zero until $2\varepsilon_F^+$ and then is non-zero again. Naturally in this approximation only one combination of **k** and **K** will contribute to this imaginary part for a given Ω, nevertheless all momenta are reached in the integral equation and for energies below $2\varepsilon_F^-$ and above $2\varepsilon_F^+$ one has to solve Eq. (5) as an integral equation which is better handled numerically in terms of an integral equation for the interaction.[13] For energies between $2\varepsilon_F^-$ and $2\varepsilon_F^+$, bound state solutions can exist with corresponding energies that emerge as the poles in Eq. (6). For energies Ω in the neighborhood of these poles one observes that due to the gap in the unperturbed spectrum, the inhomogeneous term, $g^{II}_{(0)}$, in the ladder equation, Eq. (5), can be neglected. One therefore obtains an eigenvalue equation which will give the eigenvalues $|E_0^{A\pm2}(\mathbf{K},S)-E_0|$ and the corresponding transition amplitudes z. In the nuclear case it is useful to make a decomposition of this eigenvalue equation into partial waves. This decomposition is exact in the limit **K**=0 which will now be discussed. The eigenvalue equation reads (denoting the eigenvalues by $E_\Delta^\pm$)

$$z^*_{LSJ}(k,E_\Delta^\pm) = \tfrac{1}{2} g^{II}_{(0)}(k,E_\Delta^\pm) \sum_{L'} \int \frac{dk'}{(2\pi)^3} k'^2 \langle k,LSJ|V|k',L'SJ\rangle z^*_{L'SJ}(k',E_\Delta^\pm) \quad . \tag{8}$$

In this equation $g^{II}_{(0)}$ is used without the δ–functions (see Eq. (7)) and the matrix element of the interaction V is antisymmetrized. Only physical channels are possible for which L+S+T is odd. Given the orbital angular momentum L and the total spin S, the isospin is therefore determined. The total angular momentum is denoted by J. Eq. (8) includes the possibility of bound states related to coupled channels. It was shown in Ref. 15 that for a separable interaction only the overall sign determines the occurrence of bound states. Here of course it is the overall strength of the interaction as weighted in the integral in Eq. (8) which determines this.

In addition to these bound state contributions to the two-particle propagator, one also has to deal with the continuum contributions. These are most easily obtained by solving directly the integral equation for Γ^L (see Eq.(2)) using the methods discussed in Ref. 13. Both contributions to g_L^{II} then have to be included in the calculation of the self-energy. If the interaction V produces bound states one can obtain the corresponding self-energy contributions by using the relation $\Delta\Gamma^L = V g_L^{II} V$ (see Eq. (2) and Fig. 3). The contribution from the A+2 pole yields[14]

$$\Sigma_+^\Delta(\mathbf{k},\omega) = \int \frac{d\mathbf{k}'}{(2\pi)^3} \frac{|\alpha(\mathbf{k}_r;\mathbf{K},E_\Delta^+)|^2}{\omega - E_\Delta^+ + \varepsilon(\mathbf{k}') + i\eta} \tag{9}$$

where

$$\alpha(\mathbf{k}_r;\mathbf{K},E_\Delta^+) = \int \frac{d\mathbf{q}}{(2\pi)^3} z(\mathbf{q};\mathbf{K},E_\Delta^+)\langle \mathbf{q}|V|\mathbf{k}_r\rangle \tag{10}$$

is used and $\mathbf{k}_r$ denotes the relative momentum $\tfrac{1}{2}(\mathbf{k}-\mathbf{k}')$. The transformation from partial wave representation to vector momentum representation is implied here allowing in principle for the occurrence of different bound states referring to different partial wave channels. The analogous result for the A-2 pole contribution reads

$$\Sigma_-^\Delta(\mathbf{k},\omega) = \int \frac{d\mathbf{k}'}{(2\pi)^3} \frac{|\alpha(\mathbf{k}_r;\mathbf{K},E_\Delta^-)|^2}{\omega - E_\Delta^- + \varepsilon(\mathbf{k}') - i\eta} \theta(|\mathbf{k}'| - k_F) \quad . \tag{11}$$

The contribution of the bound states will only be considered for **K**=0. This restriction is conventional and will provide the simplest method to include both the continuum and the bound state contributions to the self-consistency problem. This restriction is not necessary and should be further studied. Including only the zero total momentum contribution to Eqs. (9) and (11) leads to setting

$$|\alpha(\mathbf{k}_r;\mathbf{K},E_\Delta^\pm)|^2 = (2\pi)^3\delta(\mathbf{K})\Delta_\pm^2(\mathbf{k}_r) \,. \tag{12}$$

With this assumption one obtains pole terms in these self-energy contributions

$$\Sigma_\pm^\Delta(k,\omega) = \frac{\Delta_\pm^2(k)}{\omega - E_\Delta^\pm + \varepsilon(k) \pm i\eta} \,. \tag{13}$$

The + sign in Σ_Δ refers to momenta below k_F, the - sign to momenta above which reflects the analytic structure of the denominator. These self-energy contributions, $\Sigma_\pm^\Delta$ and Σ^C, must then be used to solve the Dyson equation (Eq. (1)). Using the on-shell approximation[13] (see Eq. (3)) for the continuum contribution one obtains

$$\Sigma(k,\omega) = \Sigma_\pm^\Delta(k,\omega) + \Sigma^C(k,\varepsilon(k)). \tag{14}$$

The solution of the Dyson equation then is reduced to a simple two-dimensional eigenvalue problem from which the conventional BCS propagator emerges.[15] This propagator has two poles for each value of k if one assumes that only one partial wave channel provides bound pair states

$$E^\pm(k) = \tfrac{1}{2}(E_\Delta - \varepsilon(k) + \frac{k^2}{2m} + \Sigma^C(k))$$
$$\pm [(\tfrac{1}{2}(E_\Delta - \varepsilon(k) - \frac{k^2}{2m} - \Sigma^C(k)))^2 + \Delta^2(k)]^{1/2} \,. \tag{15}$$

Here the $\pm$ associated with the two different poles on the right hand side of the equation have been suppressed. It is now possible to define the functions U_+^B and U_-^B (see Eq. (4)) by subtracting the kinetic energy and the continuum contribution from Eq. (15), as follows

$$U_+^B(k) = \tfrac{1}{2}(E_\Delta^- - \varepsilon(k) - \frac{k^2}{2m} - \Sigma^C(k))$$
$$+ [(\tfrac{1}{2}(E_\Delta^- - \varepsilon(k) - \frac{k^2}{2m} - \Sigma^C(k)))^2 + \Delta_-^2(k)]^{1/2} \,, \quad k > k_F \tag{16}$$

and

$$U_-^B(k) = \tfrac{1}{2}(E_\Delta^+ - \varepsilon(k) - \frac{k^2}{2m} - \Sigma^C(k))$$
$$- [(\tfrac{1}{2}(E_\Delta^+ - \varepsilon(k) - \frac{k^2}{2m} - \Sigma^C(k)))^2 + \Delta_+^2(k)]^{1/2} \,, \quad k < k_F \,, \tag{17}$$

respectively. Instead of keeping two branches for all momenta, this prescription leads back to a single sp energy for each momentum by keeping only the dominant branch. This choice is made for practical reasons since it complicates the solution of the continuum integral equation substantially when BCS sp propagators are used. Nevertheless, already the influence of the continuum leads to similar effects related to the dispersion of sp strength[13] and should ultimately be considered in a complete solution to the problem depicted in Fig. 3 in which the full dressing of the sp propagator is self-consistently determined.

At this stage one has to go back to the ladder equation and solve it again in a next iteration step towards a self-consistent solution. It is important to observe that the sp gap resulting from Eqs. (16) and (17) has been dynamically calculated and is due to the presence of the bound pair states. The functions $U_{\pm}^{B}$ are then included in the self-consistency procedure and represent the bound pair state contibution to the sp spectrum. In addition the continuum contribution must be considered until one finally arrives at a self-consistent spectrum according to Eq.(4).

III. BOUND PAIR STATES IN NUCLEAR MATTER

To demonstrate the interest of the scheme outlined in the previous section some simple calculations will be discussed which assess the importance of pairing phenomena in nuclear systems. It will be crucial to treat the nuclear interaction with its full state-dependence. Already from knowledge of the nucleon-nucleon phase-shifts one deduces that the interaction is attractive in the $^3S_1$–$^3D_1$ and $^1S_0$ channel. Because of its relevance for neutron stars also the $^3P_2$–$^3F_2$ channel is considered here. The bound pair states are obtained by solving Eq. (8). To establish the density range where such correlations play a role, the following survey is made. Using a kinetic energy spectrum without modification due to continuum contributions a gap (k-independent) at k_F is introduced by shifting the whole sp spectrum for $k < k_F$ by a constant energy in such a way that for each density the pairing instability is just avoided. This means that with this gap the eigenvalues of Eq. (8) are never complex but a slightly smaller gap would lead to two complex eigenvalues. Clearly, the larger the required gap the stronger the correlations. The resulting minimum gap is shown in Fig. 4 as a function of k_F for the three partial wave channels mentioned above.

The correlation in the $^3S_1$–$^3D_1$ is very strong, requiring a gap of slightly more than 15 MeV around $k_F \approx 1.2\text{fm}^{-1}$ to prevent the instability. The limit $k_F \rightarrow 0$ gives half the binding energy of the deuteron as expected. The $^1S_0$ channel requires a maximum separation between particle and hole sp energies of about 3 MeV around $k_F = 0.8\ \text{fm}^{-1}$. The correlations are weakest for the $^3P_2$–$^3F_2$ channel which requires at most a 0.5 MeV gap around $k_F = 2.0\ \text{fm}^{-1}$. It is important to keep in mind that these results cannot be a final estimate of the real gap. The influence of the continuum on the sp spectrum is such that it gives the most attractive contribution to the sp energy for very small momenta. This influence will enhance somewhat the difference between particle and hole energies, thereby reducing somewhat the gap. In addition the effect of the pairing on the average occupation and depletion of sp states (in terms of v^2 and u^2 factors) should be included self-consistently for a careful treatment of the pairing properties.

Nevertheless, one can consider the results shown in Fig. 4 as a kind of upper bound on the gap. It is extremely unlikely that a more complete treatment of pairing will lead to larger gaps and therefore one can expect that the real gap at k_F for each channel will lie below the curves shown in Fig. 4. It is also conceivable that the modification of the sp spectrum due to the continuum and long-range correlations[25] will change the sp spectrum such that no instability remains. This is most likely to occur at higher density where such effects are large. On the other hand it is quite unlikely that the pairing in the $^3S_1$–$^3D_1$ channel can be suppressed by such effects around normal nuclear density since the required gap there in the present calculation is extremely large. One can therefore conclude that nuclear matter around normal density must be considered as a superfluid with pairing in deuteron quantum numbers. As such it can be called a new state of nuclear matter.

It is interesting to note that the density range where pairing is likely to appear for the other two channels coincides quite nicely with other estimates for such effects for neutron matter.[26–29] Calculations as outlined in sect. II are therefore also planned for neutron matter.

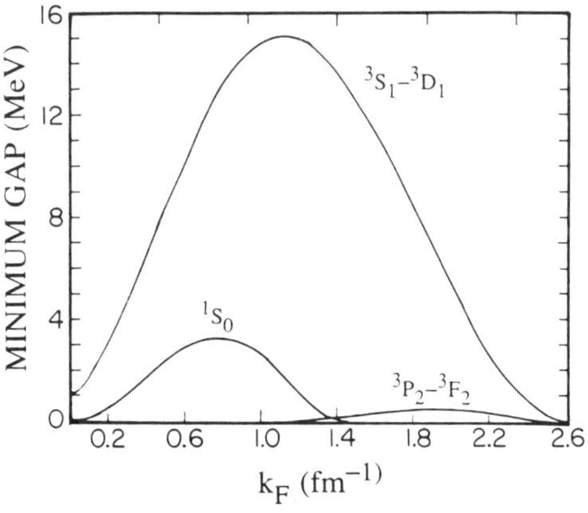

Figure 4. Minimum gap in the sp spectrum at k_F to avoid a pairing instability in the $^3S_1$–$^3D_1$, $^1S_0$, and $^3P_2$–$^3F_2$ channel of the Reid soft core interaction as a function of k_F. Except for the gap the sp spectrum included only kinetic energies.

It is now possible to use the knowledge of Fig. 4 to actually show the eigenvalues of Eq.(8) as a function of k_F. By choosing the maximum gap from Fig.4 for each channel one ensures that when this gap is used for all densities no complex eigenvalues will occur in solving Eq. (8). The resulting eigenvalues for the $^3S_1$–$^3D_1$ channels are shown in Fig. 5 as a function of k_F. The continuum for particle-particle (pp) excitations starts at the top of the figure and the continuum for hole-hole (hh) excitations ends at the bottom. The full curves show the eigenvalues of Eq. (8) including pp and hh propagation. Not always two eigenvalues will appear between the two continua. In the limit of zero density one eigenvalue will remain which corresponds to the deuteron bound state of two nucleons. This eigenvalue is therefore located at exactly the binding energy of the deuteron below the pp continuum.

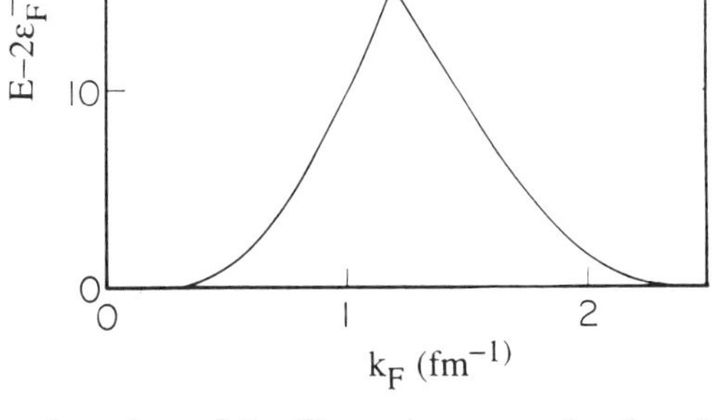

Figure 5. The eigenvalues of Eq. (8) are shown as a function of k_F using the maximum gap in the sp spectrum derived from Fig. 4 for the $^3S_1$–$^3D_1$ channel.

The coherence observed here in this channel is substantially stronger in the medium for a considerable range of densities than in free space. This can be related to the importance of the tensor force in generating this coherence. In free space the lowest energy states in Eq. (8) (which then corresponds to the Schrödinger equation in momentum space) have zero relative momentum. These states do not couple directly to the D state. In the medium the lowest energy states in Eq. (8) have momenta corresponding to k_F and can directly couple to the D state as a result of which the attraction becomes stronger in the medium. It should also be mentioned that Eq. (8) has similar properties as an RPA eigenvalue equation which also allows very collective solutions (it sometimes also leads to imaginary solutions). The procedure outlined here for the complete two-particle phase space using the bare nucleon-nucleon interaction in nuclear matter, is used in finite nuclei in restricted model spaces employing effective interactions and is referred to as ppRPA.[30] A version which directly discretizes the ladder eigenvalue problem for nuclear matter without considering the question of the gap can be found e.g., in Ref. 31.

The tensor correlations are so strong that even if one eliminates the contribution of hh propagation to Eq. (8), one still first finds increasing attraction as a function of k_F. This result is shown as the dashed curve in Fig. 5. Note that this eigenvalue curve corresponds to the eigenvalue of the Bethe-Goldstone equation with kinetic energies as sp spectrum. At low density the pp eigenvalue (full curve) obtained with inclusion of holes is close to this curve since the hh phase space is very small. The difference becomes very large, however, around $k_F = 1.2\ \text{fm}^{-1}$ emphasizing the importance of hh correlations.

Similar results are shown in Fig. 6 for the $^1S_0$ channel. Again there is substantially more collectivity generated by including hh propagation although even the Bethe-Goldstone equation supports a bound state in a considerable density range. From the perspective of RPA the Bethe-Goldstone equation only generates Tamm-Dancoff like correlations and will therefore give rise to less collective features. These bound pair states should therefore be considered as the most collective solutions of the ladder equation and therefore deserve appropriate consideration in describing the physics of nuclear matter. It is curious that these correlations have never been considered with modern interactions since they already show up in the case of the Bethe-Goldstone equation.

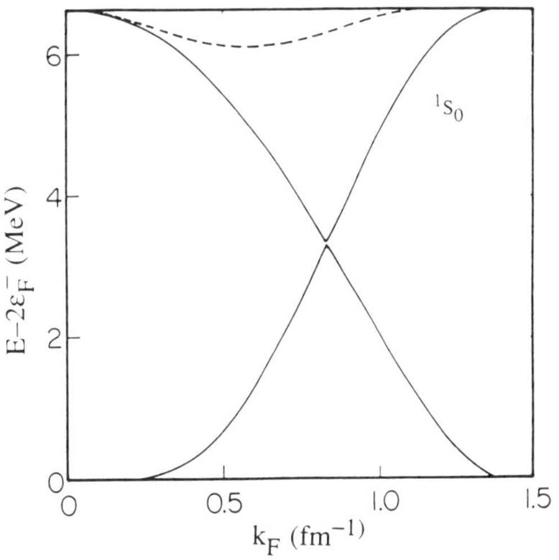

Figure 6. Same as Fig. 5 but for the $^1S_0$ channel.

The correlations associated with pairing in $^3S_1$–$^3D_1$ quantum numbers are also present in finite nuclei with N=Z. For comparison a similar plot is made in Fig. 7 as in Figs. 5 and 6 for the infinite system. Using experimental information on the neighboring odd nuclei around ^{4}He and ^{16}O one can obtain $\varepsilon_F^{p,+} + \varepsilon_F^{n,+}$ and $\varepsilon_F^{p,-} + \varepsilon_F^{n,-}$ which correspond to the lowest energy poles of the two-particle propagator as considered as a product of two dressed sp propagators (see first term on the rhs of Eq. (5)). These two energies can be compared to the top and the bottom of Figs. 5 and 6 since these correspond to the same poles in the case of the infinite system. Again, using experimental information on the ground state energies of the nuclei which are a proton-neutron pair removed from ^{4}He and ^{16}O one can obtain the position of the exact eigenvalues of the bound pair states with respect to these nuclei. The size of the correlation energy is of similar magnitude as in Fig. 5 which points out that these correlations are indeed strong in nuclei and should be carefully considered. Experimental signature for the $^1S_0$ correlations shows that the correlation energies for the N=Z nuclei considered in Fig. 7, are similar to the maximum attraction obtained in Fig. 6 at about 0.8 fm^{-1}.

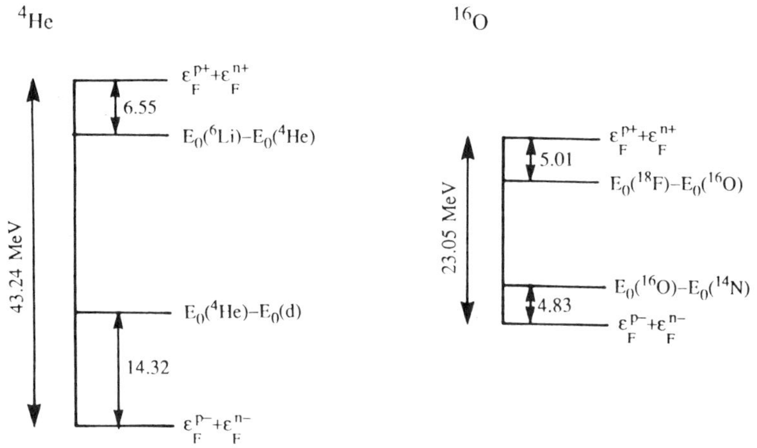

Figure 7. Position of experimental bound proton-neutron pair states with respect to sum of the proton and neutron Fermi energies for nuclei around ^{4}He and ^{16}O.

Including these bound pair correlations and the repulsive contribution to the energy due to hh terms as discussed in Ref. 16, it might be possible to get a fresh perspective on the nuclear matter saturation problem as well. In this context it is noteworthy that the saturation point for a given nucleon-nucleon interaction can be linked to the strength of its tensor component. This same tensor force also plays a very dominant role in the pairing correlations discussed above for the $^3S_1$–$^3D_1$ channel. Also the density range for which these correlations are important, is suggestive. Calculations according to the scheme outlined in sect. II are in progress. This scheme incorporates the effects of short-range correlations but at the same time is capable of dealing with strong long-range correlations in the form of pairing. First results for the self-consistent sp spectrum for the full Reid potential at $k_F = 1.36$ fm^{-1} indicate that 2–4 MeV repulsion can be expected for the energy per particle compared to a calculation in which only pp propagation in the ladder equation is considered. The former calculation leads to about 15 MeV binding energy at this density. This can be considered a promising result with regard to a possibly different saturation mechanism when self-consistent theories are considered.

REFERENCES

1. G. J. Wagner, Proc. Workshop on Nuclear Structure at High Spin, Excitation, and Momentum Transfer, Indiana University (1985), ed. H. Nann, AIP Conf. Proc. **142** (1986) 220; P. Grabmayr, S. Klein, H. Clement, K. Reiner, W. Reuter, G. J. Wagner, and G. Seegert, Phys. Lett. **164B** (1985) 15; P. Grabmayr, G. J. Wagner, H. Clement, and H. Röhm, Nucl. Phys. **A494** (1989) 244.
2. E. N. M. Quint, B. M. Barnett, A. M. van den Berg, J. F. J. van den Brand, H. Clement, R. Ent, B. Frois, D. Goutte, P. Grabmayr, J. W. A. den Herder, E. Jans, G. J. Kramer, J. B. J. M. Lanen, L. Lapikás, H. Nann, G. van der Steenhoven, G. J. Wagner, and P. K. A. de Witt Huberts, Phys. Rev. Lett. **58** (1987) 1088.
3. J. W. A den Herder, H. P. Blok, E. Jans, P. H. M. Keizer, L. Lapikás, E. N. M. Quint, G. van der Steenhoven, and P. K. A. de Witt Huberts, Nucl. Phys. **A490** (1988) 507.
4. E. N. M.Quint, Ph.D. Thesis (University of Amsterdam, 1988).
5. L. Lapikás, preprint NIKHEF (1989).
6. W. H. Dickhoff, P. P. Domitrovich, A. Polls, and A. Ramos, *Condensed Matter Theories*, vol. 5, ed. V. C. Aguilera-Navarro (Plenum, New York, 1990) in press.
7. W. H. Dickhoff, P. P. Domitrovich, K. Allaart, M. G. E. Brand, F. A. Muller, G. A. Rijsdijk, A. Polls, and A. Ramos, in *Topical Workshop on Two Nucleon Emission Reactions*, Proc. EIPC Vol. 3, O. Benhar and A. Fabrocini, eds. (Giardini, Pisa, 1990) in press.
8. P. P. Domitrovich and W. H. Dickhoff, to be published.
9. M. G. E. Brand, G. A. Rijsdijk, F. A. Muller, K. Allaart, and W. H. Dickhoff, to be published.
10. V. R. Pandharipande, C. N. Papanicolas and J. Wambach, Phys. Rev. Lett. **53** (1984) 1133.
11. M. G. E. Brand, K. Allaart, and W. H. Dickhoff, Phys. Lett. **214B** (1988) 483; Nucl. Phys. **A** in press.
12. S. Fantoni and V. R. Pandharipande, Nucl. Phys. **A427** (1984) 473.
13. A. Ramos, A. Polls and W. H. Dickhoff, Nucl. Phys. **A503** (1989)1; in *Condensed Matter Theories*, vol. 3, ed. J. S. Arponen, R. F. Bishop and M. Manninen (Plenum, New York, 1987) p. 319.
14. W. H. Dickhoff, *Condensed Matter Theories*, vol. 4, ed. J. Keller (Plenum, New York, 1989) p. 271.
15. W. H. Dickhoff, Phys. Lett.**210B** (1988) 15.
16. A. Ramos, W. H. Dickhoff, and A. Polls, Phys. Lett. **219B** (1988) 319.
17. A. A. Abrikosov, L. P. Gorkov, and I. E. Dzyaloshinski, *Methods of Quantum Field Theory in Statistical Physics* (Dover, New York, 1975).
18. K. A. Brueckner, C. A. Levinson, and H. M. Mahmoud, Phys. Rev. **95** (1954) 217.
19. B. D. Day, Phys. Rev. Lett. **47** (1981) 226; Phys. Rev. C **24** (1981) 1203.
20. B. D. Day and J. G. Zabolitzky, Nucl. Phys. **A366** (1981) 221.
21. R. V.Reid, Ann. Phys. **50** (1968) 411.
22. O. Benhar, A. Fabrocini and S. Fantoni, Nucl. Phys. **A**, in press.
23. I. E. Lagaris and V. R. Pandharipande, Nucl. Phys. **A359** (1981) 331.
24. L. D. Landau, Soviet Phys. **8** (1958) 70.
25. J. W. Clark, C.-G. Källman, C.-H. Yang, and D. A. Chakkalakal, Phys. Lett. **61B** (1976) 331.
26. C.-H. Yang and J. W. Clark, Nucl. Phys. **A174** (1971) 49; J. M. C. Chen, J. W. Clark, E. Krotscheck, and R. A. Smith, Nucl. Phys. **A451** (1986) 509.
27. M. Hoffberg, A. E. Glassgold, R. W. Richardson, and M. Ruderman, Phys. Rev. Lett. **24** (1970) 775.
28. T. Takatsuka and R. Tamagaki, Progr. Theor. Phys. **46** (1971) 114; T. Takatsuka, Progr. Theor. Phys. **48** (1972) 1517.
29 D. Pines and M. A. Alpar, Nature **316** (1985) 27.
30. P. Ring and P. Schuck, *The Nuclear Many-Body problem* (Springer, New York, 1980).
31. H. Q.Song, S. D. Yang, and T. T. S. Kuo, Nucl. Phys. **A462** (1987) 491.

THE " IDEAL " SHELL MODEL CALCULATION

A.P. Zuker

C.R.N., IN2P3-CNRS/Université Louis Pasteur

BP 20, F-67037 Strasbourg Cedex, France

The "standard" many body description of nuclei is based on non-relativistic potentials consistent with the observed nucleon-nucleon phase shifts. For over ten years the understanding of these "realistic" forces and the accuracy of calculations of ground state properties have been sufficient to establish that the standard description fails to reproduce the binding energy and density of nuclear matter, *but does not fail by much*. The contributions of S. Peiper and T. Dickhoff to this conference deal with the subject. In this talk we describe minimal phenomenological remedies for this failure and discuss their origin and consequences.

The strategy is to start from realistic forces and reach as end product detailed spectroscopic properties by taking seriously the idea that our task is to solve the Schrödinger equation exactly. A frontal attack based on purely mathematical and numerical tools is bound to fail precisely because the forces do not saturate properly. Therefore we resort to a step by step approach relying on insight offered by successful shell model phenomenology.

In section I we introduce the f^n model as an example of the literal validity of the shell model with simple configuration mixing. To account for it, an ideal calculation in 3 steps is necessary:

Ia) Define the seffconsistent field of the core and its evolution as particles are added.

Ib) Isolate a model space perturbatively.

Ic) Diagonalize the resulting effective interaction in the model space.

The definition of a model space in Ib is in general a subtle affair

Recent Progress in Many-Body Theories, Volume 2
Edited by Y. Avishai
Plenum Press, New York, 1990

and we shall discuss the possible ways in which systems may depart from extreme j-j coupling. In the process, the importance of the " monopole field " will be made evident.

Section II will start by a definition of the monopole field and of its invariant particle-hole parametrization that makes possible to isolate the terms of the Hamiltonian responsible for global (in particular, saturation) properties from the rest, responsible for spectroscopic detail. Then we list a number of facts and arguments that will help in devising corrective phenomenology. Finally some formal remarks are made on its origin.

Section III is devoted to results. First we examine how the program of minimal corrective action works. Next we present the largest diagonalization done so far (^{48}Ti) which offers an example of the success of the minimal program and the relevance of realistic forces. At the same time it is an exact model to test perturbative solutions that are seen to fare quite well. Then we explain how perturbation theory builds up quadrupole collectivity. Finally we describe an interpretation of the Lanczos algorithm as generating strength functions and apply it to magnetic dipole excitations.

Many people have collaborated over the years to the work of which this talk presents a partial and preliminary synthesis. In section II Ali Abzouzi was mainly responsible for bringing to successful completion the monopole programme started by Eduardo Pasquini, Anibal Cortès and the author some 15 years ago. The help of Etienne Caurier was determinant in the final stages of the project. He is also the author of the monumental code ANTOINE (CA89) capable of 10^6-dimensional diagonalizations: section III, to which Alfredo Poves participated actively, rests on it.

IMPORTANT : Any doubt about the meaning of a word or symbol is most likely to be resolved by the Appendix on notations.

I. The f^n model and the ideal calculation

Although nuclear physicists used to complain that they had no H atom, there is one region ($20 \leqslant N,Z \leqslant 28$) where the shell model exhibits the maximal simplicity of extreme j-j coupling. If we add a neutron to the doubly magic core of ^{40}Ca we populate preferentially four groups of states $j^\pi = 7/2^-, 3/2^-, 1/2^-$ and $5/2^-$ which correspond nicely to the orbits $f_{7/2}$, $p_{3/2}$, $p_{1/2}$ and $f_{5/2}$ expected in the extreme single particle model. The $7/2^-$ ground state carries practically all the transfer strength, which gets increasingly split as we move up but is still nicely concentrated in one dominant peak for $p_{3/2}$ (at around 2 MeV), less so around 4 MeV for $p_{1/2}$ and badly spread around 6 MeV for $f_{5/2}$ (NPA310).

If we add a further nucleon we naïvely expect that it will go to the $f_{7/2}$ orbit and indeed there is a set of states with $J = 0,2,4,6$ $(T = 1)$ and $J = 1,3,5,7$ $(T = 0)$ which dominate the low lying spectrum and square quite well with the energies calculated in a pure $f^2_{7/2}$ configuration with a δ force.

The situation suggests that $f^n_{7/2}$ configurations be used as " model space " with a Hamiltonian defined by the observed spectrum in mass 42 (i.e. $n = 2$). This type of calculations, pioneered by Bayman, Mc. Cullen and Zamick (MBZ64) are very successful in reproducing energy levels (see fig. 3 for an example).

The next step is to ask whether this very satisfactory state of affairs can be understood in terms of a more basic calculation. There is a good practical reason for doing so : the energetics of the $f^n_{7/2}$ model are good but its predictions of electromagnetic transition rates are erratic and it gives quadrupole moments that are consistently of opposite signs to those observed. Furthermore, nature seldom complies in offering spectra consistent with extreme j-j coupling for which the interaction is as readily available as in mass 42.

Conceptually the case is even stronger : what we have said about ^{41}Ca indicates we are observing dressed quasiparticles in the Landau sense rather than bare (Hartree-Fock) ones. This is corroborated by the sum rules for spectroscopic factors (e.g. $\langle ^{41}Ca|a^+_{f_{7/2}}a_{f_{7/2}}|^{41}Ca\rangle$) which are systematically smaller than 1, which is what we expect for a normal (but interacting) Fermi liquid. In other words, the naïve wavefunction $f^n_{7/2}$ has to be interpreted as a dressed " quasiconfiguration ". Schematically we have

$$\tilde{f}^n = f^n + \sum C_k |k\rangle \quad , \quad k = f^{n-\phi} h^{A_c - \chi} r^{\rho} p^{\pi} \quad , \quad \rho + \pi - \phi - \chi = 0 \qquad (1)$$

where $f \equiv f_{7/2}$; $r \equiv p_{3/2}p_{1/2}f_{5/2}$, generically ; $h =$ the occupied orbitals in ^{40}Ca (i.e. $A_c = 40$), $p =$ the free orbitals above the fp shell. (Note that $\rho,\pi,\chi \geqslant 0$ but ϕ may be negative).

The dressing involves $0\hbar\omega$ jumps (i.e. $\chi = \pi = 0$) to which only orbits in the fp shell contribute, as well as more complicated mixing, (at least $2\hbar\omega$ for parity reasons) responsible in particular for the spreading of states in ^{41}Ca. The traditional approach would consist in diagonalizing exactly in the $0\hbar\omega$ space and estimate in perturbation theory the rest of the wavefunction. What makes the f^n model so interesting is that nature suggests that perturbation theory should apply even for $0\hbar\omega$. As exact calculations in the full $(fr)^n$ space are now possible we could test the use of perturbation theory. At the same time we also expect the full diagonalizations to be accurate spectroscopic tools. Therefore the $(fr)^n$ space becomes both a soluble and a realistic model which is uncommon.

Now we come to the ideal calculation. Whatever formal framework we may want to invoke (coupled cluster, correlated basis etc.), empirical evidence should be a good guide in separating the calculations in 3 steps.

Ia) First we need a basis : in equation (1) the orbits are taken to span the full Hilbert space. They may be taken in a first approximation to be oscillator orbits but it is crucial that the correct $\hbar\omega$ be found *for each* n through a variation. In an extended system this amounts to choosing the density.

At this point we must assume that our formal framework has provided us with some *finite* forces, i.e. the strong repulsion at the origin should be screened. They may be complicated (rank higher than 2, non local, non Hermitian, and what not) but they must be density (or $\hbar\omega$!) dependent so that they saturate in agreement with the first step. (More on this at the end of section II).

Now we do something crucial : separate the resulting H in two parts

$$H = H_m + H_M \tag{2}$$

where we assume the *monopole* part H_m is the only one that participates in the variation and will be identified with the unperturbed H_o while H_M the *multipole* contribution will be called residual. In section II we show how to do this cleanly.

Ib) The next step consists in defining the model space. It is the task of H to decide what this space is, not ours. In the case H_M is weak, we can decouple perturbatively the configurations we are interested in (PZ1) and then

Ic) diagonalize the resulting effective Hamiltonian in the space that is degenerate under H_m.

This ideal situation does not occur in reality, H_M is never weak enough to define a model space such that all its eigenstates are physically identifiable. However, it does happen that the lowest levels in a spectrum are truly dominated by some configuration(s) rather than others, as the f^n model suggests. In this talk we shall assume steps Ib and Ic are carried in that order and check a posteriori in section III that it makes sense. The correct procedure in general is to manoeuver so as to recover a nondegenerate description as outlined in (ZU83).

Ib') Model spaces. Why is it that in the fp shell the $f^n_{7/2}$ space makes sense while its counterpart in the sd shell, $d^n_{5/2}$, does not ? And then, what does this mean in terms of physical behaviour ? The answer to both questions comes in what we take to be the motto of nuclear structure :

" *Multipole proposes, monopole disposes* "

In H_M we have essentially pairing, quadrupole and octupole forces that will drive the system to deform and/or superconduct. The first two may act within a single configuration (i.e. states degenerate under H_m) but in general become more efficient in producing coherence when the available space is larger. In the sd shell the $d_{5/2}$ and $s_{1/2}$ orbits are close (less than 1 MeV) and because of size effects, the two body matrix elements are larger than in the fp shell where the gap between $f_{7/2}$ and $p_{3/2}$ is 2 MeV. Empirically, rotational features are more prominent in the sd than in the f shell. It should be noticed that H_m, when restricted to a model space will always be diagonal but will not be restricted to a *single* particle field. As we shall see in next section, two body terms will modulate linearly the single particle energies as a function of n and add as well irreducibly quadratic operators. It means that as n varies, different configurations may move closer or separate so that the multipole " proposals " may be differently accepted. To fix ideas, in the fp space, $p_{3/2}$ may be high enough for low n so as to permit a perturbative treatment of f^n but then move down sufficiently to generate a well deformed spectrum. It is not the case in nature but it may well be for other sets of orbits in other regions.

In general we may expect three types of situation to obtain :

Ibi) Extreme j-j coupling as in the f^n model. Here H_m separates well all orbits.

Ibii) Configuration mixing for few orbits in the vicinity of the Fermi surface defined by H_m. It is the case in the sd shell where orbits at $0\hbar\omega$ are not well separated but the others are.

Ibiii) Breakdown of H_m everywhere.

Empirically, the most common case is Ibii) and it is associated with fairly well developped collectivity. Some experts may disagree with this point of view and claim that once deformation effects become important a spherical H_m breaks down completly. We prefer to reserve this spectacular event for superdeformation or shape isomers and keep open the possibility of strong coherence generated within finite spaces spanned by few spherical orbits.

Be as it may, one very important question is open : what do we do when we have no zeroth order Hamiltonian ? This happens even in Ibi) since we have to diagonalize in the model space. Could we replace the diagonalization by a new H_o and then do pertrubation theory ? We can certainly follow usual practice and let the central field deform but then the lowest determinantal (or BCS) solutions violate some conservation law(s) (angular momentum (J^2), number of particles (n), isospin (T^2)), and in general, perturbation theory will be of no use. The reason is that exact solutions must have good symmetries which we cannot restore perturbatively unless the starting determinant conserves them to good approximation. This is the case only for weak deformation, but then we could have perturbed by staying in the spherical description (^{48}Ti in section III provides an example). There is one exception though, of enormous interest : when the exact solutions break the symmetries. The necessary condition is that they be degenerate. Although we do not expect the ground state of any fermion system to be degenerate under H, it may well happen to be degenerate under $H-\lambda J^2$ (say), with λ suitable chosen (see ZU88 for the workings of the idea for the pairing Hamiltonian). In this case the determinant (or more generally coherent state) can be interpreted as an intrinsic state and λ as a moment of inertia.

Although we may expect an exact rotational spectrum (i.e. in J(J+1)) to appear only when some dynamical symmetry (such as Elliot's SU3) is respected by the Hamiltonian, there are several examples of very good J(J+1) behaviour when no exact symmetry exists (single large shell with quadrupole forces, $(d_{5/2}s_{1/2})^n$ or $(f_{7/2}p_{3/2})^n$ spaces with realistic forces etc.). But then there must be an approximate one which would truly deserve to be called dynamical.

The problem of determining what combination of spherical orbits can lead to well deformed spectra is a fascinating one. To be of direct empirical interest it has to obey two constraints: the forces may be schematic but at least suggested by the realistic ones and the orbits must be reasonably close to the spherical Fermi surfaces.

II. The monopole field

IIa. Definition of the monopole field

Let us assume we have a Hamiltonian H with which we can do Hartree-Fock (HF) calculations and that we have obtained a minimum of energy E_o at some spherical shell, i.e. a configuration $[N_1N_2...]$. We can always write $H = H_0+H_{res}$, so that H_{res} does not contribute to E_o while $H_0 = \sum t_i n_i + \sum V_{ij} n_i . n_j$ $(i \leqslant j)$, is the trace-equivalent operator (FR67) that reproduces the traces of configurations $[n_1...n_i...]$ at fixed n_i. Since there is only one state in the closed shell the trace coincides with E_o.

To arrive at the HF minimum we must have started from some original representation of H which we assume to be an oscillator basis. What is the most general form of H that can produce a (spherical) HF minimum ? Obviously it is sufficient to antitransform the final operators a_i^+ into the original (oscillator) ones, $a_i^+ = \sum C_{ir} b_r^+$, and it is evident that number operators n_i can only come from combinations of operators $(b_r^+ b_{r'})^0 = S_{rr'}$, coupled to zero angular momentum. Therefore, H_m in eq. (2) must have the form (ZU75)

$$H_m = \sum (\eta_{rr'} S_{rr'} + \omega_{rr'ss'} S_{rr'} S_{ss'}) \tag{3}$$

The technical problem is to separate H_m from H, originally written in normal order

$$H = \sum_{r,s} t_{rs} (b_r^+ b_s)^0 - \sum_{r\leqslant s \; t\leqslant u} (1+\delta_{rs})^{-1} (1+\delta_{tu})^{-1} (b_r^+ b_s^+)^J \cdot (b_t b_u)^J V_{rstu}^J \tag{4}$$

In principle we could recouple the operators in (4), i.e. write H in terms of $(b_r^+ b_{r'})^\lambda$ and be tempted to separate $\lambda = 0$ to construct H_m. Unfortunately the recoupling is ambiguous in some cases. We sketch the correct procedure.

For diagonal terms involving only number operators we use the trace equivalent defined above. There is no problem to extend the definition of H_m to H_{mT} so as to include isovector terms and account for traces of configurations $[n_1T_1, \ldots n_rT_r.]$ at fixed n_r, T_r.(see Appendix)

$$H_{mT} = \sum_{r\leqslant s} \left[\varepsilon_r n_r + V_{rs} n_r \cdot n_s + b_{rs} (1+\delta_{rs})^{-1} (T_r \cdot T_s - \frac{3}{4} \frac{n_r \bar{n}_r}{N_r - 1} \delta_{rs}) \right] \tag{5}$$

For non diagonal terms we assume that k orbits of fixed j may participate in a given operator in (4). The corresponding $S_{rr'}$ are then the generators of a U_k algebra ($k \leqslant 4$). Note that U_k is a subalgebra of U_K, where K is the total number of orbits allowed in the calculation. Isovector effects can be included through operators $T_{rr'} = (b_r^+ b_{r'})^{01}$ (coupled to $J = 0$, $T = 1$) that span a U_{2k} algebra.

It is now a simple matter of technique to separate from H all that is diagonal in U_k (and belongs in H_m) from what mixes representations. Details are left as an exercise.

The underlying symmetry in H_m is $U_{K_1} \times U_{K_2} \times \ldots \times U_{K\nu}$, with ν the number of different j^π values we need and K_i the arbitrary number of orbits included for each of them. For each U_K we have $U_K = U_1 \times SU_K$. We can solve H_m for an arbitrary Konfiguration $[n_1 \ldots n_i \ldots]$ since n_i is a good (U_1) quantum number.

Therefore, what H_m does for closed shells, it does with the same accuracy for all other states and it is elementary to extend HF variation to them. Of course we may solve H_m exactly but it should be sufficient to define the basis variationally and then treat together correlations of H_m and H_M. Even then the notion of central field may or may not emerge as a useful idea when we examine the results for different configurations entering a problem : the single particle state created by a_i^+ may be different for each of them. We prefer to exclude explicitly this possibility by restricting the variations. Otherwise the operators in H_M would become configuration dependent. Whatever the strategy we adopt to meet these potential difficulties, we can see the conceptual advantages of our decomposition by taking the simplest view.

Assume we want to do calculations in the (sd) shell and that we restrict our ambitious to an optimum $\hbar\omega$ for each total number of particles ($16 \leqslant A \leqslant 40$). One thing is certain: for each A we must do the full variation anew. The particles do not move in the field of the core (^{16}O) but in the field of the nucleus (^{A}X). The very notion of core is *technical* not physical. See below. For the ends of the space (^{16}O and ^{40}Ca) we have well defined closed shells, while in the middle we have to go through configuration mixing (step Ib in section I). The problem raised in the previous paragraph is equivalent to asking in ^{28}Si, say, whether we should use $d_{5/2}^{12}h^{16}$ or $d_{5/2}^{n_1}s_{1/2}^{n_2}d_{3/2}^{n_3}h^{16}$. The "core" h^{16} (see eq. (1)) will change with n but it can be dropped from the *notation*. Here two things may happen. If the choice makes no difference and leads to the same $\hbar\omega$, all to the better. If it makes a difference we should calculate carefully the state we are interested in. Then determine its average occupancy (i.e. $d_{5/2}^{n_1}s_{1/2}^{n_2}d_{3/2}^{n_3}$), do the variation for these numbers and iterate. In this case we shall predict interesting changes of radii between different states.

IIb. Trouble and cures for H_m. Les Trois Grâces

Now we turn to actual calculations in a model space and we set* (see Appendix. For λ_T refer to G2 in III)

$$H_{mT} = C(n) + \sum_{r\neq 1} \eta_r(n)\hat{\Omega}_r + \sum_{r<s} W_{rs}(n)\hat{\Omega}_{rs} + \sum_{r\leqslant s} b_{rs}(n)\hat{\tau}_{rs} \tag{6}$$

$$H_M = -\sum_{\substack{r\leqslant s\\ t\leqslant u}} W_{rstu}^{JT}(n)\lambda_T(1+\delta_{rs})^{-1}(1+\delta_{tu})^{-1}\left(b_r^+b_s^+\right)^{JT}\cdot\left(b_tb_u\right)^{JT} \tag{7}$$

We have written H_{mT} directly in terms of the invariant decomposition (FK83, ZU83) that makes it possible to split H_m in contributions that

*There is a slight abuse of language in using H_{mT} and H_M for both the full Hamiltonian and its restriction to the model space.

either depend only on n (C , the total centroid), or go as a central field (η_r the single particle energies) or are strictly of rank 2 (W_{rs}). It should be noticed that for a two body force $C = E_o + n\varepsilon + n.nW$, where E_o is the energy of the core, ε the single particle centroid and W the two body centroid. Of course C is a function of n but different from C(n) in eq. (6), in which all parameters are affected by a supplementary n dependence (i.e. $E_o \to E_o(n)$, $\varepsilon \to \varepsilon(n)$, $W \to W(n)$). Same remarks for $\eta_r = \eta_r^0 + (n-1) W_r$ from the two body force and $\eta_r(n)$, as well as all other n dependent quantities in eq. (6).

As written in (6) and (7) H is ready for calculations i.e. it assumes that steps Ia and Ib have been executed. After Ia we would have obtained exactly the same form, therefore in selecting values for the parameters we have to say what sort of renormalization we do in Ib. This is a very hard task if taken seriously (SH83, OS89, BA75) and we are going to adopt a much simpler view. The following remarks will be of use

1) We are certain that H_m, if calculated with state of the art precision will fail to saturate properly. Our prejudice is that H_M is basically correct and we want to show it by repairing phenomenologically H_m and leaving H_M untouched. Therefore we are prepared to tamper as much as necessary with the parameters in eq. (6) and as little as possible with those in eq. (7). As eq. (6) contains much that is directly observable : energies of closed shell and single particle and single hole nuclei ; we shall take directly those values from experiment. The invariant decomposition was designed precisely to make these quantities independent parameters.

2) As choice for H we have only Brueckner G-matrices calculated by various authors (KU66, KLS69, MAN88) with various potentials consistent with nucleon-nucleon phase shifts. It is for the sd shell that most calculations are in print : the agreement between different sets of matrix elements is SIMPLY AMAZING. The most recent calculations with the Paris and Bonn potentials indicate that the agreement persists (JI89).

We think these remarks provide ample justification for our choice of KLS, mainly through the availability of a 20 year old code (HCL).

3) Step Ib has to be taken seriously as we said, but we have reservations about the accuracy of calculations that ignore trouble with H_m. We have very forcibly argued that the use of proper energy denominators is crucial (PZI) in a perturbative calculation. Since we know H_m is wrong, we do not see the point in being too rigorous at this stage, although it will be of great use to have the proper techniques ready by the time we have better control of denominators. So we do something else (see 4 below)

The success of a perturbative approach hinges on a good starting point. We have often noticed in numerical work that if a basis is truncated, even severely, the overall low lying spectrum gets compressed. Of course this assumes we keep in the basis some "essential ingredients" of trouth about the wavefunctions. An example will be given in fig. 3 and table 1 (section IIIb,c). Equating the notions of "good starting point" and " essential ingredients " we conclude that step Ib must, in first approximation, provide an *overall dilation*. (This is the origin of the λ_T factors in eq. (7). See next section).

IIc. Formal comments

The mechanism that ensures saturation in selfbound systems is a strong repulsion at short distances. Here we hint in the simplest possible way how it leads to a density dependent potential in which the core effects are screened. The idea (ZU83) is to notice that Brueckner theory amounts to calculate energies in second order perturbation theory, provided the particle intermediate states are taken to be diagonal in the interaction. Then it is a matter of algebra to show that

$$G = H_R\left(1 - f(\rho)\right) \tag{11}$$

provides an effective interaction in which H_R (R = regularized) is the original potential in which the core has been set to zero, while $f(\rho)$ is a function of the density that may be taken to be linear in first approximation. For a finite system we can achieve maximum simplification by replacing ρ by ω, i.e. the oscillator frequency. This amounts to choosing "oscillator boundary conditions" and the problems raised at the end of IIa may have to be faced. It should be noted that G in eq. (11) is translationally invariant, a property we would loose by using a local density approximation. These arguments are very crude and say little about how to do things well except, perhaps, in the case of very dilute systems for which low order Brueckner theory is a good approximation. In this respect it should be said that the (MAN88) matrix elements mentioned in 2) are calculated in low order Jastrow theory. The fact they are so close to the KB and KLS values corroborates the idea they are all of the form H_R in eq. (11), although the regularization may be a bit more sophisticaced.

In next section we show that when trouble in H_m is corrected, these forces do a very decent job. As H_R is very much the same for all of them they only differ in $f(\rho)$ (the Coester line phenomenon, see figs. 5, 6 in KLZ78 for an illustration). It should be noted that $f(\rho)$ is not calculated in KB66, KLS69 or MAN88 but we are on safe grounds in asusming it would different in the three cases since they have different short range repulsion. Furthermore, KLS69 has no OPEP while the other two have it. From this we conclude that what we can test directly in spectroscopic calculations is mostly the intermediate range of the forces and that it is quite satisfactory accounted by any realistic force.

III. Results and some of their consequences

IIIa. Minimal corrective action

Now we assume we have calculated a set of G-matrix elements and implement the remarks in IIb above in 3 steps.

G1. The invariance properties of the operators Ω_r, Ω_{rs}, and τ_{rs} (see Appendix) are useful to ensure a smooth evolution of the parameters in eq. (6) by assuming we can determine them at $n = 0,1,2$ and $\bar{n} = N - n = 0,1,2$, i.e. at both ends of the shell and then " propagate " (FR67). E.g. we fix $C(n)$ at these six values and construct the minimal (5th order) polynomial in n that respects these " boundary conditions ". The procedure is the same for $\eta_r(n)$, a cubic, which we fix at $n = 1,2$; $\bar{n} = 1,2$. (Ω_r vanishes at $n = \bar{n} = 0$) and the linear $W_{rs}(n)$ and $b_{rs}(n)$ which are fixed at $n = \bar{n} = 2$ (Ω_{rs} and τ_{rs} vanish at n, $\bar{n} = 0,1$). The calculations will let us know whether further flexibility is welcome or necessary. This leaves us with many parameters : $\varepsilon, W, \eta_r^0, W_r, W_{rs}, b_{rs}$ at one end of the shell and their counterparts $\bar{\varepsilon}, \bar{W}$, etc. at the other end. What we call specifically G1 is the following remarkable result of the search for optimal parameters:

The operators $\delta W_{\pm}\left(\sum_{r<s} n_r n_s \pm \sum n_r \cdot n_r\right)$ *and* $\delta b_{\pm}\left(\sum_{r<s} \tau_{rs} \pm \sum \tau_{rr}\right)$ *are sufficient to correct monopole trouble in the sd and fp shells.*

G2. Following the remarks in 4) in section IIb we introduce two overall multiplicative parameters λ_T $(T = 0,1)$ that affect $W^{JT}_{rstu}(n)$. Work in the $N = 50$ and 82 isotones and $Z = 50$ (Sn) isotopes produce a remarkably constant $\lambda_1 = 1.5 \pm .1$. In the sd shell the same value is obtained to within uncertainties explained in G3., while $\lambda_o = 1 \pm .1$.

G3. The n dependence of W^{JT}_{rstu} must be an overall size effect. Since the matrix elements are calculated for a certain $\hbar\omega$, it is easy to relate them to the size of the nucleus if we know $W^{JT}_{rstu}(\omega)$ and $\omega(\langle r^2\rangle)$ where $\langle r^2\rangle$ is the mean square radius. To a fairly good approximation we have for the KLS matrix elements

$$W^{JT}_{rstu}(\omega) = W^{JT}_{rstu}(\omega_o)\,\frac{\omega}{\omega_o}\ , \qquad (8)$$

while ($A = A_c + n$ is the total number of particles)

$$\hbar\omega = 34.60 \frac{A^{1/3}}{\langle r^2 \rangle} \tag{9}$$

which is the classical estimate (BM64) that reduces to $\hbar\omega = 40\ A^{-1/3}$ for $\langle r^2\rangle = 0.92\ A^{2/3}$. The exact relation for the sd shell would be $40(36 + 3.5n)/(16+n)\langle r^2\rangle$, close to (9).

Putting together eqs. (8) and (9) we write

$$W^{JT}_{rstu}(n) = W^{JT}_{rstu}(\omega_0)\mu(n) \quad , \qquad \mu(n) = 34.6 \frac{A^{1/3}}{\langle r^2\rangle\omega_0} = \frac{\omega}{\omega_0} \tag{10}$$

It should be noticed that the form of eq. (7) permits only to determine $\lambda_0\mu$ (n) and λ_1/λ_0. In principle we could take $\mu(n)$ from eq. (10) by using the experimental radii when known, but we have prefered to let $\mu(n)$ vary freely assuming $\lambda_1/\lambda_0 = 1.5$ in the sd shell. Then we have played with the uncertainties and set $\lambda_0 = 1.1 - .12(n-2)/(N-2)$. By varying simultaneously the four parameters in G1. and $\mu(n)$ we obtain fig. (1) (triangles are calculated values, dots experimental, circles $(A/18)^{.3}$ (BW88)).

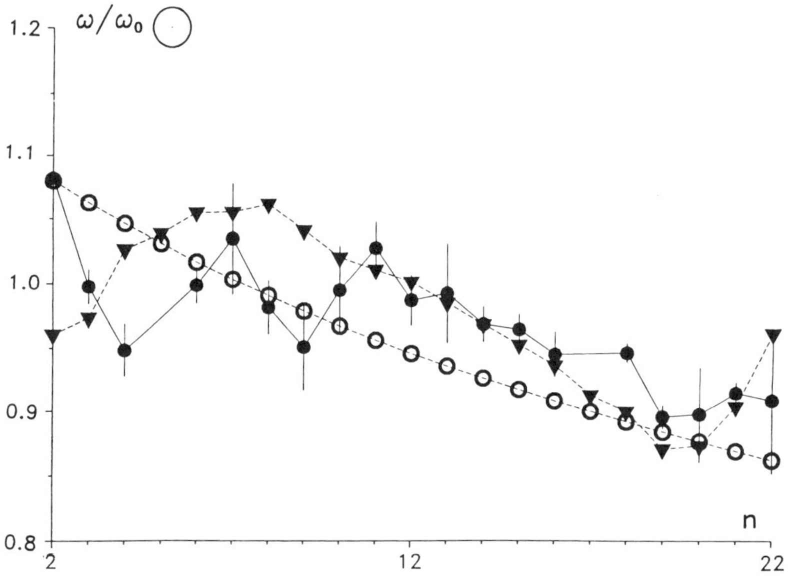

Fig. 1

The fit was made to some 180 absolute energies in the sd shell leading to root mean square deviation of 300 keV. We would like to insist on the fact that the fitted $\mu(n)$ and the experimental ω/ω_0 agree as well as could be expected. At the beginning of the shell there is such staggering (the large circle represents an informed guess of the experimental value for ^{20}O) that our parametrization can provide at most an average. At the end the agreement extends to details. In other words: we are not really " fitting ", we are checking that when left to themselves the

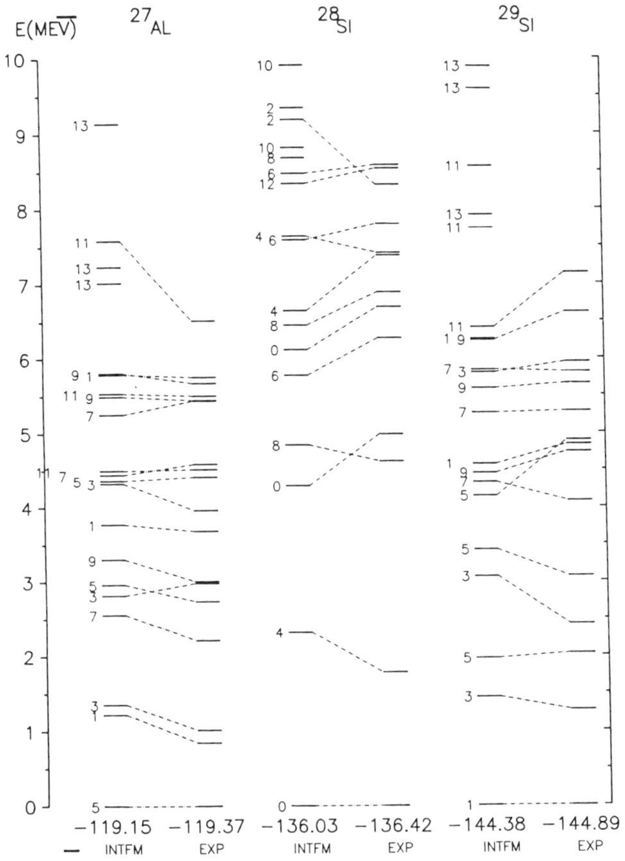

Fig. 2

spectra tell the nucleus to sit at the right oscillator well.

In fig. 2 we show the spectra obtained in the 3 nuclei at the center of the shell (none of them included in the fit) : it is definitely decent spectroscopy. Not yet of the quality of state of the art fits (BW88) in which all matrix elements are varied (63 in this case, to bring down the deviation to 180 keV for the W interaction of Wildenthal) but we vary only 4. (Numbers to the left of levels are 2J).

The sd shell has been chosen as an example because it provides the most stringent tests. All indications we have collected so far indicate that the minimal corrective program outlined above, works. From which we conclude:

"There are strong reasons to believe that trouble with the realistic interactions is confined to their monopole behaviour (H_m). Furthermore the trouble can be corrected by modifications in 4 parameters (Gl.)."

IIIb. Energetics of ^{48}Ti

Fig. 3 contains a sequence of calculations done with KB3, the Kuo-Brown interaction (KB68) minimally renormalized according to the Gl. prescription (PZ2):

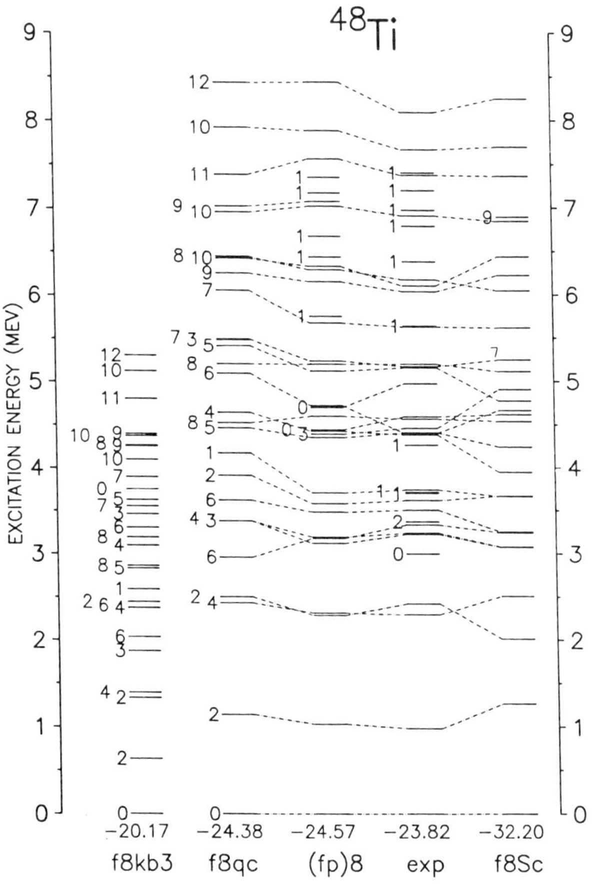

Fig. 3

f8kB3 is f^8. numbers to the left are J.

f8qc is the perturbative quasiconfiguration calculation (PZ2)

$(fp)^8$ is the exact diagonalization. Dimensionality for M = 0, $T_Z = 1$ matrix is 634744. The J labels only shown for states with no f8qc counterpart (i.e. not dominated by f^n).

exp is the experimental spectrum. J labels shown if no counterpart or dubious one (highest J = 1 peaks) with $(fp)^8$

f8Sc is the classical calculation with ^{42}Sc matrix elements.

We only note that f8qc compares well with $(fp)^8$ although there is room for refinements, $(fp)^8$ compares very well with exp, while f8Sc does not do as well but is quite good (^{48}Ti is one of the showpieces of the f^n model).

IIIc. Quadrupole behaviour and the validity of PT

Although the f8kb3 column in fig. 3 is badly compressed, there is a one to one correspondence with the exact spectrum, in line with Landau theory and with remarks in 4) section IIIb. If we examine the columns corresponding to f8Sc and f8kb3 in table 1 we see that in spite

Table 1

Quadrupole moments

J^π_n	$(fp)^8$	$f^8_{7/2}(^{42}Sc)$	$f^8_{7/2}(KB3)$	$Q_{int}(fp)^8$
2^+_1	-14.4	+3.5	+3.4	50.4
4^+_1	-11.7	+4.3	+4.4	40.9
6^+_1	+17.7	-0.5	+1.2	-61.9
6^+_2	-16.8	-1.6	-0.5	58.8
8^+_1	-13.7	-0.7	-1.1	47.9
10^+_1	-20.3	-5.8	-5.8	71.1
12^+_1	-15.9	-12.7	-12.8	55.7

$Q(2^+_1)\exp^{13} = -17.7(8)$ fm^2

of the spectral differences the wavefunctions are strikingly similar. More surprising still, the huge differences of both with the exact results. A patternless mixture of signs and sizes transforms into a well defined prolate structure as witnessed by the quasi constancy of Q_{int}, the intrinsic quadrupole moment (strictly constant for a collective rotor). Let us examine how PT can generate coherence. There are two kinds of matrix elements they play a role: W_{ffrr} renormalizes the two body force only and does not change transition operators (indeed: this is what table 1 is telling us), while W_{fffr} leads to a 3 body force and 2 body effective operators. When acting on a closed shell f^N, the former are associated with the RPA and the latter produce a *deformed* field since $j_f \neq J_r$ (refer to discussion of model spaces in Ic.). Therefore PT should inform us of precursor effects of deformation. In what follows we use α,β (i,k) for neutron (proton) states and consider the effect of a quadrupole force $-q_n.q_p$ in a large shell j. The expectation value for determinantal states $|D\rangle$ of good M projection is

$$-\langle D|q_n^{20}\cdot q_p^{20}|D\rangle = -\sum_{\alpha,i} q_i^{20}\cdot q_\alpha^{20} \cong -\sum_{\alpha,i}(3m_i^2-j^2)(3m_\alpha^2-j^2) \qquad i\leqslant i_f, \quad \alpha\leqslant\alpha_f$$

It is clear that we gain energy by choosing m_i and m_α as large as possible ($m \approx j$, oblate shape).

Now we consider the effect of a two body force $q_n.q_p' + q_p.q_n'$, where q' jumps out of the j space (i.e. $q' = (a_j^+ a_{j'})^{2\mu}$).

The resulting effective 3 body force can be calculated along the lines explained in PZ1 or PZ2 and takes the form $-q_n.Q_{np} - q_p.Q_{pn}$, where now Q_{np} and Q_{pn} are *two* body quadrupole operators. The interesting case comes from $j - j'=2$ (i.e. $f_{7/2}, p_{3/2}$ in the fp space) and leads to expectation values

$$-\langle D|q_n.Q_{np}+q_p.Q_{pn}|D\rangle \cong -\sum\left[(3m_i^2-j^2)(3m_k^2-j^2)(m_\alpha^2-j^2)^2 + p\leftrightarrow n \text{ term}\right]$$

Now it is seen that the prolate shapes ($m \approx 0$) are strongly favoured. In actual calculations the wavefunction in the model space changes only slightly from the unperturbed one (~10 %) but sufficiently to produce a smallish negative (prolate) quadrupole moment strongly boosted by the effective $Q_{np} + Q_{pn}$ operator.

IIId. Orbital enhancement and spin quenching of magnetic dipole strength. Lanczos strength functions

In general we are not interested in detailed knowledge of the spectrum beyond the lowest levels. Assume we have some operators Ω that leads from ground ($|0\rangle$) to excited states ($|E\rangle$). Let us define a sum rule state $|\Sigma\rangle$

$$|\Sigma\rangle = \Omega|0\rangle N^{-1} = N^{-1}\sum_E S(E)|E\rangle\ , \quad N^2 = \langle 0|\Omega^2|0\rangle = \sum_E S^2(E)$$

If we assume the strength function $S^2(E)$ to be sufficiently smooth, knowledge of its lowest moments should be sufficient for a valid characterization. Since $\langle\Sigma|H^n|\Sigma\rangle = N^{-2}\sum S^2(E)E^n$, we realize that each Lanczos iteration on pivot $|\Sigma\rangle$ produces two new moments of the distribution (Lanczos consists in diagonalizing in the basis $H^n|\Sigma\rangle$ after orthogonalization at each step).

In table 2 we show the values of N^2, $E_c = \langle\Sigma|H-E_o|\Sigma\rangle$ and $\omega = \left[\langle\Sigma|H^2|\Sigma\rangle - \langle\Sigma|H|\Sigma\rangle^2\right]^{1/2}$ (i.e. the sum rule, the centroid refered to the ground state energy and the width) of the strength functions for Ω, $\Omega_\ell, \Omega_\sigma$ (t, ℓ, σ in the table), where $\Omega = M1 = \Omega_\ell + \Omega_\sigma$). The exact results are compared with those of a " Tamm-Dancoff " (TD) calculation in which the ground state is f^8 and the excited states $f^8 + f_r^7$. The quenching factors $q = N^2_{exact}/N^2_{TD}$ reflect the strong erosion/(enhancement) of M1 and $\Omega_\sigma/(\Omega_\ell)$ sum rules. It is seen that quenching/(enhancement) comes with

Table 2

	t		σ		ℓ	
	TD	EXACT	TD	EXACT	TD	EXACT
N^2	15.21	10.61	16.06	10.12	.50	.98
E_c	9.14	9.43	9.72	10.09	5.14	11.72
ω	3.09	2.92	2.53	2.63	3.38	6.00

$q_t = 0.70$ $\quad q_\sigma = 0.63$ $\quad q_\ell = 1.96$ $\quad \frac{q_\ell}{q_\sigma} = 3.11$

weak/(strong) distorsion of the centroids and widths of the corresponding distributions.

We have no simple PT explanation at this stage but clearly the quenching can be incorporated as an overall renormalization of the strength of the one body operators while orbital coherence is more likely to demand mechanisms similar to those explained in the previous section.

Appendix : Notations

PT = pertubation theory

We use i,j... or r,s... to label shells of spin $j = j_i, j_j$ etc.

Configuration $[n_r, n_s ...]$: set of states with n_r particles in shell r, n_s in shell etc.

Konfiguration $[n_r, n_s ...]$: set of states with n_r particles in K shells with $j = j_r$, n_s particles in K shells with $j = j_s$ etc.

mT configuration $[n_r T_r, n_s T_s ...]$: configurations in which the isospin $T_r (T_s ...)$ of shell r(s...) is fixed.

$n = \sum n_r$ = total number of particles, T = total isospin, J = total angular momentum of system. N_r = total degeneracy of shell r (i.e. $(2j_r+1)$ or $2(2j_r+1)$ in isospin formalism).

$N = \sum n_r$ = total degeneracy of space, $\bar{n}_r = N_r - n_r$, $\bar{n} = N - n$

V^{JT}_{rstu} = two body G matrix element between normalized and antisymmetrized states. t_{rs} = kinetic energy matrix element.

$a^+_r (a_r)$ = creation (annihilation) operators

$(a^+_r a^+_s)^{JT}$ $\left((a_s a_t)^{JT}\right)$: creation (annihilation) operators for JT pairs

$n_r . n_s = n_r n_s$ if $r \neq s$; $n_r(n_r-1)/2$ if $r = s$. Otherwise . is scalar product.

$$V_{rs} = [\sum(2J+1)(2T+1)]^{-1} \sum_{JT} (2J+1)(2T+1) V^{JT}_{rsrs} \quad ,$$

$$V^T_{rs} = [\sum(2J+1)]^{-1} \sum_J (2J+1) V^{JT}_{rsrs} \quad ,$$

$$W = (N.N)^{-1} \sum N_r . N_s V_{rs} \quad , \quad W_r = (N-2)^{-1} [\sum(N_r - \delta_{rs}) V_{rs} - (N-1)W]$$

$$W_{rs} = V_{rs} - W_r - W_s - W, \quad W^{JT}_{rstu} = V^{JT}_{rstu} - \delta_{rt}\delta_{su}V^{T}_{rs}, \quad b_{rs} = V^{1}_{rs} - V^{0}_{rs} .$$

Also $\varepsilon = N^{-1}\sum N_r \varepsilon_r$, $\eta_r = \varepsilon_r - \varepsilon$,

$$\tau_{rs} = (1+\delta_{rs})^{-1}\left[T_r . T_s - \frac{3}{4}\frac{\bar{n}_r n_r}{N_r - 1}\delta_{rs}\right] ,$$

If we use the closed shell instead of the core as vacuum (particle hole transformation i.e. $a^+_r \to a_r$, $a_r \to a^+_r$ and $n_r \to \bar{n}_r$) we have $\Omega_r \to \bar{\Omega}_r = -\Omega_r$, $\Omega_{rs} \to \bar{\Omega}_{rs} = \Omega_{rs}$, $\tau_{rs} \to \bar{\tau}_{rs} = \tau_{rs}$. We call the decomposition (6) invariant because the operators go into themsleves.

References

BA75 Lectures Notes in Physics 40 (1975) (B. Barret ed., Springer).

BM64 A. Bohr and B. Mottelson, Nuclear Structure, Vol. 1 (Benjamin, 1964).

BW88 B.A. Brown and B.H. Wildenthal, Ann. Rev. Nuc. Part. Sc. 38 (1988) 29, and references therein.

CA89 E. Caurier, code ANTOINE, Strasbourg 1989.

FK83 J.B. French and V.K.B. Kota, Phys. Rev. Lett. 51 (1983) 2183.

FR67 J.B. French in Nuclear Structure, A. Hossain et al. eds. (North Holland 1967, Amsterdam).

HCL H.C. Lee code. Private communication 1969.

JI89 M.F. Jiang, R. Machleidt, D.B. Stout and T.T.S. Kuo, Phys. Rev. 40 (1989) 1857.

KB66 T.T.S. Kuo and G.E. Brown, Nucl. Phys. A85 (1966) 40.

KLS69 S. Kahana, H.C. Lee and K. Scott, Phys. Rev. 180 (1969) 956.

MAN88 J. Fiase, A. Hamoudi, J.M. Irvine and F. Yazici, J. Phys. G : Nucl. Phys. 14 (1988) 27.

MBZ64 J.D. McCullen, B.F. Bayman and L. Zamick, Phys. Rev. 134 (1964) 515.

NPA310 P.M. Endt and C. Van der Leun, Nucl. Phys. A310 (1978) 1.

OS89 M.Hjorth-Jensen and E. Osnes, Phys. Lett. 228 (1989) 281.

PZ1 A. Poves and A.P. Zuker, Physics Reports 71 (1981) 141

PZ2 A. Poves and A.P. Zuker, Physics Reports 70 (1981) 235

SH83 J. Shurpin, T.T.S. Kuo and D. Strottman, Nucl. Phys. A408 (1983)310.

ZU75 A.P. Zuker, Czech. Journal of Physics B25 (1975) 311.

ZU83 A.P. Zuker, Lecture Notes in Physics 209 (1984) (J.S. Dehesa et al., eds. Springer).

ZU88 A.P. Zuker in Annual Report of CRN, Strasbourg 1988.

NON-SPURIOUS HARMONIC OSCILLATOR STATES FOR MANY-BODY SYSTEMS

Akiva Novoselsky
Department of Nuclear Physics
The Weizmann Institute of Science, Rehovot 76100, Israel

Jacob Katriel
Department of Chemistry
Technion - Israel Institute of Technology, Haifa 32000, Israel

Abstract

An algorithm for the construction of non-spurious harmonic oscillator wave functions with arbitrary permutational symmetry is presented. The harmonic oscillator wave functions, expressed in Jacobi coordinates, are calculated recursively using a new type of harmonic oscillator coefficients of fractional parentage. These coefficients are the eigenvectors of the two-cycle class operator of the permutation group in the appropriate basis. The matrix elements of the class operators are evaluated by using a specific version of the harmonic oscillator brackets. A procedure is developed to transform the resultant harmonic oscillator states from Jacobi into single particle coordinates. The procedures proposed are expected to enhance the effectiveness of many-body computations involving harmonic oscillator basis sets.

I. Introduction

Harmonic oscillator wave functions have been widely used in computational molecular, atomic and nuclear physics, and recently also in non-relativistic quark calculations[1]. In all these applications the eigenvectors of a translationally invariant hamiltonian are evaluated in terms of harmonic oscillator eigenstates. The harmonic oscillator states used in these calculations should be constructed in such a way that the trivial center-of-mass (c.m.) motion is explicitly separated: Spurious states, in which the c.m. is excited, must be eliminated.

In order to construct non-spurious states for n identical isotropic three-dimensional harmonic oscillators we must use a set of coordinates where the c.m. is separated from the $n-1$ internal coordinates. Among the various sets of coordinates satisfying this requirement the normalized Jacobi coordinates

$$\vec{\rho}^{(n)} = \vec{R} = \frac{1}{\sqrt{n}}\sum_{j=1}^{n}\vec{r}_j \quad ; \quad \vec{\rho}_i = \sqrt{\frac{i-1}{i}}\left(\vec{r}_i - \frac{1}{i-1}\sum_{j=1}^{i-1}\vec{r}_j\right) \quad i = 2,3,\ldots,n \tag{1}$$

were found to be preferable because each internal coordinate $\vec{\rho}_i$ $i = 2,\ldots,n$ depends on the first i single particle coordinates only. This property enables the formulation of a recursive procedure for constructing the set of harmonic oscillator non-spurious states[2].

The harmonic oscillator wave functions expressed in Jacobi coordinates are naturally separated into an internal and a c.m. wave function. The c.m. wave function is totally symmetric with respect to permutations of the particle coordinates because it is expressed in terms of the totally symmetric c.m. coordinate. On the other hand, the internal coordinates do not have simple symmetry properties with respect to permutation of particle indices.

In calculations involving quantum states labeled by multiple angular momentum quantum numbers (i.e., L-S, L-S-T, L-S color and flavor), it is very useful[3] to factor the space of the states into products of subspaces, one for each angular momentum quantum number. Wave functions belonging to the various irreps of S_n should be constructed within each subspace and ultimately coupled to form a totally antisymmetric state.

A recursive algorithm was recently proposed for constructing internal wave functions belonging to irreps of S_n[2]. Suppose that all the $(n-1)$-particle internal states belonging to a given irrep of S_{n-1} are known. These states are coupled with harmonic oscillator states in terms of the n'th Jacobi coordinate, constructing n-particle internal states which are sorted into sets with common values of the resultant angular momentum and total energy. The two-cycle class operator of the permutation group is diagonalized within each such set of states. Although this operator is the sum of all the $(n-1)n/2$ transpositions (i, i') $i < i' = 2, \ldots, n$, it turns out that only the matrix element of the transposition $(n-1, n)$ has to be evaluated. After diagonalization of the appropriate matrices the eigenvalues uniquely identify the n-particle irrep of S_n, and the eigenvectors are the appropriate harmonic oscillator coefficients of fractional parentage (hocfps). A detailed description of this method was given in Ref. [2] and is reviewed in Section II.

The non-spurious harmonic oscillator states obtained in this procedure are expressed in terms of the Jacobi coordinates. On the other hand, the hamiltonian operators used in molecular, atomic and nuclear physics are conveniently expressed in terms of single particle coordinates. In order to calculate the matrix elements of these operators it is desirable to transform the non-spurious harmonic oscillator states from Jacobi into single particle coordinates.

A recursive method for carrying out this transformation of coordinates was recently introduced[4]. It starts from the n particle wave function which consists of a wave function in $n-1$ Jacobi coordinates and the (totally symmetric) c.m. motion. The wave function of the $(n-1)$'th Jacobi coordinate is separated by using the hocfps, and coupled to the c.m. wave function. These coupled wave functions can be transformed into harmonic oscillator wave functions in terms of the n'th single particle coordinate and the $(n-1)$-particle c.m. by using the appropriate harmonic oscillator brackets[5]-[8]. Repeating this procedure, recursively, one obtains after $n-1$ steps the expression for the given harmonic oscillator state in terms of n single particle coordinates. This transformation is derived in Section III.

II. Internal harmonic oscillator wave functions with arbitrary symmetry

Since the c.m. wave function is totally symmetric, the permutational symmetry of the total wave function is completely determined by that of the internal wave function. Our aim is to construct internal wave functions, consisting of $n-1$ harmonic oscillators, which belong to an irrep of the permutation group, S_n.

The permutational symmetry of an internal wave function of n-particles can be specified by a sequence of Young frames $\Gamma_2\Gamma_3 \ldots \Gamma_n$, where Γ_i is the i-particle Young frame. This sequence is equivalent to the Yamanouchi symbol Y_n[9]. All the Young frames in the sequence are good quantum numbers. The *resultant* internal angular momentum Λ_n and internal energy ($\hbar\omega\left(\epsilon_n + \frac{3}{2}(n-1)\right)$ where ϵ_n is an internal energy parameter) are good quantum numbers as well. However, the angular momenta and energies of less than n particles are not good quantum numbers. One can construct a complete set of states labeled by $|Y_n\Lambda_n\epsilon_n\alpha_n>$ where α_n is an additional label that takes care of the remaining degeneracies. For simplicity we denote the combination of quantum numbers $\Lambda_n\epsilon_n\alpha_n$ by Φ_n. $\eta^{(i)}$ stands for the individual harmonic oscillator radial and angular quantum numbers N_i and L_i corresponding to the i'th Jacobi coordinate.

The two particle internal wavefunction can be written as

$$|\Gamma_2\,\Lambda_2\,\epsilon_2\,;\vec{\rho}_2>=|\Gamma_2\,\Phi_2\,;\vec{\rho}_2>=\Psi_{N_2L_2}(\vec{\rho}_2)=|\eta^{(2)};\vec{\rho}_2> \tag{2}$$

where $\epsilon_2 = 2N_2 + L_2$, Λ_2 is the internal angular momentum and $L_2 = \Lambda_2$. Γ_2 is determined by L_2: $\Gamma_2 = [2]$ for even L_2 and $\Gamma_2 = [1^2]$ for odd L_2. The value of the z component of the angular momentum is suppressed.

To obtain three-particle internal wave functions labeled by the sequence of Young frames $\Gamma_2\Gamma_3$ and internal angular momentum Λ_3, we first form angular momentum coupled direct product states of the form

$$|\Gamma_2\Lambda_2\epsilon_2N_3L_3\Lambda_3;\vec{\rho}_2\vec{\rho}_3>=|\Gamma_2\Phi_2\eta^{(3)}\Lambda_3;\vec{\rho}_2\vec{\rho}_3> \tag{3}$$

The permutation operators transform these states within invariant subspaces, each of which is characterized by a given internal energy $\epsilon_3 = \epsilon_2 + 2N_3 + L_3$ and internal angular momentum Λ_3. The linear combinations which belong to well defined irreps of the permutation group are written in the form

$$|\Gamma_2\Gamma_3\Phi_3;\vec{\rho}_2\vec{\rho}_3> = \sum_{\substack{\Phi_2\eta^{(3)} \\ (\epsilon_3 = \epsilon_2 + 2N_3 + L_3)}} \left[\Gamma_2\Phi_2\eta^{(3)}\Lambda_3|\}\Gamma_2\Gamma_3\Phi_3\right] |\Gamma_2\Phi_2\eta^{(3)}\Lambda_3;\vec{\rho}_2\vec{\rho}_3> \quad (4)$$

Let us assume that the $(n-1)$-particle wave functions, symmetry adapted to S_{n-1}, have already been constructed. The general expression for the n-particle internal wave function, symmetry adapted to S_n, can than be written in the form

$$|Y_n\Phi_n;\vec{\rho}_2\vec{\rho}_3\ldots\vec{\rho}_n> = \sum_{\substack{\Phi_{n-1}\eta^{(n)} \\ (\epsilon_n = \epsilon_{n-1} + 2N_n + L_n)}} \left[Y_{n-1}\Phi_{n-1}\eta^{(n)}\Lambda_n|\}Y_n\Phi_n\right] |Y_{n-1}\Phi_{n-1}\eta^{(n)}\Lambda_n;\vec{\rho}_2\vec{\rho}_3\ldots\vec{\rho}_n> \quad (5)$$

where $Y_n = Y_{n-1}\Gamma_n$.

The hocfps thus defined satisfy orthogonality and completeness relations[2] similar to those satisfied by the single shell cfps for arbitrary permutational symmetry, defined in ref. [10]. On the other hand, here, the n'th particle state $|\eta^{(n)};\vec{\rho}_n> = |N_nL_n;\vec{\rho}_n>$ is not unique. We have to sum over all the different single particle states consistent with the angular momentum coupling $\vec{\Lambda}_{n-1} + \vec{L}_n = \vec{\Lambda}_n$ and the energy relation $\epsilon_n = \epsilon_{n-1} + 2N_n + L_n$, since the elements of S_n couple all those states. This is the price paid for using the Jacobi rather than single particle coordinates, the advantage of which is the explicit elimination of the c.m.

As mentioned in the Introduction, the hocfps are evaluated recursively. Starting from the set of $(n-1)$-particle internal wave functions with a given permutational symmetry, the n-particle internal states are constructed by taking angular momentum coupled direct products of the n'th harmonic oscillator wave function with these $(n-1)$-particle states. Each direct product state has well defined internal energy ϵ_n and angular momentum Λ_n. However, they are not in general symmetry adapted with respect to the permutation group.

In order to obtain n-particle states which belong to well defined irreps of the permutation group, the two-cycle class operator (the sum of all the different transpositions)

$$\mathbf{C}^2[S_n] = \sum_{i<i'}^{n}(i,i') \quad (6)$$

has to be diagonalized within sets of states having common ϵ_n, Λ_n and Y_{n-1}. The eigenvalues of this operator are[11]

$$C^2[\Gamma_n] = \frac{1}{2}\sum_i \lambda_i(\lambda_i - 2i + 1) \quad (7)$$

where the sum extends over all the rows of the Young frame Γ_n and λ_i is the number of boxes in the i'th row. This expression *uniquely* determines the various irreps of S_n obtained from a given irrep of S_{n-1} by adding one box[10]. Therefore, by diagonalization of the two-cycle class operator we can identify the various irreps from the eigenvalues; the eigenvectors are the desired hocfps.

The recurrence relation

$$\mathbf{C}^2[S_n] = \mathbf{C}^2[S_{n-1}] + \sum_{i=1}^{n-1}(i,n) \quad (8)$$

simplifies the calculation of the matrix elements of the two-cycle class operator. The first term on the right hand side is diagonal in Γ_{n-1} (the irrep of S_{n-1} that is common to all the basis states) and its value is $C^2[\Gamma_{n-1}]$ (Eq. 7)

$$< Y_{n-1}\Phi'_{n-1}\eta'^{(n)}\Lambda_n\epsilon_n|\mathbf{C}^2[S_{n-1}]|Y_{n-1}\Phi_{n-1}\eta^{(n)}\Lambda_n\epsilon_n> = C^2[\Gamma_{n-1}]\,\delta_{(\Phi'_{n-1}\eta'^{(n)}),(\Phi_{n-1}\eta^{(n)})} \quad (9)$$

where $\epsilon_n = \epsilon_{n-1} + 2N_n + L_n = \epsilon'_{n-1} + 2N'_n + L'_n$. The matrix element in Eq. (9) does not depend on the z component, M_n, of the total angular momentum Λ_n, in which it is diagonal. This allows M_n to be suppressed.

The matrix element of the second term on the right hand side of Eq. (8), between states differing by the $(n-1)$-particle Young frames, is given by

$$\frac{1}{n-1}\sum_{i=1}^{n-1} < \Gamma_{n-1}|(i,n)|\Gamma'_{n-1} >=$$
$$\delta_{\Gamma_{n-1},\Gamma'_{n-1}} \frac{1}{n_{\Gamma_{n-1}}} \sum_{Y_{n-2}} < Y_{n-2}\Gamma_{n-1}|(n-1,n)|Y_{n-2}\Gamma_{n-1} > \qquad (10)$$

where use has been made of a result derived in ref. [10]. The symbol Y_{n-2} stands for all the $(n-2)$-particle Yamanouchi symbols consistent with the Young frame Γ_{n-1}, and $n_{\Gamma_{n-1}}$ is the number of such Y_{n-2} Yamamouchi symbols, which is the degeneracy of the irrep Γ_{n-1}.

From Eq. (10) it turns out that only the matrix element of the transposition $(n-1,n)$ has to be evaluated instead of all the transpositions (i,n) $i = 1,\ldots,n-1$. Since the overall procedure involves the sequence of transpositions $(1,2),(2,3),\ldots,(n-1,n)$,it is useful to point out that these operators form a set of generators for S_n.

The $(n-1,n)$ matrix element can be evaluated by expanding the $(n-1)$-particle states in terms of the hocfps calculated previously and by using the change of coupling transformation[12] in such a way that instead of coupling the $(n-1)$-particle internal wave function to the n'th particle we couple the $(n-2)$-particle wave function to the two particle state formed by coupling the $(n-1)$'th and n'th particle angular momenta

$$\begin{aligned}
&< Y_{n-1}\Phi'_{n-1}\eta'^{(n)}\Lambda_n\epsilon_n|(n-1,n)|Y_{n-1}\Phi_{n-1}\eta^{(n)}\Lambda_n\epsilon_n >= \\
&\sum_{\substack{\Phi_{n-2}\lambda \\ \eta'^{(n-1)}\eta^{(n-1)}}} (-1)^{L'_{n-1}+L'_n+L_{n-1}+L_n}(2\lambda+1)\sqrt{(2\Lambda'_{n-1}+1)(2\Lambda_{n-1}+1)} \\
&\left\{ \begin{array}{ccc} \Lambda_{n-2} & L'_{n-1} & \Lambda'_{n-1} \\ L'_n & \Lambda_n & \lambda \end{array} \right\} \left\{ \begin{array}{ccc} \Lambda_{n-2} & L_{n-1} & \Lambda_{n-1} \\ L_n & \Lambda_n & \lambda \end{array} \right\} \\
&\left[Y_{n-2}\Phi_{n-2}\eta'^{(n-1)}\Lambda'_{n-1}|\}Y_{n-1}\Phi'_{n-1}\right] \left[Y_{n-2}\Phi_{n-2}\eta^{(n-1)}\Lambda_{n-1}|\}Y_{n-1}\Phi_{n-1}\right] \\
&< \left(\eta'^{(n-1)}\eta'^{(n)}\right)\lambda|(n-1,n)|\left(\eta^{(n-1)}\eta^{(n)}\right)\lambda > \qquad (11)
\end{aligned}$$

The summation over the $(n-2)$-particle quantum numbers and the $(n-1)$'th particle quantum numbers is limited to a small number of terms which obey the various angular momentum couplings, the relation for ϵ_n mentioned after Eq. (9), and also the following relations, resulting from the fact that the transposition $(n-1,n)$ commutes with both the $(n-2)$-particle and the n-particle hamiltonian, but not with the $(n-1)$-particle hamiltonian:

$$\begin{aligned}
\epsilon_{n-1} &= \epsilon_{n-2} + 2N_{n-1} + L_{n-1} \\
\epsilon'_{n-1} &= \epsilon_{n-2} + 2N'_{n-1} + L'_{n-1} \\
2N_{n-1} + L_{n-1} + 2N_n + L_n &= 2N'_{n-1} + L'_{n-1} + 2N'_n + L'_n
\end{aligned}$$

The last condition is actually redundent since it follows from the other ones. The z components of the angular momenta were suppressed in Eq. (11) for the same reason mentioned following Eq. (9).

The states in the matrix element on the right hand side of Eq. (11) are the wave functions corresponding to the last two Jacobi coordinates, coupled to total angular momentum λ. They can be written explicitly in the form

$$\begin{aligned}
&|(N_{n-1}L_{n-1}N_nL_n)\lambda M_\lambda >= \qquad (12) \\
&\sum_{M_{n-1}M_n} (L_{n-1}M_{n-1}L_nM_n|L_{n-1}L_n\lambda M_\lambda)\Psi_{N_{n-1}L_{n-1}M_{n-1}}(\vec{\rho}_{n-1})\Psi_{N_nL_nM_n}(\vec{\rho}_n)
\end{aligned}$$

involving angular momentum Clebsch-Gordan coefficients. M_λ, M_{n-1} and M_n are the z components of the angular momenta λ, L_{n-1} and L_n, respectively.

The Jacobi coordinates $\vec{\rho}_{n-1}$ and $\vec{\rho}_n$ (defined in Eq. (1)) are neither symmetric nor antisymmetric in the coordinates $\vec{r}_{n-1}$ and $\vec{r}_n$. Therefore, the application of the transposition $(n-1,n)$ to the wave function (12), expressed in these coordinates, is very complicated. A passage to a new set of coordinates $\vec{\tilde{\rho}}_{n-1}$ and $\vec{\tilde{\rho}}_n$, which are either symmetric or antisymmetric in the coordinates $\vec{r}_{n-1}$ and $\vec{r}_n$ is required. This passage is achieved under a rotation by an angle β

$$\begin{aligned}\vec{\tilde{\rho}}_{n-1} &= \vec{\rho}_{n-1}cos\beta - \vec{\rho}_n sin\beta \\ \vec{\tilde{\rho}}_n &= \vec{\rho}_{n-1}sin\beta + \vec{\rho}_n cos\beta \end{aligned} \tag{13}$$

where $cos\beta = \sqrt{\frac{n-2}{2(n-1)}}$ and $sin\beta = \sqrt{\frac{n}{2(n-1)}}$. Since $n > 1$ the denominator is greater than zero. Using the expressions for $\vec{\rho}_{n-1}$ and $\vec{\rho}_n$ in terms of the original particle coordinates (Eq. (1)) we find that

$$\begin{aligned}\vec{\tilde{\rho}}_{n-1} &= \frac{1}{\sqrt{2}}(\vec{r}_{n-1} - \vec{r}_n) \\ \vec{\tilde{\rho}}_n &= \sqrt{\frac{n-2}{2n}}\left[(\vec{r}_{n-1} + \vec{r}_n) - \frac{2}{n-2}\sum_{i=1}^{n-2}\vec{r}_i\right]\end{aligned} \tag{14}$$

The new coordinate $\vec{\tilde{\rho}}_{n-1}$ is antisymmetric in the coordinates $\vec{r}_{n-1}$ and $\vec{r}_n$ whereas $\vec{\tilde{\rho}}_n$ is symmetric: application of the $(n-1,n)$ operator on these coordinates yields

$$\begin{aligned}(n-1,n)\vec{\tilde{\rho}}_{n-1} &= -\vec{\tilde{\rho}}_{n-1} \\ (n-1,n)\vec{\tilde{\rho}}_n &= \vec{\tilde{\rho}}_n\end{aligned} \tag{15}$$

To exploit the symmetry properties of the new coordinates we have to express the wave function (12) in terms of wave functions in the new coordinates

$$\begin{aligned}&|\,(N_{n-1}L_{n-1}N_nL_n)\,\lambda M_\lambda; \vec{\rho}_{n-1}\vec{\rho}_n >= \\ &\sum_{n_{n-1}l_{n-1}n_nl_n} < (n_{n-1}l_{n-1}n_nl_n)\,\lambda|\,(N_{n-1}L_{n-1}N_nL_n)\,\lambda > |\,(n_{n-1}l_{n-1}n_nl_n)\,\lambda M_\lambda; \vec{\tilde{\rho}}_{n-1}\vec{\tilde{\rho}}_n >\end{aligned} \tag{16}$$

The coefficients in this expansion are the harmonic oscillator brackets for two particles having different masses. These brackets were first introduced by Talmi[5] and were then calculated explicitly for equal masses[6]-[7] and for different masses[8]. The harmonic oscillator brackets are independent of the M values. The sum in Eq. (16) is finite because of energy conservation, i.e., $2N_{n-1} + L_{n-1} + 2N_n + L_n = 2n_{n-1} + l_{n-1} + 2n_n + l_n$ and the angular momentum coupling condition $\vec{l}_{n-1} + \vec{l}_n = \vec{\lambda}$.

Applying the transformation (16) to both the bra and the ket of the matrix element on the right hand side of Eq. (11), we get a matrix element between states expressed in the new coordinates (14). Because of the symmetry properties of these new coordinates, the appropriate matrix element is trivial:

$$< (n'_{n-1}l'_{n-1}n'_nl'_n)\lambda|(n-1,n)|(n_{n-1}l_{n-1}n_nl_n)\lambda >= (-1)^{l_{n-1}}\delta_{(n'_{n-1}l'_{n-1}n'_nl'_n),(n_{n-1}l_{n-1}n_nl_n)} \tag{17}$$

The phase $(-1)^{l_{n-1}}$ is due to the antisymmetric property of the coordinate $\vec{\tilde{\rho}}_{n-1}$.

In the case where an n-particle Young frame is obtained from two (or more) $(n-1)$-particle Young frames there is no relation between the phases of the resulting hocfps, since the diagonalization is performed separately for each $(n-1)$-particle Young frame. However, from the theory of the permutation group[9] such states should be related as follows:

$$< Y_{n-2}\Gamma'_{n-1}\Gamma_n\Lambda_n\epsilon_n\alpha_n|(n-1,n)|Y_{n-2}\Gamma_{n-1}\Gamma_n\Lambda_n\epsilon_n\alpha_n >= \frac{|\sigma|}{\sqrt{\sigma^2-1}} \tag{18}$$

where $\sigma = a_r - a_s + s - r$, r and s are the row indices in which the particles n and $n-1$ are placed in the Yamanouchi symbol Y_n, and a_r and a_s are the lengths of these rows in Γ_n. By expanding both sides of the last matrix element in terms of the hocfps and using the orthogonality of the hocfps we obtain

$$\begin{aligned}&[Y_{n-2}\Gamma'_{n-1}\Phi'_{n-1}\eta'^{(n)}\Lambda_n|\}Y_{n-2}\Gamma'_{n-1}\Gamma_n\Phi_n] = \\ &\frac{|\sigma|}{\sqrt{\sigma^2-1}}\sum_{\Phi_{n-1}\eta^{(n-1)}}\left[Y_{n-2}\Gamma_{n-1}\Phi_{n-1}\eta^{(n)}\Lambda_n|\}Y_{n-2}\Gamma_{n-1}\Gamma_n\Phi_n\right] \\ &< Y_{n-2}\Gamma'_{n-1}\Phi'_{n-1}\eta'^{(n)}\Lambda_n\epsilon_n|(n-1,n)|Y_{n-2}\Gamma_{n-1}\Phi_{n-1}\eta^{(n)}\Lambda_n\epsilon_n >\end{aligned} \tag{19}$$

where $\epsilon_n = \epsilon_{n-1} + 2N_n + L_n = \epsilon'_{n-1} + 2N'_n + L'_n$. This relation yields the hocfps from Γ'_{n-1} to Γ_n in terms of the hocfps from Γ_{n-1} to Γ_n. It involves a dependence on the hocfps from Γ_{n-2} to Γ_{n-1}

and Γ'_{n-1}, which were calculated in the preceding cycle. For an irrep Γ_n which originates from more than one $(n-1)$-particle irrep, we have to calculate one set of hocfps by the diagonalization method, described above, and the other sets from Eq. (19).

III. Transformation of the harmonic oscillator states from Jacobi into single particle coordinate

The internal harmonic oscillator states with arbitrary symmetry, derived in the previous section, are expressed in terms of the normalized Jacobi coordinates (Eq. (1)). However, in many calculations in atomic and nuclear physics it is desirable to have expressions for the wave functions in terms of the single particle coordinates. This is particularly important when the harmonic oscillator states are used as a basis set in a calculation involving non-harmonic potentials, which are not easily expressible in Jacobi coordinates.

The total n-particle harmonic oscillator wave function is obtained by coupling the c.m. wave function to the internal wave function, obtaining

$$|Y_n \Phi_n c^{(n)} \mathcal{L}_n; \vec{\rho}_2 \vec{\rho}_3 \ldots \vec{\rho}_n \vec{\rho}^{(n)} > \tag{20}$$

where $\mathcal{L}_n$ is the total angular momentum and $c^{(n)}$ stands for the n-particle c.m. quantum mumbers $N^{(n)}$ and $L^{(n)}$. The permutational symmetry of this coupled state is determined by that of the internal state, as was shown in the previous section. For n-particle non-spurious states the c.m. wave function is always in the ground state and therefore $L^{(n)} = 0$ and $\mathcal{L}_n = \Lambda_n$.

We now show how to transform an n-particle harmonic oscillator state with a well defined permutational symmetry from Jacobi into single particle coordinates. Starting from the n'th particle, we use the hocfps for decoupling its wave function from the internal wave function and then we couple it to the n-particle c.m. wave function. We obtain the coupled wave function

$$|(Y_{n-1}\Phi_{n-1})(\eta^{(n)}c^{(n)})\lambda\mathcal{L}_n; (\vec{\rho}_2\vec{\rho}_3 \ldots \vec{\rho}_{n-1})(\vec{\rho}_n \vec{\rho}^{(n)}) > \tag{21}$$

where $\vec{\lambda} = \vec{L}_n + \vec{L}^{(n)}$ and $\vec{\Lambda}_{n-1} + \vec{\lambda} = \vec{\mathcal{L}}_n$.

The two harmonic oscillator functions coupled to a resultant angular momentum $\vec{\lambda}$, are expressed in terms of the coordinates $\vec{\rho}_n$ and $\vec{\rho}^{(n)}$ where

$$\begin{aligned} \vec{\rho}^{(n)} &= \frac{1}{\sqrt{n}}(\vec{r}_1 + \ldots + \vec{r}_n) \\ \vec{\rho}_n &= \sqrt{\frac{n-1}{n}}\left(\vec{r}_n - \frac{1}{n-1}(\vec{r}_1 + \ldots + \vec{r}_{n-1})\right) \end{aligned} \tag{22}$$

These coordinates can be rewritten in terms of the first $(n-1)$-particle c.m. coordinate, $\rho^{(n-1)} = \frac{1}{\sqrt{n-1}}(\vec{r}_1 + \ldots + \vec{r}_{n-1})$, and the n'th particle coordinate $\vec{r}_n$ as

$$\begin{aligned} \vec{\rho}^{(n)} &= \sqrt{\frac{n-1}{n}}\vec{\rho}^{(n-1)} + \frac{1}{\sqrt{n}}\vec{r}_n \\ \vec{\rho}_n &= -\frac{1}{\sqrt{n}}\vec{\rho}^{(n-1)} + \sqrt{\frac{n-1}{n}}\vec{r}_n \end{aligned} \tag{23}$$

This transformation is actually a rotation of the coordinates by an angle β where $cos\beta = \sqrt{\frac{n-1}{n}}$ and $sin\beta = \frac{1}{\sqrt{n}}$. Therefore, the rotation of the coordinates $\vec{\rho}^{(n)}$ and $\vec{\rho}_n$ by the angle $-\beta$ yields the coordinates

$$\begin{aligned} \vec{\rho}^{(n-1)} &= \vec{\rho}^{(n)} cos\beta - \vec{\rho}_n sin\beta \\ \vec{r}_n &= \vec{\rho}^{(n)} sin\beta + \vec{\rho}_n cos\beta \end{aligned} \tag{24}$$

The harmonic oscillator wave functions expressed in terms of the coordinates $\vec{\rho}_n$ and $\vec{\rho}^{(n)}$ in Eq. (21) can be transformed into harmonic oscillator wave functions expressed in terms of the coordintes $\vec{\rho}^{(n-1)}$ and $\vec{r}_n$ by using the harmonic oscillator brackets for different masses[8]

$$|\eta^{(n)}c^{(n)}\lambda; \vec{\rho}_n\vec{\rho}^{(n)} >= \sum_{c^{(n-1)}h^{(n)}} < (c^{(n-1)}h^{(n)})\lambda|(\eta^{(n)}c^{(n)})\lambda > |c^{(n-1)}h^{(n)}\lambda; \vec{\rho}^{(n-1)}\vec{r}_n > \tag{25}$$

where $h^{(n)}$ stands for the quantum numbers $\tilde{N}_n$, $\tilde{L}_n$ of the harmonic oscillator wave function in the single particle coordinate $\vec{r}_n$.

After performing the steps described above and separating the wave function of the last coordinate ($\vec{r}_n$) by using a change of coupling transformation[12], we obtain

$$|Y_n\Phi_n c^{(n)}\mathcal{L}_n; \vec{\rho}_2\vec{\rho}_3\ldots\vec{\rho}_n\vec{\rho}^{(n)}> = \sum_{\Phi_{n-1}c^{(n-1)}\mathcal{L}_{n-1}h^{(n)}} C^{Y_n\Phi_n c^{(n)}\mathcal{L}_n}_{\Phi_{n-1}c^{(n-1)}\mathcal{L}_{n-1}h^{(n)}}$$
$$|Y_{n-1}\Phi_{n-1}c^{(n-1)}\mathcal{L}_{n-1}h^{(n)}\mathcal{L}_n; \vec{\rho}_2\vec{\rho}_3\ldots\vec{\rho}_{n-1}\vec{\rho}^{(n-1)}\vec{r}_n> \tag{26}$$

where the coefficients are

$$C^{Y_n\Phi_n c^{(n)}\mathcal{L}_n}_{\Phi_{n-1}c^{(n-1)}\mathcal{L}_{n-1}h^{(n)}} = \sqrt{(2\Lambda_n+1)(2\mathcal{L}_n+1)}\sum_{\eta^{(n)}}\left[Y_{n-1}\Phi_{n-1}\eta^{(n)}\Lambda_n|\}Y_n\Phi_n\right]\sum_{\lambda}(2\lambda+1)$$
$$\begin{Bmatrix} \Lambda_{n-1} & \mathcal{L}_n & \lambda \\ L^{(n)} & L_n & \Lambda_n \end{Bmatrix}\begin{Bmatrix} \Lambda_{n-1} & \mathcal{L}_n & \lambda \\ \tilde{L}_n & L^{(n-1)} & \mathcal{L}_{n-1} \end{Bmatrix} < (c^{(n-1)}h^{(n)})\lambda|(\eta^{(n)}c^{(n)})\lambda> \tag{27}$$

The summation over the quantum numbers N_n and L_n (denoted by $\eta^{(n)}$) is restricted by the condition $\epsilon_n = \epsilon_{n-1} + 2N_n + L_n$, where ϵ_n and ϵ_{n-1} are specified by Φ_n and Φ_{n-1}.

Application of Eq. (26) to two particles harmonic oscillator wave function yields

$$|Y_2\Phi_2 c^{(2)}\mathcal{L}_2; \vec{\rho}_2\vec{\rho}^{(2)}> = \sum_{\Phi_1 c^{(1)}\mathcal{L}_1 h^{(2)}} C^{Y_2\Phi_2 c^{(2)}\mathcal{L}_2}_{\Phi_1 c^{(1)}\mathcal{L}_1 h^{(2)}}|Y_1\Phi_1 c^{(1)}\mathcal{L}_1 h^{(2)}\mathcal{L}_2; \vec{\rho}^{(1)}\vec{r}_2> \tag{28}$$

Note that Y_1 is the (trivial) one particle Yamanouchi symbol, $\vec{\rho}^{(1)} = \vec{r}_1$, $\Phi_1 = 0$, $L^{(1)} = \mathcal{L}_1$ and $c^{(1)} = h^{(1)}$. Therefore, in this case we completely transform from the Jacobi into the single particle coordinates in a single step. Eq. (28) can be written in the form

$$|Y_2\Phi_2 c^{(2)}\mathcal{L}_2; \vec{\rho}_2\vec{\rho}^{(2)}> = \sum_{h^{(1)}h^{(2)}} C^{Y_2\Phi_2 c^{(2)}\mathcal{L}_2}_{h^{(1)}h^{(2)}}|h^{(1)}h^{(2)}; \vec{r}_1\vec{r}_2> \tag{29}$$

For three particles we have to apply the transformation in Eq. (26) twice in order to obtain the harmonic oscillator states in single particle coordinates, obtaining

$$|Y_3\Phi_3 c^{(3)}\mathcal{L}_3; \vec{\rho}_2\vec{\rho}_3\vec{\rho}^{(3)}> = \sum_{\Phi_2 c^{(2)}\mathcal{L}_2 h^{(3)}} C^{Y_3\Phi_3 c^{(3)}\mathcal{L}_3}_{\Phi_2 c^{(2)}\mathcal{L}_2 h^{(3)}}|Y_2\Phi_2 c^{(2)}\mathcal{L}_2 h^{(3)}\mathcal{L}_3; \vec{\rho}_2\vec{\rho}^{(2)}\vec{r}_3> =$$
$$\sum_{\substack{\Phi_2 c^{(2)}\mathcal{L}_2 h^{(3)} \\ h^{(1)}h^{(2)}}} C^{Y_3\Phi_3 c^{(3)}\mathcal{L}_3}_{\Phi_2 c^{(2)}\mathcal{L}_2 h^{(3)}} C^{Y_2\Phi_2 c^{(2)}\mathcal{L}_2}_{h^{(1)}h^{(2)}}|h^{(1)}h^{(2)}\mathcal{L}_2 h^{(3)}\mathcal{L}_3; \vec{r}_1\vec{r}_2\vec{r}_3> \tag{30}$$

For the n-particle harmonic oscillator wave function we have to apply Eq. (26) recursively $(n-1)$ times in order to transform completely to the single particle coordinates:

$$|Y_n\Phi_n c^{(n)}\mathcal{L}_n; \vec{\rho}_2\vec{\rho}_3\ldots\vec{\rho}_n\vec{\rho}^{(n)}> = \sum_{\begin{pmatrix} \Phi_{n-1}c^{(n-1)}\mathcal{L}_{n-1}h^{(n)} \\ \vdots \\ \Phi_2 c^{(2)}\mathcal{L}_2 h^{(3)} \\ h^{(1)}h^{(2)} \end{pmatrix}}$$
$$C^{Y_n\Phi_n c^{(n)}\mathcal{L}_n}_{\Phi_{n-1}c^{(n-1)}\mathcal{L}_{n-1}h^{(n)}}\cdots C^{Y_3\Phi_3 c^{(3)}\mathcal{L}_3}_{\Phi_2 c^{(2)}\mathcal{L}_2 h^{(3)}} C^{Y_2\Phi_2 c^{(2)}\mathcal{L}_2}_{h^{(1)}h^{(2)}}|h^{(1)}h^{(2)}\mathcal{L}_2 h^{(3)}\mathcal{L}_3\ldots h^{(n)}\mathcal{L}_n; \vec{r}_1\vec{r}_2\ldots\vec{r}_n> =$$
$$\sum_{\{\Phi_{i-1}c^{(i-1)}\mathcal{L}_{i-1}h^{(i)};i=2,3,\ldots,n\}}\prod_{j=2}^{n} C^{Y_j\Phi_j c^{(j)}\mathcal{L}_j}_{\Phi_{j-1}c^{(j-1)}\mathcal{L}_{j-1}h^{(j)}}|h^{(1)}h^{(2)}\mathcal{L}_2 h^{(3)}\mathcal{L}_3\ldots h^{(n)}\mathcal{L}_n; \vec{r}_1\vec{r}_2\ldots\vec{r}_n> \tag{31}$$

Note that the Yamanouchi symbol Y_n determines all the Yamanouchi symbols Y_i ; $i = 1, 2, \ldots, n-1$.

Conclusions

Harmonic oscillator wave functions are widely used as basis states in molecular, atomic and nuclear computations. A straightforward construction of an n-particle basis from these wave functions generates a large number of spurious states, involving center-of-mass motion. A common device employed to eliminate these states is the addition of an appropriate operator with a relatively large coefficient to the hamiltonian, in order to push them up in energy. An obvious drawback is that a huge basis set is employed, a substantial part of which is totally ineffective.

In the present article we introduce a computationally efficient procedure for constructing non-spurious harmonic oscillator states with arbitrary symmetry. The explicit elimination of the spurious states results in a very significant reduction in the size of the basis employed. This is a very crucial point for calculations involving a very large number of states. Moreover, the states constructed using this method have a definite permutational symmetry. This property is essential for calculations involving multiple angular momentum quantum numbers, such as non-relativistic quark calculations.

References

[1] M. Oka and K. Yazaki, *Baryon Baryon Interaction from Quark Model Viewpoint* in *Quarks and Nuclei*, W. Weise, Ed. (World Scientific, Singapore, 1985).

[2] A. Novoselsky and J. Katriel, *Non-Spurious Harmonic Oscillator States with Arbitrary Symmetry*, Ann. Phys. (N. Y.) (in press).

[3] J.-Q. Chen, A. Novoselsky, M. Vallières and R. Gilmore, Phys. Rev. **C 39**, 1088 (1989).

[4] A. Novoselsky and J. Katriel, *Non-Spurious Harmonic Oscillator States in Single Particle Coordinates*, J. Math. Phys. (in press).

[5] I. Talmi, Helv. Phys. Acta **25**, 185 (1952).

[6] M. Moshinsky, Nucl. Phys. **13**, 104 (1959).

[7] T. A. Brody and M. Moshinsky, *Tables of Transformation Brackets* (Gordon and Breach, 1967).

[8] A. Gal, Ann. Phys. (N. Y.) **49**, 341 (1968).

[9] M. Hamermesh, *Group Theory* (Addison-Wesley, MA, 1959).

[10] A. Novoselsky, J. Katriel and R. Gilmore, *J. Math. Phys.* **29** (1988), 1368.

[11] A. Partensky, *J. Math. Phys.* **13** (1972), 1503.

[12] A. De-Shalit and I. Talmi, *Nuclear Shell Theory* (Academic Press, New York, 1963).

INHOMOGENEOUS QUANTUM LIQUIDS: MANY-BODY THEORY ON THE TESTBENCH

E. Krotscheck

Center for Theoretical Physics, Department of Physics

Texas A&M University, College Station, Texas 77843

This contribution reports on recent progress in the development of a uniform picture for non-uniform quantum many-body systems. It is shown how the optimized Hypernetted-Chain (HNC/EL) theory is applied to a number of non-uniform quantum systems of physical interest: Films of ^{4}He atoms adsorbed to a substrate, quantum solids, and electrons in atoms and ions.

At one of the last Many-Body conferences I was accused of talking about *many-body systems*, not *many-body theories*. I would, therefore, like to make the interplay between the methodological and the empirical aspect of many-body theory clear from the outset: I am interested in the theoretical description of *real systems* that occur in nature. I am asking what minimal criteria a many-body theory should satisfy in order to allow me to study such real system. I am *not* interested in "models" that have been designed, in the absence of sufficiently powerful tools to describe the real world, in order to illuminate specific aspects of many-body systems.

The basic structure of the problem is most clearly illuminated in ^{4}He at zero temperature. The simplicity of the system makes the implementation of the theory technically quite straightforward and the physical interpretation most transparent. ^{4}He is also attractive since a large number of very clean experimental data are available. The methods to be described here have also been successfully applied to electrons in the metal surface and in atoms and to model nuclei.

We start the microscopic description of a many-body system in an external potential with an empirical Hamiltonian

$$H = \sum_{i=1}^{N} \left\{ -\frac{\hbar^2}{2m} \nabla_i^2 + U_{ext}(\mathbf{r}_i) \right\} + \frac{1}{2} \sum_{1 \leq i < j \leq N} v(|\mathbf{r}_i - \mathbf{r}_j|), \tag{1}$$

where $U_{ext}(\mathbf{r})$ is an external potential, and $v(|\mathbf{r}_i - \mathbf{r}_j|)$ is the interaction between individual particles.

Our goal is to predict the behavior of N of particles in a given volume Ω assuming no other knowledge but the fundamental interactions. ^{4}He is one of the systems where this problem can partly be solved by large-scale numerical simulations. But there is a large number of experimental data on, for example, low-lying excitated states, whose energetics, structure, and lifetime is most adequately discussed in the lan;;uage of conventional many-body theory.

Recent Progress in Many-Body Theories, Volume 2
Edited by Y. Avishai
Plenum Press, New York, 1990

A microscopic theory for many strongly interacting particles must be able to deal with the strong, short-ranged repulsion between its constituents *and* make valid predictions for the macroscopic structure of the system. If we start with no other assumption but that we have N particles in a given volume Ω, the theory must also be able to *predict* to some extent the geometric properties of the resulting system. For a certain particle number, the volume may be filled by a uniform liquid. If the particle number is decreased, there may be a regime of average densities N/Ω where *both* a uniform and a non-uniform state are possible. However, if the uniform phase is diluted to the density point where the compressibility becomes negative, the theory for the uniform phase should cease to have solutions. A similar effect should occur if the particle number is increased: The liquid should eventually solidify. The stability regime of each of these phases is determined by the response of the system to external perturbations.

From the above qualitative arguments, one concludes that an adequate treatment of both short- and long-ranged correlations is essential for a qualitatively correct microscopic description of a self-bound many-body system. Such a treatment is provided by the optimized hypernetted chain theory; I will spend the first part of my contribution describing how the theory is applied for non-uniform systems. One starts[1] with a variational *ansatz* for the ground-state wave function of a system of N particles with coordinates $\mathbf{r}_1, \ldots, \mathbf{r}_N$:

$$\Psi_0(\mathbf{r}_1,\ldots,\mathbf{r}_N) = \exp\frac{1}{2}\left\{\sum_{1\le i\le N} u_1(\mathbf{r}_i) + \sum_{1\le i<j\le N} u_2(\mathbf{r}_i,\mathbf{r}_j) + \ldots\right\}\Phi_0(\mathbf{r}_1,\ldots,\mathbf{r}_N). \tag{2}$$

The one-body function $u_1(\mathbf{r})$ describes the spatial structure of the system, the two-body function $u_2(\mathbf{r}_i,\mathbf{r}_j)$ describes the short- and long-ranged correlations between pairs of particles, and $\Phi_0(\mathbf{r}_1,\ldots,\mathbf{r}_N)$ is the ground state of a weakly interacting model system; its main purpose is to guarantee antisymmetry of the wave function for Fermions. The correlation functions and the model wave function are determined by minimization of the ground-state energy-expectation value E_0.

$$\frac{\delta E_0}{\delta u_1(\mathbf{r})} = 0, \qquad \frac{\delta E_0}{\delta u_2(\mathbf{r}_i,\mathbf{r}_j)} = 0. \tag{3}$$

If one uses approximations for the energy functional, it is essential to use only such approximations that lead to approximate Euler equations which have the same structure as the Euler equations of the exact problem; the variational problems (3) may otherwise be meaningless. The existence of physically meaningful solutions of the variational problems (3) has been shown to be intimately related to an adequate, self-consistent treatment of both short- and long ranged correlations. Again, it turns out that the HNC hierarchy of approximations has such meaningful variational minima at each level of its implementation[2]; it includes both the right physics and the appropriate mathematical structure.

To show how short- and long ranged correlations are treated in the HNC/EL theory, we start with the basic equations for the simplest case, the homogeneous Bose liquid. These equations are most conveniently formulated in terms of the pair correlation function $g(r)$, and the static structure function $S(k) = 1 + \rho\int d^3r[g(r) - 1]e^{i\mathbf{k}\cdot\mathbf{r}}$. One finds[2]

$$S(k) = \left\{\sqrt{1 + \frac{4m\rho}{\hbar^2k^2}\tilde{V}_{ph}(k)}\right\}^{-\frac{1}{2}} \tag{4}$$

where

$$V_{ph}(r) = g(r)v(r) + \frac{\hbar^2}{m}|\nabla\sqrt{g(r)}|^2 + [g(r) - 1]w_I(r). \tag{5}$$

is the "particle-hole interaction", note that

$$V_{ph}(|\mathbf{r} - \mathbf{r}'|) = \frac{\delta^2 E_0}{\delta\rho_1(\mathbf{r})\delta\rho_1(\mathbf{r}')}. \tag{6}$$

$w_I(r)$ is the so-called "induced interaction", its analytic form is irrelevant for our further discussions. In Eq. (4) we recover the RPA expression for the static form factor. The HNC/EL theory supplements the RPA with a microscopic theory of the particle- hole interaction $V_{ph}(r)$. Some algebraic manipulations also permit to write the equations in a "Bethe-Goldstone" like form.

We see from Eq. (4) that the theory has, as required, no *uniform* solution with negative compressibility, $\tilde{V}_{p-h}(0+) < 0$, c.f. Eq. (6). The theory has also a unique signature for symmetry breaking with a finite wave number: Then, the term under the square-root Eq. (4) becomes negative at *finite* wave numbers. If an instability of either kind occurs, one must allow for the appropriate symmetry breaking in the wave function (1) by including a one-body function $u_1(\mathbf{r})$ and breaking the translational invariance of $u_2(\mathbf{r}_i, \mathbf{r}_j)$. The theory then has solutions if the system is, *within the assumed geometry*, stable against small perturbations.

The next task is to derive an algorithm for the numerical solution of the generalizations of the equations (4-6) in geometries with broken symmetry. The generalization of the PPA equation (3) for inhomogeneous systems is written in the form[3]

$$\left[S^{-1} * H_1 * S^{-1} - S_F^{-1} * H_1 * S_F^{-1}\right](\mathbf{r}_1, \mathbf{r}_2) = 2\tilde{V}_{p-h}(\mathbf{r}_1, \mathbf{r}_2), \tag{7}$$

where

$$H_1 \equiv -\frac{\hbar^2}{2m}\frac{1}{\sqrt{\rho_1(\mathbf{r})}}\nabla\rho_1(\mathbf{r})\cdot\nabla\frac{1}{\sqrt{\rho_1(\mathbf{r})}}. \tag{8}$$

The asterisk denotes the convolution integral, and S^{-1} is the inverse of the static structure function in the sense of the convolution product. With the tilde-version of a given function we denote the corresponding "density-weighted" function, i.e. $\tilde{V}_{p-h}(\mathbf{r}_1, \mathbf{r}_2) \equiv \sqrt{\rho_1(z_1)}V_{p-h}(\mathbf{r}_1, \mathbf{r}_2)\sqrt{\rho_1(z_2)}$. $S_F(\mathbf{r}_1, \mathbf{r}_2)$ in Eq. (7) is the static structure function of the non-interacting Fermi system defined by the model ground state $\Phi_0(\mathbf{r}_1, \ldots, \mathbf{r}_N)$, it is one for a Bose system.

A normal mode decomposition is used to turn the two-body equation (7) into a set of independent one-body equations. For this purpose, we consider the generalized eigenvalue problem

$$H_1 * \left[S_F^{-1} * H_1 * S_F^{-1} + 2\tilde{V}_{p-h}\right]\psi^{(\ell)} = \hbar^2\omega_\ell^2\psi^{(\ell)}. \tag{9}$$

With the normalization $(\psi^{(\ell)}|H_1^{-1}|\psi^{(\ell')}) = \delta_{\ell,\ell'}$, the solution of the PPA equation (7) can be represented as

$$S(\mathbf{r}_1, \mathbf{r}_2) = \sum_\ell \frac{\psi^{(\ell)}(\mathbf{r}_1)\psi^{(\ell)}(\mathbf{r}_2)}{\hbar\omega_\ell}. \tag{10}$$

The low-lying eigenfrequencies $\hbar\omega_\ell$ can be identified with the low-lying excitation energies of the system[3].

The particle-hole interaction for the inhomogeneous case is a straightforward generalization of Eq. (5). Given a solution of Eqs. (9), (10), the variational energy expectation value for Bosons can be written as

$$E = \frac{\hbar^2}{2m}\int d^3r|\nabla\sqrt{\rho_1(\mathbf{r})}|^2 + \frac{1}{2}\int d^3r_1 d^3r_2 \rho_1(\mathbf{r}_1)V_{p-h}(\mathbf{r}_1,\mathbf{r}_2)\rho_1(\mathbf{r}_2) \\ -\frac{1}{2}\int d^3r_1 d^3r_2 d^3r_3 \tilde{V}_{p-h}(\mathbf{r}_1,\mathbf{r}_2)\tilde{h}(\mathbf{r}_2,\mathbf{r}_3)\tilde{h}(\mathbf{r}_3,\mathbf{r}_1) \tag{11}$$

with $h(\mathbf{r}_1,\mathbf{r}_2) \equiv g(\mathbf{r}_1,\mathbf{r}_2) - 1$. The simplest symmetry breaking that can be treated with reasonable computational effort is the plane surface geometry. In this case, Eq. (9) can be decoupled as a function of the momentum parallel to the surface, i.e. the eigenfunctions $\psi^{(\ell)}(\mathbf{r})$ have the form $\psi^{(\ell)}(\mathbf{r}) = \psi_{n,\mathbf{q}}(z)e^{i\mathbf{q}\cdot\mathbf{r}_{\parallel}}$, where $\mathbf{q}$ is the momentum parallel to the surface, and z the distance of a particle from the surface. The index n labels the quantum number related to the motion in the z direction. For systems with spherical symmetry, the eigenvalue equation (9) is separated in spherical harmonics. The remaining manipulations, which are, for example, used in atoms[4], are very similar to the plain surface geometry.

The periodic density in a lattice[5] can be expanded as

$$\sqrt{\rho_1(\mathbf{r})} = \sum_m C_m e^{i\mathbf{Q}_m\cdot\mathbf{r}}, \tag{12}$$

where the $\mathbf{Q}_m$ are the reciprocal lattice vectors. The PPA equation decouples in the momentum within the first elementary cell, i.e. we have

$$\sum_n \left\{H_1(\mathbf{k}) * \left[H_1(\mathbf{k}) + 2\tilde{V}(\mathbf{k})\right]\right\}_{mn} \psi_n^{(\ell)}(\mathbf{k}) = \hbar^2\omega_\ell^2(\mathbf{k})\psi_m^{(\ell)}(\mathbf{k}), \tag{13}$$

where the operators $H_1(\mathbf{k})$ and $\tilde{V}_{p-h}(\mathbf{k})$ are considered as matrices with indices corresponding to the reciprocal lattice vectors. The pair distribution function takes the form

$$\tilde{h}(\mathbf{r}_1,\mathbf{r}_2) = \sum_{m,n} e^{i(\mathbf{Q}_m\cdot\mathbf{r}_1 - \mathbf{Q}_n\cdot\mathbf{r}_2)}\tilde{h}_{mn}(\mathbf{r}_1 - \mathbf{r}_2) \tag{14}$$

with

$$\tilde{h}_{mn}(\mathbf{r}) = \int \frac{d^3k}{(2\pi)^3} e^{i\mathbf{k}\cdot\mathbf{r}} \left\{\sum_\ell \frac{\psi_m^{(\ell)}(\mathbf{k})\psi_n^{(\ell)}(\mathbf{k})}{\hbar\omega_\ell(\mathbf{k})} - \delta_{mn}\right\}. \tag{15}$$

From the structure of Eqs (14), (15), it is clear that $h(\mathbf{r}_1,\mathbf{r}_2)$ can be expanded as

$$h(\mathbf{r}_1,\mathbf{r}_2) = \sum_m e^{i\mathbf{Q}_m\cdot\mathbf{R}} h_m(\mathbf{r}), \tag{16}$$

where $\mathbf{R}$ is the center of mass vector, and $\mathbf{r} = \mathbf{r}_1 - \mathbf{r}_2$. A systematic nearest-neighbor expansion scheme is then based on the expansion (16).

The first physical system considered is our application of the basic formalism is a number of ^{4}He atoms adsorbed to a substrate which is described by an external field $U_{ext}(z)$. The substrate potential $U_{ext}(z)$ is obtained by averaging Lennard-Jones interactions over a half space. Given the substrate potential, the two-body interaction, and the surface coverage $n = \int dz\rho_1(z)$, the physical problem is completely defined. We have solved the HNC/EL equations for the ^{4}He numerically[6] for a sequence of surface coverages n. Some typical examples of density profiles are shown in Fig. 1. Most remarkable are the strong density-oscillations of the ^{4}He film, which are due to the geometric core-exclusion between the ^{4}He particles.

Fig. 1. The density $\rho_1(z)$ of the background film of ^4He atoms (dotted lines) and of the ^3He impurity, $\rho_1^I(z)$ (solid lines) are shown for surface coverages of n = 0.15, 0.20, 0.25, 0.30, and 0.35 ^4He atoms/Å^2. The ^4He densities of all films shown are virtually identical within the first layer, whose density maximum is about 0.08 atoms/Å^3. The ^3He impurity density is normalized such that $\int dz \rho_1^I(z) = 1$. The figure is taken from Ref. 6

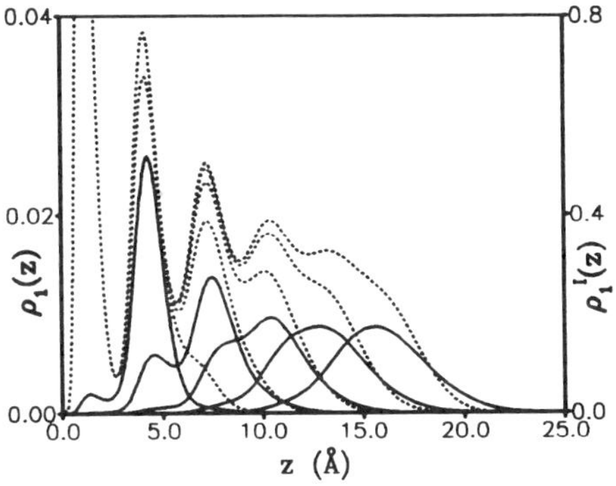

At this point one has to ask the question on how one might be able to probe the layer structure of such adsorbed films. Unfortunately, quantities that can be calculated easily are often hard to measure and vice versa. Ground-state energy measurements are difficult, it also turns out that the energy per particle does not depend strongly on the layer structure of the system[6], and the chemical potential has only a weak modulation. Neutron scattering experiments to probe the liquid surface are even more difficult since the penetration depth of the neuton is about 100 Å, whereas the surface width of ^4He is of the order of 10 Å. If one is interested in exploring the layer structure, one has to look for sensitive experiments.

A handy probe of the surface structure is a ^3He impurity since it can be distinguished from the ^4He background by both its magnetic moment and its statistics. Experimental efforts focus consequently on the following areas:

(a) Measurements of the binding energy and the specific heat of ^3He impurities[7], and

(b) Measurement of magnetic properties of ^3He impurities adsorbed to ^4He surfaces[8].

Besides probing the structure of the underlying ^4He film, these experiments also provide the exciting possibility to study quasi-twodimensional Fermi systems. The experimental possibilities require the development of three levels of theoretical tools:

(a) Single-impurity theories aim at the calculation of impurity binding-energy and effective mass. The static correlations between particles determine if and how many bound states the impurity particle can have at the surface, and whether it penetrates to the substrate. The calculation of the (complex) self-energy of the impurity particle gives access to the specific heat and the mobility.

(b) Mixture films in the dilute limit correspond to a system of *two* static impurities[9]. To the extent that the concentration of the impurities is small, one can ignore all higher-order correlations between the impurity particles and possible dynamic effects due to momentum-dependent correlations.

(c) A theory for two-component systems is needed for large "impurity" concentrations. This enables us to study the structure of quantum-liquid interfaces.

Formally, impurities may be added to the system by extending the wave function to include impurity-background and impurity-impurity correlation functions $u_1^I(\mathbf{r}^I)$, $u_2^{IB}(\mathbf{r}_i^I, \mathbf{r}_j)$, and $u_2^{II}(\mathbf{r}_i^I, \mathbf{r}_j^I)$. For Fermion impurities, the wave function is multiplied with a Slater determinant to ensure the antisymmetry with respect to exchange. To include a momentum dependence, one may either add a "backflow" function to the one-impurity correlations, or calculate the self-energy in CBF perturbation theory. The new impurity-background, impurity-impurity, and backflow correlation functions are again determined by solving Euler equations.

Let us turn to the results of our calculations (see Refs. 10-12). We have carried out extensive calculations for the static ground state of one ^{3}He impurity on a family of ^{4}He backgrounds ranging from a double layer ($n = 0.12\text{Å}^{-2}$) to a system of about five helium layers ($n = 0.35\text{Å}^{-2}$). The solid lines in Fig. 1 show the impurity density in comparison with the background density. We see that the ^{3}He particle is, as expected, inside the outermost ^{4}He layer for our calculation of a double-layer system. As the thickness of the background increases, the ^{3}He particle is pushed outward into the low-density regime of the film.

The interaction of the ^{3}He impurity with the ^{4}He background gives rise to the "hydrodynamic effective mass". This contribution to the total effective mass has been calculated using second-order CBF perturbation theory for the impurity self-energy in terms of the emission and re-adsorption of ripplons and phonons. Our results are compared in Fig. 2 with the experimental results of Refs. 7. The calculated hydrodynamic mass is somewhat lower than the experimental mass. However, the experiment has been done for a finite ^{3}He coverage of 0.3 layers. Therefore we must include the effects of the quasiparticle interaction between the ^{3}He atoms.

The interaction between ^{3}He impurities is intelligently discussed in terms of Landau's Fermi-Liquid theory in two dimensions[13]. As it is the case for the background film above, the variational theory provides a microscopic, *local* effective interaction ("quasiparticle interaction") between two impurities, which includes both the direct Van der Waals force, local screening, and the exchange of ripplons and phonons through the ^{4}He background. In the case of a dilute system of impurities, we may take the impurity-impurity interaction as an effective interaction between any two ^{3}He atoms for a *finite* impurity population. The quasiparticle interaction depends only on the relative momentum and spin of the two impurity atoms:

$$f(i,j) = f_c(q_{\|}) + f_\sigma(q_{\|})\sigma_1 \cdot \sigma_2 \tag{17}$$

Given the interaction (17), we can calculate the Fermi-Liquid parameters of the two-dimensional Fermi liquid,[13]

$$\begin{aligned} N(0)f_c(|\mathbf{q}_1 - \mathbf{q}_2|) &= \sum_{m=0}^{\infty} F_m^s \cos(m\phi) \\ N(0)f_\sigma(|\mathbf{q}_1 - \mathbf{q}_2|) &= \sum_{m=0}^{\infty} F_m^a \cos(m\phi), \end{aligned} \tag{18}$$

where ϕ is the angle between $\mathbf{q}_1$ and $\mathbf{q}_2$, and $N(0) = m^*/\pi\hbar^2$ the density of states at the Fermi surface. From these, we obtain the magnetic susceptibility $\chi(0)$, in units of the susceptibility of the free two-dimensional Fermi gas, χ_{30},

$$\chi(0)/\chi_{30} = (m_H/m_3)(1 + F_1^s/2)/(1 + F_0^a) \tag{19}$$

and the total effective mass

$$m^* = m_H(1 + F_1^s/2). \tag{20}$$

Figure 2 (solid line) shows the total effective mass m^* calculated from Eq. (20). For one active layer of ^{4}He, the hydrodynamic mass is about 1.8 m_3, in reasonable agreement with both the data of Refs. 7 and the conclusion of Ref. 8 from their thinnest ^{4}He film. In this regime the correction due to the quasiparticle interaction becomes quite sizable with increasing ^{3}He concentration, but it is difficult to make a quantitative statement due to both the high density of the background and the rapid variation of the effective mass.

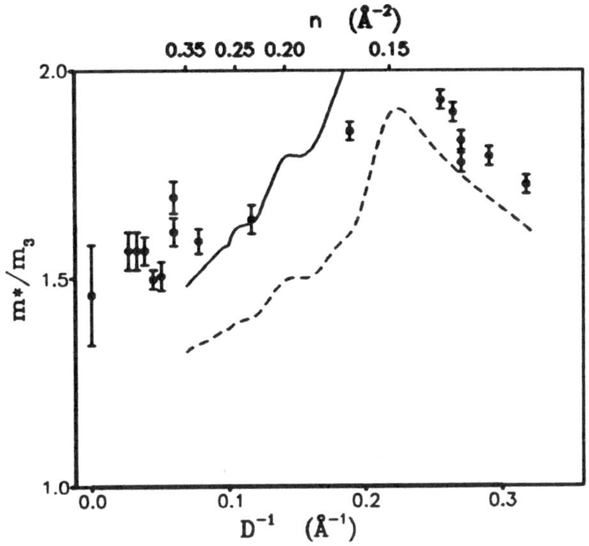

Fig. 2. The calculated effective mass m^* is shown in units of the bare ^{3}He mass m_3 as a function of the *inverse film thickness* D^{-1} (dashed line). The circles with error bars are the experimental data of Refs. 7. The upper scale shows the surface coverage corresponding to the film thickness D. The dashed line shows the hydrodynamic effective mass m_H for 0.3 layers of ^{3}He. This figure is taken from Ref. 12.

For all cases with thicker ^{4}He films, the accuracy of our theoretical prediction is quite satisfactory. The microscopic theory predicts a hydrodynamic mass that is consistent with the estimate $m_H/m_3 = 1.26 \pm 0.15$ given in Ref. 8, but below the results of Refs. 7. The agreement with the latter data is improved when the corrections due to the quasiparticle interaction are included. The quasiparticle interaction between ^{3}He atoms in the surface gives a density-dependent correction to the effective mass. The contribution is about 10% for the case of 0.3 layers of ^{3}He.

A comparison of the theoretical and the experimental magnetic susceptibility ratio $\chi(0)/\chi_{30}$ is shown in Fig. 3. Here, we have taken a hydrodynamic mass of $1.26m_3$, which gives the best fit to experimental data[8]. This choice is consistent with our calculation of the effective mass since the experimental data refer to a much smaller ^{3}He coverage. We find a quite satisfactory agreement between theory and experiment. The agreement is not quite so good for smaller ^{4}He coverages, where our theoretical description is intrinsically less accurate.

The good agreement between the experimental results and our calculation has been obtained with a *static, momentum-dependent* quasiparticle interaction. The calculation of the magnetic susceptibility involves phase-space integrals of the quasiparticle interaction over momenta between $q = 0$ and $q = 2k_F$.

Therefore, with increasing impurity density one probes the *momentum-dependence* of the quasiparticle interaction. Measurements of the type of Ref. 8 are extemely useful to enhance our theoretical understanding of the quasiparticle interaction in liquid ^{3}He.

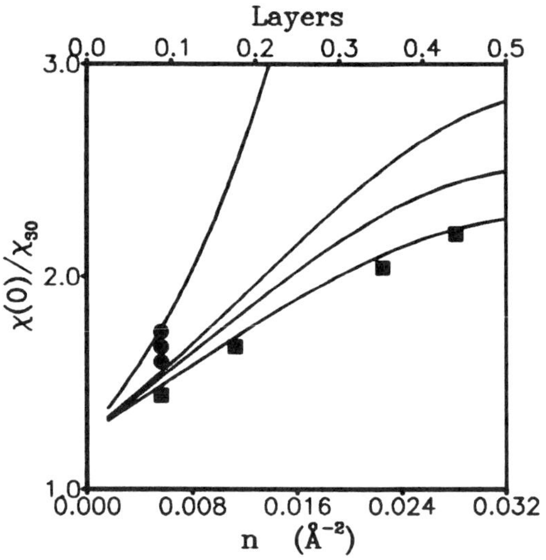

Fig. 3. Magnetic susceptibility χ/χ_{30} of the ^{3}He film as a function of areal density n in atoms/Å^2, for ^{4}He coverages between 0.15 atoms/Å^2 (uppermost curve), and 0.30 atoms/Å^2 (lowest curve). Solid squares: experimental data[7] for a surface coverage of 9.5 ^{4}He layers. Circles: experimental data[7] for a ^{3}He density of 0.088 layers, for ^{4}He coverages of 2.8, 3.4, 4.75, and 5.24 atoms/Å^2. This figure is taken from Ref. 11

The application of HNC-variational methods to the quantum solid[5] is technically similar, but emphazises an entirely different aspect of the theory. Our theoretical description of surface phenomena was quite successful partly because the interesting physics happens in the low-density regime of the surface. The liquid-solid phase transition in ^{4}He happens at a rather high density where the simplest version of the HNC theory has serious deficiencies. One does therefore not expect more than qualtitative statements. In this respect, the HNC/EL theory for the liquid-solid phase transition is mostly an exercise in obtaining the necessary *know-how* for studying electron correlations in a real metal. An interesting formal aspect of the solidification problem is that one is, in particular in the HNC approximation, far away from a *local* instability. This means that any *small* change in the ground state wave function will not be sufficient to predict a phase transition.

Preliminary numerical results in the HNC approximation indicate a liquid-solid phase transition at a density of about 0.028 Å^{-3}. The first iteration of the PPA procedure yields to some overbinding which is quite expected since the lattice structure should induce a more pronounced structure of the short-ranged correlations, which lead in turn to an increase of the kinetic energy.

Let us finally turn to our treatment of electron correlations in atoms. Apart from the necessity to include Fermi statistics, electronic systems are formally treated very similarly to ^{4}He. In ^{4}He, the inclusion of multi-particle correlations and elementary diagrams produces increasingly better descriptions of the ground state structure without generating much additional physical insight. In electronic systems, one can obtain a good description of the ground state within the simplest approximation, FHNC//0, The next higher order corrections generate more or less random contributions to the ground state energy of very slowly decreasing magnitude.

The Hamiltonian of the electons is the one spelled out in Eq. (1), with the external potential $U_{ext}(\mathbf{r})$ being the Coulomb potential of the nucleus, $-Ze^2/r$. The optimal $u_1(\mathbf{r})$ depends solely on the choice of the single-particle orbitals ψ_i which are solutions of a correlated Hartree-Fock equation[14]:

$$[-\frac{1}{2}\nabla^2 + U_{ext}(\mathbf{r}) + V_H(r)]\psi_i(r) + \int d^3r' V_g(\mathbf{r},\mathbf{r}')\rho_1(\mathbf{r},\mathbf{r}')\psi_i(\mathbf{r}') = \epsilon_i\psi_i(\mathbf{r}). \tag{21}$$

Here, $V_H(\mathbf{r})$ is the generalized Hartree potential, $V_g(\mathbf{r},\mathbf{r}')$ is the exchange/correlation interaction, and $\rho_1(\mathbf{r},\mathbf{r}')$ is the one body density matrix of the *non-interacting* system.

The energy of atomic systems is almost entirely accounted for by Hartree Fock. Correlations in electronic systems are significant only in the regimes of low electron density, which means the outer region of the atom. The *correlation energy* E_c is defined as the difference between the total energy E_{TOT} and the Hartree Fock energy E_{HF}:

$$E_c = E_{TOT} - E_{HF} \tag{22}$$

This correlation energy consists of two parts: a positive contribution which comes from the fact that the single particle orbitals ψ_i are solutions of the *correlated* Hartree-Fock equation, and thus are not the best *uncorrelated* Hartree-Fock orbitals; and a generic correlation contribution. This latter negative quantity exceeds the positive Hartree Fock shift, giving a total negative value for the correlation energy.

The correlation energy is of the order of 1% or less of the total energy. Clementi and Veillard[15] have determined this energy from experiments on four and ten electron atomic systems. Their results, shown in the Table, have been corrected for center of mass energy and relativistic effects, and thus can be compared directly to our results for non-relativistic atoms with an infinite mass nucleus. It is seen from the table that our results are in good agreement with these experiments.

	a	*b*	*c*	*d*	*e*
Be	-0.0944	-0.096	-0.224	-0.099	-0.094
Ne^{+6}	-0.18	-0.10	-0.333	-0.026	-0.136
Ne	-0.39	-0.33	-0.74	-0.41	-0.39
Mg^{+2}	-0.40	-0.37			
Ca^{+10}	-0.463	-0.41			

Correlation energies for various atoms and ions in atomic units. a: experimental data[15]; b: FHNC[4]; c: LSD results[16]; d: generalized gradient expansion of Langreth and Mehl[17]; and e: generalized gradient expansion of Perdew[18].

Our data for four-electron systems should be regarded with some caution since the (Fermi)-HNC theory has been designed to treat systems with large particle number and leaves out terms of relative magnitude $1/N^2$. A calculation of these corrections would require additional effort.

It is abundantly clear that the local density approximation is inadequate for the correlation energies. This was expected by those who introduced this approximation[19]. It is also seen there that adjusted gradient expansions[17,18] give agreement with experiment which is comparable to our theory. However it was necessary in each of these gradient expansions to fit the data either at one[17] atom or to several atoms[18].

Acknowledgements

This contribution is a review of work performed in collaboration with C. E. Campbell, J. L. Epstein, T. Pang, G. N. Parker, and M. Saarela. The work on atoms was carried out during a visit of the Theoretical Physics Institute at the University of Minnesota in Minneapolis, whose support and hospitality is gratefully acknowledged. The work was supported, in part, by the National Science Foundation under Contract PHY-8806265 and the Robert A. Welch foundation under Grant A-1111.

References

1. E. Feenberg, *Theory of Quantum Fluids* (Academic, New York, 1969).
2. C. E. Campbell and E. Feenberg, Phys. Rev. **188**, 396 (1969).
3. E. Krotscheck, G.-X. Qian, and W. Kohn, Phys. Rev. **B31**, 4245 (1985).
4. Tao Pang, C. E. Campbell, and E. Krotscheck, Chem. Phys. Lett (in press) and J. Chem. Phys. (to be published).
5. E. Krotscheck, G. N. Parker, and M. Popović, to be published.
6. J. L. Epstein and E. Krotscheck, Phys. Rev. **B37**, 1666 (1988).
7. X. Wang and F. M. Gasparini, Phys. Rev. **B38**, 11245 (1988); B. K. Bhattacharyya, M. J. DiPirro, and F. M. Gasparini, Phys. Rev. **B30**, 5029 (1984).
8. J. M. Valles, Jr., R. H. Higley, R. B. Johnson, and R. B. Hallock, Phys. Rev. Lett. **60**, 428 (1988).
9. J. C. Owen, Phys. Rev. Lett. **47**, 586 (1981).
10. E. Krotscheck, M. Saarela, and J. L. Epstein, Phys. Rev. **B38**, 111 (1988).
11. E. Krotscheck, M. Saarela, and J. L. Epstein, Phys. Rev. Lett. **61**, 1728 (1988).
12. J. L. Epstein, E. Krotscheck, and M. Saarela, preprint (1989).
13. S. M. Havens-Sacco and A. Widom, J. Low Temp. Phys. **40**, 357 (1980).
14. E. Krotscheck and W. Kohn, Phys. Rev. Lett. **57** (1986) 862; E. Krotscheck, G.-X. Qian, and W. Kohn, Phys. Rev. **B32** (1985) 5693.
15. E. Clementi, J. Chem. Phys. **38** (1963) 2248; A. Veillard and E. Clementi, J. Chem. Phys. **45** (1968) 2415.
16. Y. Tong and L. J. Sham, Phys. Rev. **144** (1966) 1.
17. D. C. Langreth and M. J. Mehl, Phys. Rev. **B28** (1983) 1809.
18. J. Perdew, Phys. Rev. **B33** (1986) 8822.
19. W. Kohn and P. Vashishta, in *Theory of the inhomogeneous electron gas*, p. 79 ff, edited by S. Lundqvist and N. H. March (Plemum, N.Y. 1983).

COUPLED CLUSTER PARAMETRIZATIONS OF MODEL FIELD THEORIES AND THEIR BARGMANN-SPACE REPRESENTATIONS

J. S. Arponen* and R. F. Bishop†

*Department of Theoretical Physics, University of Helsinki
Siltavuorenpenger 20C, SF-00170 Helsinki, Finland

†Department of Mathematics
University of Manchester Institute of Science and Technology
P O Box 88, Manchester M60 1QD, England

1. INTRODUCTION

Coupled cluster techniques[1-7] have by now been very successfully applied to numerous quantum systems of strongly-interacting particles and fields.[2,8-12] One of the key features of the whole coupled cluster method (CCM) is its ability to incorporate rather naturally and at a very fundamental level, such unifying concepts as supercoherent states, generalized many-body mean fields and generalized order parameters, and exact mappings onto corresponding multilocal classical field theories. This is particularly true of the most recent version of the theory, the so-called extended coupled cluster method (ECCM).[4,6,7] In common with its predecessor, the normal coupled cluster method (NCCM) of Coester and Kümmel,[1] the essence of the formalism is its intrinsic universality in being able to be applied to any system governed by some underlying Schrödinger dynamics. Furthermore, the methods are both exact in principle, and capable of being systematically implemented at various levels of approximation in practice. In its most general form the CCM provides a complete dynamical description of a many-body system by formulating it in terms of a dynamical variational principle for the action.[4,7]

The relationships between the ECCM, the NCCM and the more primitive configuration-interaction (CI) method,[13] and the way that they form a very natural hierarchy of formulations, have been explored in some detail elsewhere.[7] By focusing on the interpretation of each method in terms of time-independent perturbation theory, and their decompositions in terms of Goldstone diagrams, it was possible to formulate each of the three methods in terms of suitably-defined generalized tree diagrams. Their connectivity properties are intimately related to the linked-cluster properties of the different cluster or correlation amplitudes which characterize each of the three methods. Furthermore, these linked-cluster properties are important with regard to the problem of the size-extensivity or size-consistency (i.e, full separability) of the resulting many-body description.[14] Whereas the CI method suffers from well-known deficiencies in this respect, coupled cluster methods in principle take good care of these separability problems, insofar as the relevant correlation amplitudes which describe the interacting many-body system and its properties, obey the cluster property.

Recent Progress in Many-Body Theories, Volume 2
Edited by Y. Avishai
Plenum Press, New York, 1990

In particular, in the ECCM (and only in the ECCM) *all* of the corresponding amplitudes which fully characterize (for example, the ground state of) the many-body system, obey the exact cluster property. Since the ECCM is therefore capable, in principle, of describing such phenomena as phase transitions, spontaneous symmetry breaking, and states of topological excitation, as well as nonequilibrium properties, its potential range of applicability is very wide. Particular applications to date include gauge-field descriptions of both a charged impurity in a polarizable medium (e.g., positron annihilation in metals),[15] and the zero-temperature quantum hydrodynamics of a strongly-interacting condensed Bose fluid.[11] In view of its promise for further use, it seems particularly appropriate to investigate as rigorously as possible such fundamental features of the ECCM as the existence and convergence properties of the method at its various levels of approximation. To this end, we compare in this paper all three parametrizations (*viz.*, CI, NCCM, and ECCM), via their holomorphic representations in the Bargmann Hilbert space.[16] Each method is illustrated by applying it to various quantum anharmonic oscillators, chosen both as an important class of model field theories in themselves, and as an especially stringent test of the ECCM in particular.

One of the most important outcomes of the present work is that by utilizing the holomorphic representation of the various amplitudes, we are able to algebraize the various formulations completely. In particular the topological linking or connectivity requirements associated with the structure of the vertices in the emergent tree-diagram structures, which are normally analyzed diagrammatically, now have considerable new light shed upon them. Indeed, we know of no other comparable example where the asymptotic analytic behaviour of the various amplitudes which fully and exactly parametrize the system, has been so fully analyzed. As an example, we show how certain formally divergent series may be given precise - but generally non-unique - interpretations. We stress from the outset that our final results, which take the form of a generating function for the expectation values of arbitrary operators, are rather general and not simply restricted to the illustrative case of the anharmonic oscillators.

The outline of the remainder of the paper is as follows. Sections 2 and 3 are devoted to describing the basic elements of respectively the three fundamental methods (CI, NCCM and ECCM) and their parametrizations of the wavefunction, and the Bargmann Hilbert space and the associated coherent states which underpin the holomorphic representation. The Schrödinger and Fock representations of the wavefunction and their inter-relationships are described in Sec. 4, where we also discuss the holomorphic representation of all three (CI and CCM) parametrizations for the case of simple field theories with a single bosonic mode. These techniques are illustrated in more detail in Sec. 5 by specific reference to the anharmonic oscillator problem. After describing in Sec. 6 how general expectation values of arbitrary operators may be generated wholly algebraically, the results are discussed in Sec. 7.

2. BASIC ELEMENTS; PARAMETRIZATIONS OF THE WAVEFUNCTION

We consider the bra and ket ground-state wavefunctions $\langle\tilde{\psi}|$ and $|\psi\rangle$ respectively, of an arbitrary many-body system,

$$H|\psi\rangle = E_0|\psi\rangle, \ \langle\tilde{\psi}|H = E_0\langle\tilde{\psi}| \quad , \tag{1}$$

where we do not necessarily assume that the Hamiltonian H is hermitian. For ease of presentation we restrict ourselves to "closed-shell systems" for which each of the parametrizations that we consider may be referred to some suitable single model or reference state $|0\rangle$. This state can be chosen rather freely, so long as it is a *cyclic vector* in the sense that the algebra of all possible operators in the many-body Hilbert space is spanned by the two Abelian sub-algebras of creation and destruction operators defined with respect to it. In this way we may define suitable complete sets of (multiconfigurational) orthonormal creation operators $\{C_k^\dagger\}$, and their hermitian adjoint destruction counterparts $\{C_k\}$, where each many-body configuration-space index k represents some appropriate *set* of discrete or continuous (single-particle) labels. Their choice depends upon the specific system under consideration. Examples have been given elsewhere.[6]

The three particular parametrizations of the ground-state wavefunction, namely the CI method, the NCCM, and the ECCM, are now specified respectively as follows:

CI

$$|\psi\rangle \equiv F|0\rangle \;,\; \langle\tilde{\psi}| \equiv \langle 0|\tilde{F} \;, \qquad F \equiv \sum_k F_k C_k^\dagger \;,\; \tilde{F} \equiv \sum_k \tilde{F}_k C_k \;; \tag{2a}$$

NCCM

$$|\psi\rangle \equiv e^S|0\rangle \;,\; \langle\tilde{\psi}| \equiv \langle 0|\Omega e^{-S} \;, \qquad S \equiv \sum_k{}' S_k C_k^\dagger \;,\; \Omega \equiv \sum_k \Omega_k C_k = 1 + \sum_k{}' \Omega_k C_k \;; \tag{2b}$$

ECCM

$$|\psi\rangle \equiv e^S|0\rangle \;,\; \langle\tilde{\psi}| \equiv \langle 0|e^{\tilde{\Sigma}} e^{-S} \;, \qquad \tilde{\Sigma} \equiv \sum_k{}' \tilde{\sigma}_k C_k \;; \tag{2c}$$

where a prime on a sum over the set-index k indicates that the term k = 0 is excluded, where, by definition $C_0^\dagger \equiv I$, the identity operator. The three methods are then completely specified by the respective pairs of creation and destruction operators $\{F,\tilde{F}\}$, $\{S,\Omega\}$, and $\{\Sigma,\tilde{\Sigma}\}$, where in the latter (ECCM) case,

$$\Sigma|0\rangle \equiv (I - |0\rangle\langle 0|)\, e^{\tilde{\Sigma}} S|0\rangle \;,\; \Sigma \equiv \sum_k{}' \sigma_k C_k^\dagger \;, \tag{3}$$

or, equivalently, by the complete sets of correlation amplitudes $\{F_k,\tilde{F}_k\}$, $\{S_k,\Omega_k\}$, and $\{\sigma_k,\tilde{\sigma}_k\}$, in terms of which they may be decomposed.

If we denote these three sets of parameters generically as $\{x_k,y_k\}$, we have shown elsewhere[7] that in each case the exact ground-state energy expectation value, $\bar{H} \equiv \bar{H}[x_k,y_k] \equiv \langle\tilde{\psi}|H|\psi\rangle/\langle\tilde{\psi}|\psi\rangle$, is stationary with respect to each member of the set,

$$\partial\bar{H}/\partial x_k = 0 = \partial\bar{H}/\partial y_k \;. \tag{4}$$

These coupled sets of equations (4) determine the ground-state equilibrium values of the respective amplitudes. Only in the case of the CI method are the resulting equations linear. Both CCM parametrizations provide intrinsically nonlinear decompositions of the underlying time-independent Schrödinger equation (1). Furthermore, in a full dynamical description via the time-dependent Schrödinger equation, in which the energy eigenstates $|\psi\rangle$ and $\langle\tilde{\psi}|$ are replaced by their non-stationary counterparts, the equations of motion in all three parametrizations have been shown[7] to have the identical form,

$$i dx_k/dt = \partial\bar{H}/\partial y_k \ , \ -i dy_k/dt = \partial\bar{H}/\partial x_k \ . \tag{5}$$

These are just the canonical equations of motion in a classical (multi-body) phase space, in which the role of canonical co-ordinates is thus played by the multiconfigurational correlation amplitudes themselves.

The primary difference between the three parametrizations lies in their separability properties. Whereas the exact wave operators F and $\tilde{F}$, and the correlation operator Ω are *multiplicatively* separable, the various cluster operators S, Σ, and $\tilde{\Sigma}$ are *additively* separable. The important physical properties of the system are extensive quantities, and these are clearly most directly formulated in terms of parameters which are additively separable even when the full description is approximated by truncation in the multiconfigurational phase space. Furthermore, this additive separability is intrinsically related to the connectivity properties of the associated amplitudes. For example, the coefficients $\{S_k\}$ are just the Hubbard linked-cluster amplitudes which represent the sums of complete sets of connected (or linked) open-ended Goldstone diagrams for the ground state, whereas the comparable coefficients $\{F_k\}$ also contain disconnected (or unlinked) terms. In fact, all of the amplitudes $\{S_k\}$, $\{\sigma_k\}$ and $\{\tilde{\sigma}_k\}$ are linked, and hence obey the cluster property, whereas the amplitudes $\{F_k\}$, $\{\tilde{F}_k\}$ and $\{\Omega_k\}$ contain unlinked pieces. We see, in particular, that only the ECCM thus comprises a parametrization in which *all* of the amplitudes are linked.

The classical phase space for all three parametrizations is in principle of equal complexity. Nevertheless, the increasing amount of connectivity in the chain CI → NCCM → ECCM (which, incidentally, is also related to the various "generalized time-ordering" properties[4,7] associated with the tree-diagram structures which emerge from each method in terms of their time-independent perturbation-theoretic content), is expected to bring about a similar increase in the effective compactification of the respective phase space. In particular, the ECCM phase space is a complex differentiable manifold with a symplectic structure induced by a generalized classical Poisson bracket form between the amplitudes,[6] each of which is now *multilocal* in the sense of obeying the cluster property in co-ordinate space. The associated compactification of the phase space may be expected to allow the physically important region to be described in terms of an effective classical mean (multilocal) field theory of much reduced dimensionality and, hence, complexity.

The above discussion opens the intriguing possibility that a further exploration of the various parametrizations, and the ECCM in particular, might shed new light on the relationships between classical and quantum mechanics and on the quantization of classical systems. Despite the obvious intuitive physical appeal of the ECCM, many open questions remain. Thus, the very existence of the ECCM phase space is far from clear for infinite Hilbert spaces. The related practical problems of the convergence properties of the method (and the associated need for appropriate regularization procedures), and how to devise useful approximation schemes, are also of fundamental importance.

In order to address these issues we examine here the one-body anharmonic oscillator as a much-studied system with infinite-dimensional Hilbert space.

As a model interacting field theory, the anharmonic oscillator is highly singular in the sense of being maximally *nonlocal*. It thus provides a most stringent test for a method such as the ECCM whose linked-cluster properties suit it particularly to physical systems with normal locality and separability properties. Furthermore, it is well-known that ordinary perturbation theory fails to converge for this system, however weak the anharmonicity.[17,18] Nevertheless, the Fock space associated with such one-body systems can be described in terms of (functions of) a single creation operator, $a^\dagger$, and its hermitian adjoint, a.

3. BARGMANN REPRESENTATION; COHERENT STATES

For a one-body problem in one dimension, with Hamiltonian

$$H = \tfrac{1}{2}p^2 + V(x) \ , \tag{6}$$

the position operator x and its canonically conjugate momentum operator $p = -id/dx$, are conventionally mapped into their canonical Fock-space creation and destruction operator counterparts by,

$$x = 2^{-\frac{1}{2}}(a^\dagger + a) \ , \ p = 2^{-\frac{1}{2}}i(a^\dagger - a) \ , \tag{7}$$

which obey the usual bosonic canonical commutation relation (CCR), $[a,a^\dagger] = I$. The corresponding (normalized) vacuum state $|0\rangle$, $a|0\rangle = 0$, which now plays the role of model state or cyclic vector, is simply the ground state of the harmonic oscillator Hamiltonian, $H^0 = \frac{1}{2}p^2 + \frac{1}{2}x^2$. It has the usual coordinate-space representation, $\langle x|0\rangle \equiv \psi^0(x) = \pi^{-\frac{1}{4}} \exp(-\frac{1}{2}x^2)$. Arbitrary ket and bra states in the Fock space can now be represented in the form,

$$|g\rangle \equiv g(a^\dagger)|0\rangle \ , \ \langle f^*| \equiv \langle 0|f(a) = [f^*(a^\dagger)|0\rangle]^\dagger = (|f^*\rangle)^\dagger \ ;$$

$$g(z) \equiv \sum_{n=0}^{\infty} g_n z^n \ , \ f(z) = \sum_{n=0}^{\infty} f_n z^n \ . \tag{8}$$

The normalized eigenstates of the operator a are the usual Glauber coherent states $|z\rangle$,

$$a|z\rangle = z|z\rangle \ ; \ |z\rangle = e^{(za^\dagger - z^*a)}|0\rangle = e^{-\frac{1}{2}|z|^2}e^{za^\dagger}|0\rangle \ , \tag{9}$$

where z is an arbitrary complex number. Their co-ordinate-space representation is readily seen to be given as,

$$\langle x|z\rangle = \pi^{-\frac{1}{4}} \exp(-\tfrac{1}{2}|z|^2 - \tfrac{1}{2}z^2 + \sqrt{2}xz - \tfrac{1}{2}x^2) \ . \tag{10}$$

They obey the well-known completeness and overlap relations,

$$I = \pi^{-1}\int d^2z \ |z\rangle\langle z| = \int_{-\infty}^{\infty} dx \ |x\rangle\langle x| \ , \tag{11}$$

$$\langle z|z'\rangle = \exp(-\tfrac{1}{2}|z|^2 - \tfrac{1}{2}|z'|^2 + z^*z') \ . \tag{12}$$

In the Bargmann space the CCR algebra is represented by the algebra,

$$a^\dagger \rightarrow z \ , \ a \rightarrow d/dz \tag{13}$$

of the complex variable z and its derivative. Arbitrary states $f(a)g(a^\dagger)|0\rangle$ in the Fock space now have the holomorphic Bargmann representation

$f(d/dz)g(z)$. In this way the bosonic quantum field theory in a Hilbert space may be mapped into the corresponding (classical) field theory of a complex function in a particular normed space, the so-called Bargmann Hilbert space.[16] Thus, the ground-state Schrödinger equation (1) with the Hamiltonian of Eq. (6) becomes the following ordinary differential equation in the complex plane,

$$-\tfrac{1}{4}(d/dz - z)^2F(z) + V[(d/dz + z)/\sqrt{2}]F(z) = E_0F(z) , \tag{14}$$

in terms of the CI representation $|\psi\rangle = F(a^\dagger)|0\rangle$ of the ground-state wavefunction.

It is easy to check, using the CCR and Eqs. (9) and (11), that the scalar product of two such arbitrary states as in Eq.(8) can be expressed in any of the following forms,

$$\langle f^* | g\rangle = \pi^{-1}\int d^2z\ e^{-|z|^2} f(z)g(z^*) = \pi^{-1}\int d^2z\ e^{-|z|^2} f(z^*)g(z) \tag{15a}$$

$$= f(d/dz)g(z)\big|_{z=0} \tag{15b}$$

$$= \sum_{n=0}^{\infty} n!f_n g_n . \tag{15c}$$

If the two states are normalizable, both (holomorphic) functions $f(z)$ and $g(z)$ will be entire functions of order $\rho \le 2$, and each of the relations (14a) - (14c) yields the same (convergent) result. Differences will arise only when the Hilbert space of normalizable wavefunctions is extended to a more general linear vector space which may encompass states that are not normalizable in the standard metrics. It is just such an extension which is necessitated by the CCM parametrizations of the state vectors.

4. HOLOMORPHIC REPRESENTATIONS OF THE CI AND CCM PARAMETRIZATIONS FOR SIMPLE BOSONIC FIELD THEORIES

In the case of simple bosonic field theories with a single mode, created by the operator $a^\dagger$, we may therefore now use the Bargmann-space concepts of Sec. 3 to give a holomorphic representation of the various operators defined generally in Eqs. (2a) - (2c) that define our three basic (CI, NCCM and ECCM) parametrizations of Sec. 2. In the first place we consider the detailed mapping between the Schrödinger and Bargmann representations of an *arbitrary* wavefunction,

$$|\psi\rangle = F(a^\dagger)|0\rangle , \ F(z) \equiv \sum_{n=0}^{\infty} F_n z^n , \tag{16}$$

written in its CI form [c.f., Eq. (2a)]. By considering the overlap $\langle z^*|\psi\rangle$, inserting a complete set of position eigenstates $|x\rangle$, and using Eq. (10), it is straightforward to derive the Fourier-like mapping,

$$F(z) = \pi^{-\frac{1}{4}} \exp(-\tfrac{1}{2}z^2)\int_{-\infty}^{\infty} dx\ \exp(\sqrt{2}zx - \tfrac{1}{2}x^2)\psi(x) . \tag{17}$$

By considering the Fourier transform of Eq. (17), or by likewise considering the overlap $\langle x|\psi\rangle$ and inserting a complete set of coherent states as in Eq. (11), one may also derive the inverse relations,

$$\psi(x) = 2^{-\frac{1}{2}}\pi^{-\frac{3}{4}}\exp(\tfrac{1}{2}x^2)\int_{-\infty}^{\infty} d\eta\, \exp(-i\sqrt{2}x\eta - \tfrac{1}{2}\eta^2)\, F(i\eta) \tag{18a}$$

$$= \pi^{-5/4}\, \exp(-\tfrac{1}{2}x^2)\int d^2z\, \exp(-|z|^2 - \tfrac{1}{2}z^{*2} + \sqrt{2}xz^*)\, F(z)\ . \tag{18b}$$

For the further CCM parametrizations it is conventional to impose the so-called *intermediate normalization condition* on $|\psi\rangle$, namely $\langle 0|\psi\rangle = \int dx\psi^0(x)\psi(x) = 1$, which implies that $F_0 = 1$ in Eq. (16). Furthermore, the CCM representations of Eqs. (2b) - (2c) imply that $\langle\tilde{\psi}|\psi\rangle = 1$. In the case of a hermitian Hamiltonian this simply gives $\langle\tilde{\psi}| = N^{-2}\langle\psi|$, where $N^2 \equiv \langle\psi|\psi\rangle = \int dx|\psi(x)|^2 = \langle 0|F^*(a)F(a^\dagger)|0\rangle$, and hence $\tilde{F}(a) = N^{-2}F^*(a)$.

In the CCM, the wave operator $F(a^\dagger)$ is now parametrized in the exponential form,

$$F(a^\dagger) = \exp\, S(a^\dagger)\ ;\ S(z) \equiv \sum_{n=1}^{\infty} S_n z^n\ . \tag{19}$$

The remaining NCCM operator $\Omega(a)$, where $\langle\tilde{\psi}| \equiv \langle 0|\Omega(a)\exp[-S(a^\dagger)]$, is then simply given as,

$$\langle 0|\Omega(a) = N^{-2}\langle 0|F^*(a)F(a^\dagger)\ ;\ \Omega(z) \equiv \sum_{n=0}^{\infty} \Omega_n z^n\ , \tag{20}$$

with $\Omega_0 = 1$. If we define the average value of an arbitrary operator $\mathfrak{O} = \mathfrak{O}(a^\dagger,a)$ as $\langle\mathfrak{O}\rangle \equiv \bar{\mathfrak{O}} \equiv \langle\tilde{\psi}|\mathfrak{O}|\psi\rangle = N^{-2}\langle 0|F^*(a)\mathfrak{O}(a^\dagger,a)F(a^\dagger)|0\rangle$, it is immediately clear from Eq. (20) that the function $\Omega(z)$ is the moment-generating function for the powers of the creation operator $a^\dagger$,

$$\Omega_n = \langle(a^\dagger)^n\rangle/n!\ ,\ \Omega(z) = \langle\exp(za^\dagger)\rangle. \tag{21}$$

By inserting a complete set of co-ordinate states $|x\rangle$ into the representation of Eq. (21) and making use of Eq. (7), we readily find the expression,

$$\Omega(z) = N^{-2}\int_{-\infty}^{\infty} dx\ \psi^*(x)\ \exp[2^{-\frac{1}{2}}z(x - d/dx)]\psi(x)\ . \tag{22}$$

The exponential operator is easily transformed by use of the Baker-Campbell-Hausdorff theorem to yield the explicit representation,

$$\Omega(z) = N^{-2}\ \exp(-\tfrac{1}{4}z^2)\int_{-\infty}^{\infty} dx\ \exp(xz/\sqrt{2})\psi^*(x)\psi(x-z/\sqrt{2})\ , \tag{23}$$

which is bilinear in the Schrödinger wavefunction.

In the ECCM parametrization, the function $\tilde{\Sigma}(z)$ of Eq. (2c) is given by $\Omega(z) = \exp\,\tilde{\Sigma}(z)$, and is hence simply the associated cumulant function whose coefficients provide a measure of the *connected averages*,

$$\tilde{\Sigma}(z) \equiv \sum_{n=1}^{\infty}\tilde{\sigma}_n z^n\ ,\ \tilde{\sigma}_n = \frac{1}{n!}\langle(a^\dagger)^n\rangle_{\text{conn.}}\ . \tag{24}$$

The relationship between the coefficients Ω_n and $\tilde{\sigma}_n$ is thus just the usual one between the moments and cumulants of a probability distribution. Finally, the remaining ECCM function $\Sigma(a^\dagger)$ given by Eq. (3), is readily shown to have the form,

$$\Sigma(z) \equiv \sum_{n=1}^{\infty} \sigma_n z^n = \langle 0|\Omega(a)(e^{za} - 1)S(a^\dagger)|0\rangle,$$

$$\sigma_n = \sum_{m=0}^{\infty} \frac{(m+n)!}{n!} \Omega_m S_{m+n} = \frac{1}{n!} \frac{d^n}{dz^n} \Omega(d/dz)S(z)\Big|_{z=0} . \tag{25}$$

If an arbitrary operator $A(a^\dagger,a)$ and the functions $\exp[\pm S(a^\dagger)]$ are now expressed in their Bargmann form, we readily find, for example, that the NCCM form of the average-value functional $\bar{A}$ has the holomorphic representation,

$$\bar{A} = \Omega(d/dz)A(z,d/dz + S'(z))\Big|_{z=0} , \tag{26}$$

where $S'(z) \equiv dS/dz$. When $|\psi\rangle$ and $\langle\tilde{\psi}|$ are the exact ground eigenstates of $H = H(a^\dagger,a)$, the usual coupled set of nonlinear NCCM equations for the coefficients $\{S_n\}$,

$$\langle 0|a^n e^{-S} H e^{S}|0\rangle = 0 \ ; \ n = 1,2,\ldots , \tag{27}$$

which follows immediately from the Schrödinger equation (1), now simply correspond to the stationarity condition [c.f., Eq. (4)],

$$\partial\bar{H}/\partial\Omega_n = (d/dz)^n \, H(z,d/dz + S'(z))\Big|_{z=0} = 0 \ ; \ n \geq 1 , \tag{28}$$

in terms of the NCCM average-value functional $\bar{H} = \bar{H}(S_n,\Omega_n)$. Similarly, the remaining NCCM equations for the coefficients $\{\Omega_n\}$,

$$\langle 0|\Omega e^{-S}[H,(a^\dagger)^n]e^{S}|0\rangle = 0 \ ; \ n = 1,2,\ldots , \tag{29}$$

correspond to the stationarity condition,

$$\partial\bar{H}/\partial S_n = \Omega(d/dz)[H(z,d/dz + S'(z)),z^n]\Big|_{z=0} = 0 \ ; \ n \geq 1 . \tag{30}$$

The CI and ECCM stationarity equations may be comparably expressed. It is clear from the above discussion that it will be useful to study the analytic properties of the various functions $F(z)$, $S(z)$, $\Omega(z)$, $\Sigma(z)$ and $\tilde{\Sigma}(z)$ at the exact ground-state stationary point and in its vicinity. In order to make further progress it is simplest to treat a specific example.

5. EXAMPLE: THE ANHARMONIC OSCILLATOR

We now consider the anharmonic oscillator model, corresponding to Eqs. (1) and (6) with

$$V(x) = \tfrac{1}{2}\lambda x^{2K} , \ K = 2,3,\ldots . \tag{31}$$

If we write $F(z) \equiv \exp(-\tfrac{1}{2}z^2)f(z)$, Eq. (14) shows that the function $f(z)$ satisfies the differential equation,

$$\left(\frac{d}{dz}\right)^{2K} f(z) = \frac{2^{K+1}}{\lambda} \left[\left(\frac{1}{2}\frac{d}{dz} - z\right)^2 + E_0\right] f(z) . \tag{32}$$

By using Eq. (17) and the fact that the ground-state wavefunction $\psi(x)$ is real, positive, and of even parity, we see easily that $f(z)$ is an even entire function of z which is real and positive on the real axis. Equation (32) shows that in the asymptotic region where $r = |z| \to \infty$, there are 2K qualitatively different solutions,

$$f_{(n)}(z) \to \exp(b\varepsilon^n z^\nu) \; ; \; n = -K+1, \ldots, K ,$$

$$\nu \equiv \frac{K+1}{K} , \; \varepsilon \equiv \exp\left(\frac{i\pi}{K}\right) , \; b \equiv \frac{1}{\nu} 2^{\nu/2} \lambda^{-(\nu-1)/2} , \tag{33}$$

and hence that f(z) is an entire function of order ν. The normalizability of the ground-state wavefunction $\psi(x)$ implies from Eqs. (15a) and (16) that the product $\exp(-|z|^2)|F(z)|^2$ is integrable over the whole complex plane. Hence, for real x and y, we have $|f(x + iy)| \to o(1)$ as $y \to \pm\infty$. Thus, for example, in the first quadrant, $0 \le \arg(z) < \frac{1}{2}\pi$, of the complex plane, only the K asymptotic branches of Eq. (33) with n = 0,1, ..., K−1 are permitted, and in the fourth quadrant, $-\frac{1}{2}\pi < \arg(z) \le 0$, only the K branches with n = 0,−1, ..., −K+1. By matching the various branches, using the usual WKB arguments, one can show that the actual asymptotic behaviour is given by,

$$f(z) \xrightarrow[r \to \infty]{} A \exp[b(\pm z)^\nu] , \; |\arg(\pm z)| < \tfrac{1}{2}\pi . \tag{34}$$

In order to investigate the analytic behaviour of the CCM function S(z) = ln F(z), we are also interested in the zeros z_m of F(z), $F(z_m) = 0$. Similar WKB arguments to those used above show that on the imaginary axis the asymptotic solutions of Eq. (34) in the left and right half-planes are of equal magnitude and can generate an infinite sequence of zeros on the imaginary axis, with asymptotic positions at $z_m = \pm iy_m$, where

$$y_m \xrightarrow[m \to \infty]{} \left[\frac{\pi}{b} \sec\left(\frac{\pi}{2K}\right)\right]^{1/\nu} m^{1/\nu} . \tag{35}$$

It also seems probable that *all* of the zeros lie on the imaginary axis.

In order to examine the analytic properties of the NCCM function $\Omega(z)$, we may use the representation of Eq. (23). It is easy to show that $\psi(x)$ can be analytically continued into the complex plane where it is an entire function of order K+1. One may then use its asymptotic behaviour in Eq. (23) to show that $\Omega(z)$ is also an entire function of order K+1, with asymptotic behaviour $\Omega(x) \sim \exp(-k\sqrt{\lambda}|x|^{K+1})$, where k is a constant, as $|x| \to \infty$. Equation (23) also shows that $\Omega(z)$ has no zeros on the real axis.

We may now use the Hadamard decomposition theorem and the fact that f(z) is an entire function of order $1 < \nu < 2$, to yield the representations,

$$f(z) = \prod_m (1-z/z_m)\exp(z/z_m) , \; S(z) = -\tfrac{1}{2}z^2 + \sum_m [\ln(1-z/z_m) + z/z_m] , \tag{36}$$

where the product and sum run over the infinite sequence $\{z_m\}$ of zeros of f(z). It is now immediately clear, for example, that the state $S(a^\dagger)|0\rangle$ is *not normalizable* within the Hilbert space. One consequence is that the CCM operator $S(a^\dagger)$ cannot simply be reckoned to be an operator which generates "small" correlations in some suitably defined perturbative sense, even when one might naively intuit that this is possible (e.g., in the case of small coupling constants, λ). By comparing Eqs. (19) and (36), and recalling that f(z) = f(−z), and hence that S(z) = S(−z), we find

$$S_{2n} = -\tfrac{1}{2}\delta_{n,1} - \frac{1}{2n}\sum_m z_m^{-2n} , \; n \ge 1 . \tag{37}$$

Equation (35) shows that the sum in Eq. (37) is convergent for all $n \ge 1$, and that the coefficients for high n are determined only by the two zeros, $\pm z_1$

say, closest to the origin. Thus, if $z_1 = i\rho$, the sequence becomes asymptotically geometric,

$$S_{2n} \xrightarrow[n \to \infty]{} \frac{(-1)^{n-1}}{n} \rho^{-2n} . \tag{38}$$

Turning now to the ECCM function $\Sigma(z)$, another consequence of the non-normalizability of the function $S(a^\dagger)|0\rangle$ is that formal expressions such as those in Eq. (25) must be handled with extreme caution. For example, the infinite sum for σ_n is easily seen to be formally divergent. Among the several techniques that may be applied to give an interpretation to these divergent or ill-defined (non-unique) expressions, we mention that Fourier and Borel resummation methods may respectively be applied to the definitions (15b) and (15c) of the general scalar product of the arbitrary functions $\langle f^*|$ and $|g\rangle$ of Eq. (8). For example, in the latter case we may write

$$\langle 0|f(a)g(a^\dagger)|0\rangle = \int_0^{\infty.\eta} dw\ e^{-w}\langle 0|f(a)g_B(wa^\dagger)|0\rangle , \tag{39}$$

where the integration variable w is in the direction of the unit vector $\eta \equiv \exp(i\varphi)$ with $|\varphi| < \pi/2$, and where

$$g_B(z) \equiv \sum_{n=0}^{\infty} \frac{1}{n!} g_n z^n \tag{40}$$

is the Borel transform of g(z) given by Eq. (8).

Such techniques may now be employed to evaluate the ECCM function $\Sigma(z)$, calculated in terms of its first derivative from Eq. (25) as,

$$\Sigma'(z) = \langle 0|\Omega(a)e^{za}S'(a^\dagger)|0\rangle = \Omega(d/du)S'(u+z)|_{u=0} , \tag{41}$$

in which the multi-valued function S(z) has been replaced by its single-valued first derivative S′(z). Thus, given that the Borel transform of the function $g(z) = 1/(z-c)$ is $g_B(z) = -c^{-1}\exp(z/c)$, and that for an arbitrary entire function f(z) one has the simple relation,

$$f(a)\exp(\xi a^\dagger) = \exp(\xi a^\dagger)f(a+\xi) , \tag{42}$$

where ξ is an arbitrary constant, we find the important relation,

$$\langle 0|f(a)\frac{1}{(a^\dagger-c)}|0\rangle = -\frac{1}{c}\int_0^{\infty.\eta} dw\ e^{-w}f(w/c) , \tag{43}$$

for a class of functions f(z). The direction of integration must be chosen so that the integral converges. Within the half-plane $|\arg(\eta)| < \pi/2$ there may be several disjoint sectors in which this occurs, *in which case the scalar product would not be uniquely defined.*

We may now apply Eq. (43) to Eq. (41), with S(z) given by the representation of Eq. (36). The asymptotic behaviour of the entire function $\Omega(z)$ and the distribution of the poles of S′(z) [and see Eq. (35)] together imply that the function $\Sigma(z)$ does not have a unique representation. Upon expansion, each such solution does however lead to precisely the same (divergent) sum of Eq. (25). The existence of such a multiplicity of solutions must ultimately be caused by the nonlinear CCM decomposition of the original linear Schrödinger eigenvalue problem. On the other hand, it is still possible to impose such physical constraints as that the ground-state wavefunction $\psi(x)$ be real and of even parity (which imply in turn that in Eq. (25) the coefficients σ_n are both real and non-zero only for n even), in order to construct a unique $\Sigma(z)$.

We omit the somewhat involved details of the remaining derivation and quote only the result, in terms of the odd function R(x) defined, for $x \in \mathbb{R}$, as

$$R(x) \equiv \delta'(x) - \tfrac{1}{2}\mathrm{sgn}(x)\sum_m \exp(-xz_m) \,, \tag{44}$$

where $\delta(x)$ is the Dirac delta. From the asymptotic distribution of the zeros $\{z_m\}$ of F(z) from Eq. (35), it is clear that R(x) is to be understood as a generalized function or distribution, in the usual Schwartz sense. Thus, the sum in Eq. (44) is formally divergent and needs to be interpreted. The sum is first partitioned into terms which are analytic in either the upper or lower complex x-plane, and the function R(x) is then defined in terms of the boundary values of these sectionally holomorphic functions on the real axis. In this way, we derive our final expressions for S(z) and $\Sigma(z)$ in terms of their first derivatives as,

$$S'(z) = \int_{-\infty}^{\infty} dx\; R(x)\,\sinh(xz) \;,\; \Sigma'(z) = \int_{-\infty}^{\infty} dx\; \rho(x)\,\sinh(xz) \,, \tag{45}$$

where $\rho(x) \equiv R(x)\Omega(x)$. Equivalently, we have that S(z) and $\Sigma(z)$ are even functions with non-vanishing coefficients,

$$S_{2n} = \frac{1}{(2n)!}\int_{-\infty}^{\infty} dx\; R(x)x^{2n-1} \;,\; \sigma_{2n} = \frac{1}{(2n)!}\int_{-\infty}^{\infty} dx\; \rho(x)x^{2n-1} \,, \tag{46}$$

given in terms of the moments of the respective distributions R(x) and $\rho(x)$. It is clear that whereas $\Sigma(z)$ is defined via Eq. (45) as an entire function, S(z) is only similarly defined in an infinitesimal region around the origin and on the imaginary axis, from where it may, however, be analytically continued.

6. ARBITRARY EXPECTATION VALUES: A GENERATING FUNCTION

We turn finally to the question of calculating the exact ground-state expectation value of an arbitrary operator in terms of either set of complete CCM coefficients, namely $\{\Omega_n, S_n\}$ for the NCCM and $\{\sigma_n, \tilde{\sigma}_n\}$ for the ECCM. To be specific, let us consider an arbitrary operator *in normal-ordered form*, $\mathfrak{N} = \mathfrak{N}(a^\dagger, a) \equiv :\mathfrak{N}(a^\dagger, a):$. Its expectation value,

$$\bar{\mathfrak{N}} \equiv \langle \mathfrak{N}(a^\dagger, a)\rangle = \langle 0|\Omega(a)e^{-S(a^\dagger)}\mathfrak{N}(a^\dagger, a)e^{S(a^\dagger)}|0\rangle \,, \tag{47}$$

may then be given in terms of a generating function A(u,v) as follows,

$$\bar{\mathfrak{N}} = \mathfrak{N}\left(\frac{\partial}{\partial u}, \frac{\partial}{\partial v}\right) A(u,v)\Big|_{u=v=0} \;;\; A(u,v) \equiv \langle e^{ua^\dagger} e^{va}\rangle \,. \tag{48}$$

By making use of Eq. (42), we readily derive the relation,

$$A(u,v) \equiv \langle 0|\Omega(a+u)\,\exp\left[\int_0^v dw S'(a+w)\right]|0\rangle \,. \tag{49}$$

By expanding the exponential as a power series, by using the explicit representation of Eq. (45), and by using the relation (42), it is then a simple matter to prove the respective NCCM and ECCM parametrizations of the expectation-value generating function,

$$A(u,v) = \sum_{n=0}^{\infty} \frac{1}{n!} \prod_{i=1}^{n} \left[\int_{-\infty}^{\infty} dx_i \; R(x_i) \left(\frac{e^{vx_i}-1}{x_i}\right)\right] \Omega(u + x_1 + \ldots + x_n) \;, \tag{50a}$$

$$= \sum_{n=0}^{\infty} \frac{1}{n!} \prod_{i=1}^{n} \left[\int_{-\infty}^{\infty} dx_i \; \rho(x_i) \left(\frac{e^{vx_i}-1}{x_i}\right)\right]$$

$$\times \exp\left[\tilde{\Sigma}(u + x_1 + \ldots + x_n) - \tilde{\Sigma}(x_1) - \ldots - \tilde{\Sigma}(x_n)\right] . \tag{50b}$$

By using the expansions (20) and (24) together with Eq. (46), it is clear that Eqs. (48) and (50) now permit an expansion of the average-value functional in either the NCCM form $\bar{\mathfrak{A}} = \bar{\mathfrak{A}}(S_n,\Omega_n)$ or the ECCM form $\bar{\mathfrak{A}} = \bar{\mathfrak{A}}(\sigma_n,\tilde{\sigma}_n)$. In both cases we have infinite-order multinomials of the respective sets of coefficients. On the other hand Eqs. (50a) and (50b) represent closed-form expressions for $\bar{\mathfrak{A}}$ as a functional of only two functions, namely $\bar{\mathfrak{A}} = \bar{\mathfrak{A}}[R(x),\Omega(x)]$ for the NCCM and $\bar{\mathfrak{A}} = \bar{\mathfrak{A}}[\rho(x),\tilde{\Sigma}(x)]$ for the ECCM. For some proper classes of these two pairs of functions, the integral expressions from Eqs. (48) and (50) are clearly convergent, whereas by contrast the convergence properties of the expansions in terms of the respective coefficient are far from clear. On the other hand, these expansions may be given a diagrammatic representation in the usual way, and it is clear that their precise linking and connectivity properties are a direct reflection of the more well-known linked (L) and double-linked (DL) expansions for the NCCM and ECCM average-value functionals of an arbitrary operator $\mathfrak{O}$, that have been described more fully elsewhere,[2,4,6] and which may be given in the respective operator forms,

$$\bar{\mathfrak{O}} = \sum_{n=0}^{\infty} \frac{1}{n!} \langle 0 | \Omega (\mathfrak{O} S^n)_L | 0 \rangle \tag{51a}$$

$$= \sum_{n=0}^{\infty} \frac{1}{n!} \langle 0 | e^{\tilde{\Sigma}} (\mathfrak{O} \Sigma^n)_L | 0 \rangle_{DL} \; . \tag{51b}$$

As a specific illustration we consider, for example, the terms proportional to $\sigma_2\sigma_4\tilde{\sigma}_2\tilde{\sigma}_4$ in the ECCM expansion of the expectation value $\langle (a^\dagger)^2 a^2 \rangle$, which we denote by $\mathcal{T}$, say. From Eq. (50) we have

$$\langle e^{ua^\dagger} a^2 \rangle = \int_{-\infty}^{\infty} dx_1 \; x_1 \rho(x_1) e^{\tilde{\Sigma}(u+x_1)-\tilde{\Sigma}(x_1)}$$

$$+ \int_{-\infty}^{\infty} dx_1 \int_{-\infty}^{\infty} dx_2 \; \rho(x_1)\rho(x_2) e^{\tilde{\Sigma}(u+x_1+x_2)-\tilde{\Sigma}(x_1)-\tilde{\Sigma}(x_2)} \; . \tag{52}$$

Obviously the terms $\mathcal{T}$ arise from the second term in Eq. (52). By expanding out the exponential in the integrand, and keeping terms of order u^2, we find that the terms $\mathcal{T}$ of interest arise from the term,

$$\langle (a^\dagger)^2 a^2 \rangle_{\mathcal{T}'} = 2\tilde{\sigma}_2\tilde{\sigma}_4 \int_{-\infty}^{\infty} dx_1 \int_{-\infty}^{\infty} dx_2 \; \rho(x_1)\rho(x_2) [\{(x_1+x_2)^4 - x_1^{\,4} - x_2^{\,4}\}$$

$$+ \{6(x_1+x_2)^2[(x_1+x_2)^2 - x_1^{\,2} - x_2^{\,2}]\} + \{8(x_1+x_2)^4\}] . \tag{53}$$

In particular, Eq. (46) shows that the terms $\mathcal{T}$ arise from the terms in the integrand of Eq. (53) proportional to $x_1^3 x_2$ or $x_1 x_2^3$. The three terms in braces in Eq. (53) lead to three contributions labelled (a), (b) and (c) respectively with specific numerical values $96k_i\sigma_2\sigma_4\tilde{\sigma}_2\tilde{\sigma}_4$, where $k_a = 8$, $k_b = 24$, $k_c = 64$. Equation (53) also shows that the connectivity properties of the associated diagrams are as shown in Figs. (1a) – (1c) respectively, where the diagram (c) should also include an (unshown) "exchange contribution" to itself obtained by reconnecting the line joining $\tilde{\sigma}_2$ to σ_2 so that it connects $\tilde{\sigma}_2$ to σ_4 and simultaneously reconnecting any one of the lines joining $\tilde{\sigma}_4$ to σ_4 so that it connects $\tilde{\sigma}_4$ to σ_2.

7. DISCUSSION

It should be clear from the above discussion that we have succeeded in achieving a *complete algebraization* of the CI and CCM parametrizations and their various important properties, by utilising the holomorphic representations for our model field theories in their respective Bargmann Hilbert spaces. Most particularly, the abstract topological linking requirements related to the vertex structure of the corresponding tree diagrams associated with their respective expansions, have been especially illuminated. It should be emphasized that although we have illustrated the techniques by reference to the anharmonic oscillator problem, many of the final results of Secs. 5 and 6 are much more general. This is especially true of the form of the holomorphic representation of the CCM amplitudes in terms of the distribution of zeros for the CI wavefunction F(z) [and see Eq. (45)], and the corresponding form of the generating function for arbitrary expectation values [and see Eq. (50)].

We have shown how the linked-cluster amplitudes $\{S_n\}$ and $\{\sigma_n\}$ may be represented as the moments of certain Schwartz distributions, which themselves depend on the distribution of the zeros of the holomorphic wavefunction F(z).

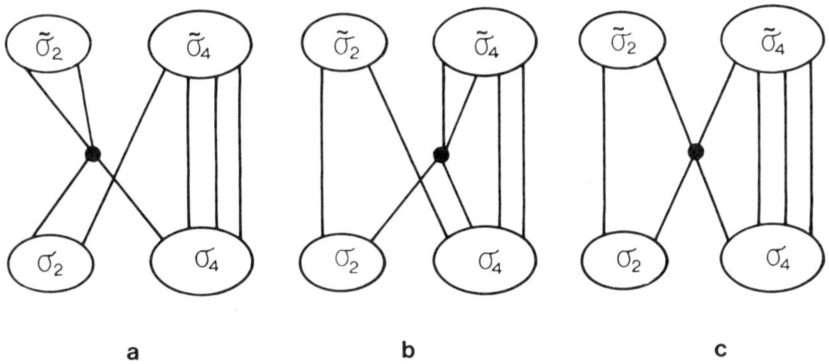

Fig. 1. Diagrammatic representation of some terms in the ECCM expansion of the expectation value $\langle (a^\dagger)^2 a^2 \rangle$.

Whereas the amplitudes $\{S_n\}$ decrease asymptotically as a geometric series, the amplitudes $\{\sigma_n\}$ decrease much more rapidly in the same large-n limit. This has the consequence that their practical calculation is extremely difficult. In turn, this difference in the asymptotic behaviour of the two sets of amplitudes is itself a consequence of the fact that the ECCM function $\Sigma(z)$ is a holomorphic function, whereas its NCCM counterpart $S(z)$ is nonanalytic since it has an infinite number of branch points.

In practical implementations of the CI method or either version of the CCM, one must approximate. In each case, the natural truncation scheme is the so-called SUB(N) hierarchy in which the respective infinite amplitude expansions (in the appropriate configuration space) of the fundamental operators described in Eqs. (2a) - (2c), are truncated at the level where the set-index k contains no more than N single-particle (or single pairs of particle-hole) labels. For our simple model field theories considered here, the otherwise infinite expansions of Eqs. (16), (19), (20), (24) and (25) are simply curtailed as finite N^{th}-order polynomials. The potentially hazardous nature of these truncations is now easy to see. Thus, we know that as $|z| \to \infty$, the asymptotic behaviour of $F(z)$ is as given in Eq. (34), where $\nu < 2$ for the anharmonic oscillator models. Nevertheless, in the NCCM representation, for example, $F(z)$ is replaced in the SUB(N) approximation by a function $F_{(N)}(z)$ whose asymptotic behaviour is proportional to $\exp\ (S_N z^N)$ as $|z| \to \infty$. Clearly, the corresponding approximated wavefunctions $F_{(N)}(a^{\dagger})|0\rangle$ are no longer normalizable if $N > 2$. Hence, such NCCM approximation schemes automatically involve excursions out of the usual Hilbert space.

Similarly, whereas the exact NCCM parametrization requires the function $S(z)$ to satisfy Eq. (28) for all n, namely to satisfy the condition that the function $\tilde{H}(z) \equiv H(z, d/dz + S'(z))$ be a constant $(= E_0)$ everywhere, the SUB(N) approximation merely requires that its first N derivatives vanish at the origin. Thus, $\tilde{H}(z)$ is ultimately represented in the SUB(N) approximation by a finite-order polynomial, and it is clear that although the behaviour near the origin may be very well represented, the asymptotic behaviour is seriously in error. Since the physically important region is in the immediate vicinity of the origin, we may thus easily understand both why the SUB(N) scheme suffers from such seemingly serious drawbacks as lack of normalizability, and yet can be extremely accurate in practice, even at very low levels of the truncation index N.

In conclusion, it is interesting to enquire how the formalism as developed here for simple model field theories with a single creation operator $a^{\dagger}$, might be extended to local continuum field theories where $a^{\dagger} \to a^{\dagger}(x)$. In such cases the complex variable z must presumably be replaced by a complex field $z(x)$, and the holomorphic wavefunction $F(z)$ and its zeros by a corresponding wavefunctional $F[z(x)]$ and its zero functions. Whether a practical formulation of such concepts can be realized, remains a goal for the future. In any case, we believe that the present work has already paved the way to a deeper understanding of both formulations of the CCM.

ACKNOWLEDGEMENT

One of us (RFB) gratefully acknowledges support for this work in the form of a research grant from the Science and Engineering Research Council of Great Britain.

REFERENCES

1. F Coester, Nucl. Phys. 7: 421 (1958);
   F Coester and H G Kümmel, Nucl. Phys. 17: 477 (1960).
2. H G Kümmel, K H Lührmann and J G Zabolitzky, Phys. Rep. 36C: 1 (1978).
3. R F Bishop and K H Lührmann, Phys. Rev. B 17: 3757 (1978).
4. J Arponen, Ann. Phys. (NY) 151: 311 (1983).
5. R F Bishop and H G Kümmel, Phys. Today 40 (No. 3): 52 (1987).
6. J S Arponen, R F Bishop and E Pajanne, Phys. Rev. A 36: 2519 (1987); *ibid.* 2539 (1987).
7. J Arponen, R F Bishop and E Pajanne, in: "Condensed Matter Theories", Vol. 2, P Vashishta, R K Kalia and R F Bishop (eds.), Plenum, New York (1987), p. 357.
8. R F Bishop and K H Lührmann, Phys. Rev. B 26: 5523 (1982).
9. H G Kümmel, in: "Nucleon-Nucleon Interaction and Nuclear Many-Body Problems", S S Wu and T T S Kuo (eds.), World Scientific, Singapore (1984), p. 46.
10. H Kümmel, in: "Recent Progress in Many-Body Theories", Vol. 1, A J Kallio, E Pajanne and R F Bishop (eds.), Plenum, New York (1988), p. 265.
11. J Arponen, R F Bishop, E Pajanne and N I Robinson, Phys. Rev. A 37: 1065 (1988).
12. J Noga, S A Kucharski and R J Bartlett, J. Chem. Phys. 90: 3399 (1989).
13. R K Nesbet, Phys. Rev. 109: 1632 (1958).
14. R Chowdhuri, D Mukherjee and M D Prasad, in: "Aspects of Many-Body Effects in Molecules and Extended Systems", D Mukherjee (ed.), Lecture Notes in Chemistry Vol. 50, Springer, Berlin (1989), p. 3.
15. J Arponen, R F Bishop and E Pajanne, in: "Condensed Matter Theories", Vol. 2, P Vashishta, R K Kalia and R F Bishop (eds.), Plenum, New York (1987), p. 373.
16. V Bargmann, Comm. Pure Appl. Math. 14: 180, 187, (1961); *ibid.* 20: 1 (1967); Rep. Math. Phys. 2: 221 (1971).
17. C M Bender and T T S Wu, Phys. Rev. 184: 1231 (1969); Phys. Rev. D 7: 1620 (1973).
18. B Simon, Ann. Phys. (NY) 58: 76 (1970).

FERMION MONTE CARLO ALGORITHMS FOR QUANTUM FLUIDS

R. M. Panoff

Kinard Laboratory of Physics
Clemson University
Clemson, SC 29634-1911

I. INTRODUCTION

Accurate first-principles calculations of the macroscopic properties of quantum fluids from microscopic interactions remain a key focus of many-body theorists.[1] Our approach can be presented in summary fashion by saying that we seek to solve the Schrödinger equation:

$$H\psi_0 = E\psi_0, \tag{1}$$

to obtain the ground-state wave function ψ_0 for a given Hamiltonian H whose energy eigenvalue is E. For anyone familiar with many-body approaches to this problem, it is clear that we are presented with a problem of shocking proportions: we are seeking to solve one equation which has (at least) four unknowns:

1. We generally do not know the correct Hamiltonian which corresponds to material reality.
2. We generally do not know in closed form the ground-state wave function.
3. We generally do not know the energy eigenvalue for the exact Hamiltonian.
4. We generally do not know how to proceed with the calculation of the many-dimensional integrals inherent in the problem.

The history of where to begin and how to proceed with narrowing the number of unknowns in the problem, then, traces the history of many-body theory over the past half century.[2]

The particular quantum systems of interest have changed somewhat over the years. Considerable early effort went into attempting to calculate the saturation properties of nuclear and neutron matter before it became clear that there was sufficient uncertainty in the description of the microscopic interaction to render discrepancies between theory and experiment irreconcilable.[3] As computational methods were devised – especially in the field of clever approximations – it became clear that some unifying assumptions had to be made about the interaction in order to allow for comparison between competing calculations. Even after agreement was reached among calculational schemes, there remained the question of the comparison between theory and experiment.

Recent Progress in Many-Body Theories, Volume 2
Edited by Y. Avishai
Plenum Press, New York, 1990

Since many of the approximations introduced into the calculations arose because of the state-dependence of the nuclear interaction, some theorists moved their attention to the terrestial quantum fluids,[4-7] with liquid Helium-3 and Helium-4 as the archetypal fermion and boson systems, respectively. By such a move, at least the uncertainty in the Hamiltonian could be greatly reduced. In such a context, describing the ground-state properties of the helium liquids, quantum Monte Carlo first proved itself as one of the most powerful tools of the trade. The work of Malvin H. Kalos and co-workers[6-28] in the development and refinement of both variational and Green's Function Monte Carlo methods, and their application to the helium systems, stand as major achievements and as an important demonstration of the ability of large-scale computational science to impact the directions of theoretical physics.

Application of quantum Monte Carlo methods to boson systems such as ^{4}He, though requiring significant care and sophistication, can be considered to be rather straightforward compared to the last twenty years of attempts to apply these methods to fermion systems. Since only a positive probability distribution can be represented as a list of walker coordinates – the key results of a quntum Monte Carlo calculation are sets of configurations of particle coordinates drawn from a specified distribution – difficulties arise which are associated with the antisymmetric nature of the fermion wave function. These difficulties are well known in the many areas of physics which have applied quantum Monte Carlo, from atomic and condensed matter physics back to the nuclear matter problem which gave rise to the many-body discipline. Variational Monte Carlo (VMC) avoids the problem by dealing with distributions of particle/walker coordinates drawn from a distribution proportional to the *square* of the wave function. In Green's Function Monte Carlo (GFMC) calculations, however, the wave function itself must be sampled, and the the statistical error increases exponentially as the calculation proceeds. In this brief review, since a desired goal is to return to the nuclear roots of our field, I shall concentrate on displaying a comparison of methods,[19,27] and results from their application to liquid ^{3}He, which have been designed recently to reduce or eliminate this so-called fermion problem.

II. GOING BEYOND GROUND-STATE ENERGY CALCULATIONS

Before proceeding to explore fermion Monte Carlo algorithms, perhaps it would be useful to motivate the large-scale, computationally intensive effort by listing our goals and taking the simplest theoretical and calculational path which will realize them. Certainly on the list of qualities of a "good" theoretical description of a system, bulk atomic liquid ^{3}He, for example, we would find the following:

- Agrees well with experiment
- No adjustable parameters
- Computationally compact
- No uncontrolled approximations
- Remains a variational upperbound
- Model independent

Suppose this were a complete description of an ideal theoretical treatment, and we only wanted to calculate the ground-state energy as the measure of how well our theory does. Then I would propose the following theory as one which certainly qualifies and is infinitely simpler than the supercomputer implementation of any competing many-body calculational scheme:

Nature has fixed the ground-state energy per particle of liquid ^{3}He to be:

$$E_{nature} = -\frac{\pi^2}{4}\text{K} = -2.465\text{K}$$

which, when compared to the experimental value of -2.47 K, can be seen to fit each of the criteria on the list above. Kelvin turn out to be merely the appropriate units at equilibrium density for such a system. So what is wrong with such a simple theory? It lacks *insight*. Being able to calculate the ground-state energy of a system, then, isn't much of an accomplishment unless there is also some understanding as to *why* the theory agrees (or fails to agree) with experimental results.

It is that understanding, which starts from and returns to the microscopic description of the interacting system, which we seek, and failing to find it in the simple theory above, or in certain phenomenological calculations, or with methods which have an undesired level of uncontrolled approximation, we are led to adapt quantum Monte Carlo methods to generate our description of the ground-state properties of a many-fermion system.

III. THE MICROSCOPIC MODEL

A non-relativistic system of N helium atoms which interacts via static two-body forces only has a Hamiltonian of the form

$$H = \sum_{i=1}^{N} -\frac{\hbar^2}{2m}\nabla_i^2 + \sum_{1\leq i<j\leq N} v(r_{ij}). \tag{2}$$

The pair interaction $v(r_{ij})$ for the helium systems is taken to be the HFDHE2 interaction of Aziz, *et al.* [29] which for ^{4}He yields excellent agreement with experiment.[13] For two reasons, accurate variational calculations yielding a good ground-state trial function are an essential preliminary to practical and reliable GFMC calculations. In the first place, improved trial wave functions significantly reduce the statistical error by initiating the iterative calculation nearer the ground state. Moreover, an improved importance function acts to suppress population fluctuations. A good trial wave function is especially crucial to a successful attack on the fermion problem.

A physically motivated trial wave function of modified Feenberg form[2] is:

$$\Psi_T = \psi_3 \det(\exp[i\vec{k}_i \cdot (\vec{r}_j + \sum_{l\neq j} \eta(r_{lj})\vec{r}_{lj})]), \tag{3}$$

where

$$\psi_3 = \exp\left[-\frac{1}{2}\sum_{i<j} \tilde{u}(r_{ij}) - \frac{\lambda_T}{4}\sum_{l}\sum_{i<j\neq l} \xi(r_{li})\xi(r_{lj})\vec{r}_{li}\cdot\vec{r}_{lj}\right], \tag{4}$$

and

$$\tilde{u}(r) = u(r) - \lambda_T \xi^2(r) r^2. \tag{5}$$

This *ansatz* includes explicit two- and three-body correlations; the state-dependent correlations incorporate "backflow" effects.[10–12] The two-body correlation $u(r_{ij})$ is an appropriately scaled solution of an Euler-Lagrange equation obtained from Fermi hypernetted chain methods[30] or the solution of a parametrized two-body Schrödinger-like equation.[31] The triplet correlations are parametrized as

$$\xi(r) = \exp\left[-\left(\frac{r - r_T}{w_T}\right)^2\right] \tag{6}$$

and the momentum-dependent correlations are introduced through

$$\eta(r) = \lambda_B \exp\left[-\left(\frac{r - r_B}{w_B}\right)^2\right] + \frac{\lambda'_B}{r^3}. \tag{7}$$

The wave function is required to be periodic, so all correlations should smooth to zero at a distance equal to half the side of the simulation volume, r_{max}. One way of accomplishing this is to make the replacement:

$$f(r) \rightarrow f(r) + f(2r_{max} - r) - 2f(r_{max}) \tag{8}$$

for each of the correlation functions. Since the correlations which minimize the variational energy for liquid ^{3}He are relatively short ranged, this finite-size adaptation has no real effect on the magnitude of the correlations. Longer range correlations may be studied with large numbers of particles in increasing larger periodic boxes, but the calculations which we have done to date have concentrated on 54 particles. Finite-size effects, though not neglible, are believed to be small enough that the fundamental physics may be displayed by these quantum simulations.

In variational calculations, the expectation value of the Hamiltonian, an upper bound to the ground-state energy E_0, may be evaluated as

$$\begin{aligned} E_0 &\leq \frac{\int \psi_T^*(R) H \psi_T(R) dR}{\int \psi_T^*(R) \psi_T(R) dR} \\ &= \int \frac{H \psi_T^*(R)}{\psi_T^*(R)} \frac{\psi_T^*(R) \psi_T(R)}{\int \psi_T^*(R') \psi_T(R') dR'} dR \end{aligned} . \tag{9}$$

The second form of the integral is easily calculated using the method of Metropolis *et al.* [32], usually with a low variance answer since $H\psi_T^*(R)/\psi_T^*(R)$ is nearly constant if $\psi_T(R)$ is close to the true ground-state wave function. A good trial wave function minimizes Eq. (9) and its variance.

The optimum set of variational parameters[27] obtained for a system of 54 particles at the experimental equilibrium density for the trial function above are: $\lambda_T = -1.80$, $r_T = 0.66\sigma$, $w_T = 0.50\sigma$, $\lambda_B = 0.14$, $r_B = 0.74\sigma$, $w_B = 0.54\sigma$, and $\lambda'_B = 0.15\sigma^3$. Although the functional forms of the correlation functions are similar to those used in integral-equation studies, there are significant differences between the optimum parameters obtained with the two methods.[23,33] In the Monte Carlo calculations, no approximations are necessary in evaluating the variational upper bound, subject to the *caveat* that the calculation refers to a finite number of particles with periodic boundary conditions.

IV. A MIRROR POTENTIAL FRAMEWORK

Employing the accurate variational wave function as an importance function, the Green's function Monte Carlo method[4,6,8,9,13,17,21,22,25] is used to calculate ground-state properties of the system. The GFMC method is based upon the fact that the Schrödinger equation in imaginary time is equivalent to a diffusion equation in real time. The GFMC algorithm is implemented by choosing a set of points $\{R\}$ in configuration space and iterating the equation:

$$\Psi^{n+1}(R) = E_T \int G(R, R')\Psi^n(R')dR'. \tag{10}$$

Although the Green's function is not known analytically, it can be sampled exactly through the use of an ancillary random walk. The GFMC method has been applied to macroscopic bosonic systems, such as liquid ^{4}He, with great success. It has also been applied to few-body fermion systems where elaborate sampling schemes are possible because the dimensionality of the system is small.

Exact GFMC treatment of many-body fermion systems, however, has proven to be very elusive. The antisymmetry requirement is a global property that is difficult to incorporate successfully into the diffusion algorithm, which is local in character. The fact that the wave function is not positive definite introduces a statistical error which grows exponentially as Eq. (10) is iterated. Fermion algorithms typically introduce two positive populations, the difference of which corresponds to the desired antisymmetric wave function:

$$\Psi^A = \Psi^+ - \Psi^-. \tag{11}$$

One class of fermion algorithms introduces pair-wise filtering of Ψ^+'s and Ψ^-'s to reduce the variance by controlling the growth of the symmetric noise.[15] A wider set of fermion GFMC algorithms can be viewed in the framework of mirror potentials as shown by Carlson and Kalos.[19] By introducing two coupled equations,

$$[H(R) + c(R)\Psi^+(R)]\Psi^-(R) = E\Psi^-(R), \tag{12a}$$

$$[H(R) + c(R)\Psi^-(R)]\Psi^+(R) = E\Psi^+(R), \tag{12b}$$

and making an appropriate choice of the mirror potentials $c(R)\Psi^\pm(R)$, it is possible, in principle, to retain two stable and distinct positive populations for Ψ^+ and Ψ^- such that the difference is a solution of the original Schrödinger equation. Two very useful fermion methods were introduced before the mirror potential framework was realized. The fixed-node approximation[16,34] and transient estimation [11,34] turn out to be limiting cases of Eqs. (12). The *fixed-node* approximation corresponds to the limit $c(R) \to \infty$ in Eqs. (12), whereby the diffusion of the two populations is restricted to two isolated regions in configuration space. This prescription yields an upper bound to the true ground-state energy, and leads to the lowest-energy wave function with the same nodes as the trial function. Alternatively, *transient estimation* corresponds to the limit $c(R) \to 0$ in Eqs. (12) and results in a series of decreasing upper bounds with exponentially increasing statistical error. Accurate trial functions are extremely important for this second method, since the goal is to achieve convergence to the true ground state before the growing statistical error dominates.

Recently, an intermediate choice for $c(R)$ has been studied,[27]

$$c(R)\Psi^{\pm} = \frac{[E^* - H]\Psi_T^S}{\Psi_T^{\mp}} \tag{13}$$

with

$$\Psi_T^{\pm} = \Psi_T^S \pm \Psi_T^A \tag{14}$$

and

$$\Psi_T^S = \sqrt{(\Psi_T^A)^2 + (\beta\Psi_T^B)^2} \tag{15}$$

wherein Ψ_T^A is the antisymmetric trial wave function of Eq. (3). The "bosonic" wave function Ψ_T^B is taken as the product of the two- and three-body correlations ψ_3 of Eq. (4) only, and is therefore symmetric. The quantities E^* and β are parameters of the approximate mirror potential. Although the mirror potential itself is arbitrary, the use of Ψ_T rather than Ψ on the right hand side of Eq. (13) gives only an approximate cancellation in Eqs. (12). Nevertheless, we have found that this approximation allows for an accurate description of the ground state. In the limit that the antisymmetric trial function Ψ_T^A is exact, the mirror potential equations yield an exact solution independent of E^* and β.

Table 1. Results of Monte Carlo calculations for 54 ^{3}He atoms in a periodic box at $\rho\sigma^3 = 0.273$, where $\sigma = 2.556$ Å. All energies in K per particle. MP = Mirror Potential, FN = Fixed-Node approximation, and TE = Transient Estimation.

Method	E^*	β	Energy	$\langle T\rangle$	$\langle V\rangle$
Variational	—	—	-2.13±0.02	12.22±0.03	-14.35±0.02
GFMC-MP	-2.30	60.0	-2.24±0.04	12.33±0.14	-14.57±0.14
GFMC-MP	-2.35	70.0	-2.32±0.07	12.22±0.20	-14.54±0.20
GFMC-MP	-2.40	70.0	-2.27±0.03	12.42±0.16	-14.69±0.16
GFMC-MP	-2.35	80.0	-2.30±0.04	12.34±0.14	-14.64±0.14
GFMC-FN	—	—	-2.37±0.01	12.28±0.04	-14.65±0.03
GFMC-TE	—	—	-2.44±0.04	12.40±0.10	-14.84±0.10

In Table 1, results obtained at the experimental equilibrium density of ^{3}He are collected. So that they may be taken in context, mirror potential results are presented in the table along with results of variational, transient estimate, and fixed-node studies. The optimally determined trial function, as determined by the variational calculation, has been used in each of the GFMC studies. This wave function determines the antisymmetric part of the initial distribution in all cases, the antisymmetric part of the mirror potential, and the nodal surfaces in the fixed-node calculation. To date, calculations for the intermediate choice of mirror potential in Eq. (13) give energies between -2.24 ± 0.04 and -2.32 ± 0.07 K per particle, depending upon the parameters. While significantly lower than the variational result, the mirror potential results in general are not as low as the fixed-node (-2.37 ± 0.01). However, and this is the principal motivation behind the intermediate choice, the mirror potential procedure yields accurate energies *without the necessity of fixing the nodal surface of the wave function.* Consequently, it should have great applicability where the nodal surface is hard or even impossible to specify. Mirror potentials also allow for even more accurate higher-order treatments.[19]

In our calculations, the rapid increase in the statistical error made it hard to determine if the transient estimate calculation has converged to the ground state; nevertheless, the upper bound of -2.44 ± 0.04K/particle is the lowest obtained to date and is very close to the experimental equilibrium energy of -2.47 K/particle. For this reason we have concluded that the Aziz potential serves as well for ^{3}He as for ^{4}He as an effective many-body interaction. With a greatly expanded energy scale, Figure 1 shows the significant improvement of the fixed-node energy equation of state over the pure variational wave function. At equilibrium density, the transient estimate result spans both the experimental curve and the fixed-node curve within its small statistical error.

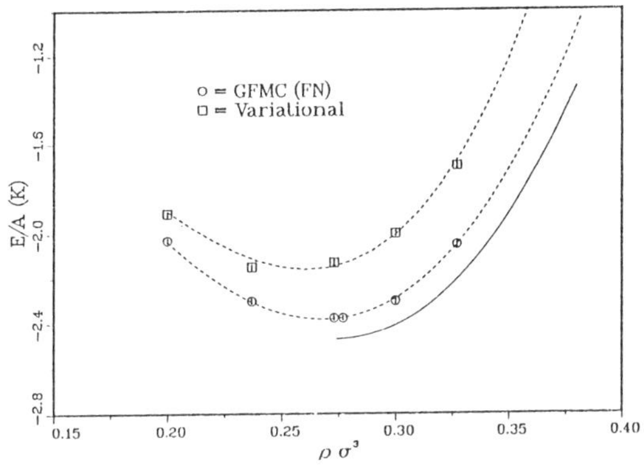

Fig. 1. Energy per particle vs. density for ^{3}He at zero temperature. The dashed lines are fits to the variational and fixed-node GFMC calculations, while the solid line gives the experimental equation of state

V. EXTRAPOLATION OF GROUND-STATE PROPERTIES

One detail which is often overlooked in describing the general method of calculating ground-state properties from Monte carlo calculations arises because importance sampling has been incorporated into the algorithm to control the variance. In such a case, GFMC simulations actually yield configurations – again, lists of particle coordinates – chosen from the asymptotic distribution $\psi_T \psi_0$. A "mixed estimation" may be defined for any operator F by

$$\langle F \rangle_M = \frac{\int \psi_0(R) F \psi_T(R) dR}{\int \psi_0(R)\psi_T(R) dR}. \tag{16}$$

The calculation of the total energy in the previous section is precisely such a mixed estimation. In that case, ψ_0 is an eigenfunction of the Hamiltonian operator and so the mixed estimator is exact, *i.e.*, the total energy calculated with the mixed estimator asymptotically approaches[25] the true ground state E_0. On the other hand, to form ground-state expectation values for other operators which do not commute with the Hamiltonian, corrections must be introduced. Assuming that the trial function is close to the ground-state wave function, $\psi_T = \psi_0 + \delta\psi$, a linear extrapolation is made to obtain $\langle F \rangle_0$,

$$\langle F \rangle_0 = 2\langle F \rangle_M - \langle F \rangle_V \tag{17}$$

where $\langle F \rangle_V$ is the expectation value from the variational calculation. The extrapolated expectation value gives $\langle F \rangle_0$ to order δ^2. While other methods[8,25] exist for obtaining $\langle F \rangle_0$, they have higher variance for small δ.

Particular choices for the functional forms of Jastrow or triplet or backflow correlations, along with the values for the parameters contained therein, may be quite different while yielding very similar variational energies. However, other properties of the system, at the variational level may be very different.[23,24] Nevertheless, for our choice of importance function and employing the fixed-node approximation of GFMC for fermions, the extrapolated ground-state properties ought to be quite reliable. We associate high reliability in our calculations with small extrapolations.

The output from both variational and GFMC calculations are configurations composed of particle coordinates. These configurations are analyzed to determine other properties of the systems. For instance, Table 1 above also gives the kinetic energy $\langle T \rangle$ and potential energy $\langle V \rangle$ at equilibrium density. For every case, the kinetic energy per particle is between 12 and 12.5 K, significantly larger than some previous interpretations of experimental data[20,35,36]. One striking feature, then, of Table 1 is that for a wide variety of wave functions, there is actually very little change in the kinetic or potential energies calculated separately. As the wave function is improved, with an assumed greater degree of overlap with the ground state as higher-order correlations are included, there remains the sensitive cancellation of quantities of roughly the same magnitude. This may explain the wide variation in results from approximate calculations which may generate the two contributions to the total energy separately, even with relatively small errors, but which are far from the ground state when added together.

These same configurations may be analyzed[18,24,28] to give $n(\vec{r})$ and $n(\vec{k})$ at k-vectors contained within the simulation box. The one-body density matrix, $n(\vec{r})$, is the Fourier transform of the momentum distribution,

$$\begin{aligned} n(\vec{r}) &= \int e^{i\vec{k}\vec{r}} n(\vec{k}) d\vec{k} \\ &= \left\langle \frac{\psi(r_1, r_2, \cdots, r_i + \vec{r}, \cdots, r_n}{\psi(r_1, r_2, \cdots, r_i, \cdots, r_n)} \right\rangle \end{aligned} \tag{18}$$

and is a measure of the change in the wave function for a given displacement r. The expectation value in Eq. (18) is evaluated simply by moving particle i by an amount r and averaging the ratio of the wave functions over an ensemble of configurations. In this analysis, one may use two different methods, one of which computes $n(r)$ by moving particle i in the configurations a random amount, giving n_0 accurately and the best values for $n(k)$ at specific k-vectors. Alternatively, one determines $n(r)$ by moving particle i increments of dr along a prescribed direction yielding precise values of $n(r)$ at small r and therefore better values for the kinetic energy. In a fermion system, $n(r)$ fluctuates about zero for large r, reflecting the antisymmetry of the wave function.

As the calculation of $n(r)$ proceeds using the first of the two methods above, $n(\vec{k})$ is determined directly for those wave vectors contained in the simulation box. That is,

$$n(\vec{k}) = \left\langle \frac{e^{i\vec{k}\cdot(\vec{r}_i+\vec{r})}\psi(\vec{r}_i + \vec{r})}{\psi(\vec{r}_i)} \right\rangle . \tag{19}$$

In general, a sphericalized $n(k)$ could be obtained as the Fourier transform of the calculated $n(r)$,

$$n(k) = \frac{1}{2\pi^2} \int e^{-ikr} n(r) r^2 dr. \tag{20}$$

In the fermion case, due to the discontinuity at the fermi surface, transforming a sphericalized $n(r)$ is not computationally appropriate.

Given a reliable $n(k)$ the average kinetic energy per particle can be calculated from its second moment:

$$\langle KE \rangle = \frac{\hbar^2}{2m} \left\langle k^2 \right\rangle = \frac{\int_0^\infty k^4 n(k) dk}{\int_0^\infty k^2 n(k) dk}. \tag{21}$$

To carry out the integrals in Eq. (21), the large-k behavior of $n(k)$ must be known. For this purpose, an exponential tail can be fit to $n(k)$. While there is no proof that the exact large-k behavior is exponential, the near-tail behavior of $n(k)$ suggests that this is at least a reasonable approximation. No great effort is required to produce a fit which is continuous, with a continuous first derivative, and for which the integral in Eq. (21) reproduces within statistics the value for the kinetic energy from the eigenvalue calculation.

Alternatively, the kinetic energy may be determined from the curvature of $n(r)$ as $r \to 0$, using

$$\left\langle k^2 \right\rangle = -\frac{3}{n(r)} \frac{d^2 n(r)}{dr^2}\bigg|_{r=0} . \tag{22}$$

Again, as a numerical check on the internal consistency of our calculations, the value of the kinetic energy obtained by this second method agrees within statistics with that from the eigenvalue calculation as given in Table 1. These alternative evaluations form the basis for the confidence we have in our work: the confidence comes, not primarily from our gratifying agreement with new experimental analyses, but from the efforts we have made to control or eliminate approximations in our calculations, achieving an internal consistency in the numerical evaluation of distinct quantities which are formally equivalent.

An example of how close the variational and extrapolated values are for the ground-state properties of ^{3}He, $n(r)$ in a 54-particle ^{3}He system at equilibrium density is given in Figure 2, while the direct calculation of $n(k)$ is given in Figure 3.

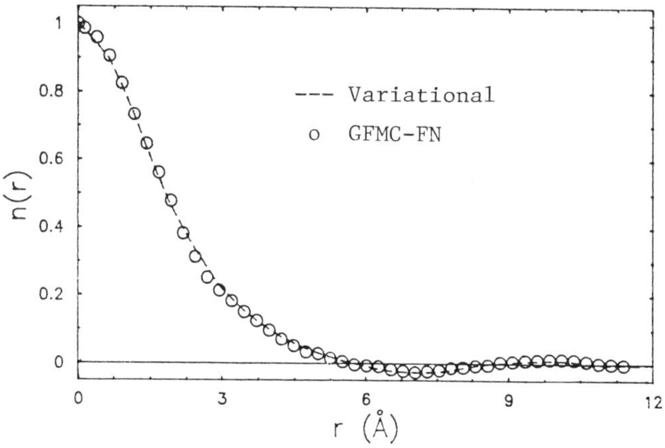

Fig. 2. The one-body density matrix $n(r)$ for 54-particle calculation of ^{3}He at equilibrium density from both variational and extrapolated GFMC calculations.

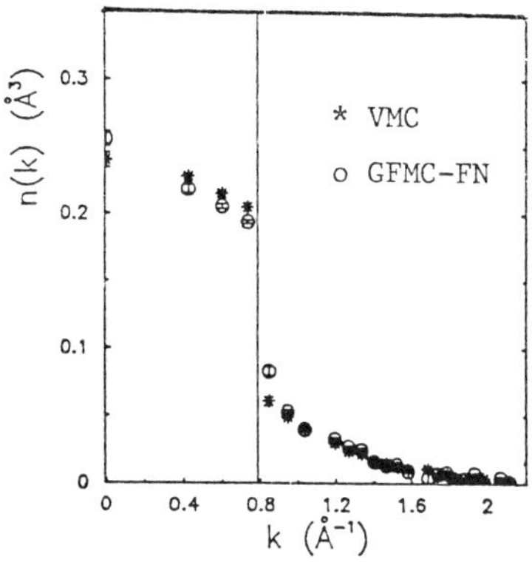

Fig. 3. The momentum distribution $n(k)$ for ^{3}He at equilibrium density from both variational and extrapolated GFMC calculations.

As indicated above, while calculations with different variational wave functions produce a wide variation in results for many ground-state properties other than the energy per particle, the extrapolated results have proven to be quite reliable.

VI. CONCLUSIONS

Attacking the fermion problem has proven to be an exhausting, yet exhilarating effort. Progress in understanding the range of algorithms and approximations incorporated in the mirror potential framework has made it possible to focus on a reasonable comparison between theory and its results and experiment. The uncertainty in any remaining discrepancy may be fixed by adjusting the model interaction[37] or going beyond a pair interaction; finite size effects under further study may also account for the remaining discrepancy. Given continued access to advanced computing resources, we anticipate a return to the nuclear regime and its problems with state-dependent interactions.

I would like to acknowledge the seminal ideas and work, as well as the outstanding leadership of the Courant Institute's quantum Monte Carlo group, of Malvin H. Kalos, honored as the newest Eugene Feenberg Medalist. In addition, the significant contributions by Joe Carlson and Paula Whitlock to the work reviewed here is gratefully acknowledged, as are the ideas, suggestions, contributions, and encouragement of D. M. Ceperley, J. M. C. Chen, G. V. Chester, Michael A. Lee, and Kevin E. Schmidt. The computations reported here were made possible by supercomputer access supported by the Nuclear Physics Program, U. S. Department of Energy, at the National Energy Research Supercomputer Center.

REFERENCES

1. See other contributions in this volume and earlier volumes of the series *Recent Progress in Many Mody Theory* and the *Proceedings of the Conference on Frontiers of Quantum Monte Carlo*, J. Stat. Phys. 43 (1986).
2. Eugene Feenberg, *Theory of Quantum Fluids,* (Academic Press, New York, 1969).
3. J. W. Clark, "The Crisis in Nuclear Matter Theory," 1976 (unpublished).
4. J. Zabolitzky, in *Progress in Particle and Nuclear Physics*, A. Faessler, ed; vol. 16 (Pergamon, Oxford, 1986).
5. W. L. McMillan, Phys. Rev. **138**, 442 (1965).
6. M. H. Kalos, D. Levesque, and L. Verlet, Phys. Rev. **A9**, 2178 (1974).
7. D. M. Ceperley, G. V. Chester, and M. H. Kalos, Phys. Rev. **B** 1977.
8. D. M. Ceperley and M. H. Kalos, in *Monte Carlo Methods in Statistical Physics*, K. Binder, ed; Topics in Current Physics, vol. 7 (Springer, Berlin, Heidelberg, New York, 1979) Chap. 4.
9. P. A. Whitlock, D. M. Ceperley, G. V. Chester and M. H. Kalos, Phys. Rev. **B19**, 5598 (1979).
10. K. E. Schmidt, M. H. Kalos, Michael A. Lee and G. V. Chester, Phys. Rev. Lett. **45**, 573 (1980).
11. M. A. Lee, K. E. Schmidt, M. H. Kalos and G. V. Chester, Phys. Rev. Lett. **46**, 728 (1981).
12. K. E. Schmidt, M. A. Lee, M. H. Kalos and G. V. Chester, Phys. Rev. Lett. **47**, 807 (1981).
13. M. H. Kalos, Michael A. Lee, P. A. Whitlock and G. V. Chester, Phys. Rev. **B24**, 115 (1981).
14. R. M. Panoff, J. W. Clark, Michael A. Lee, K. E. Schmidt, M. H. Kalos, G. V. Chester, Phys. Rev. Lett. **48**, 1675 (1982).
15. D. M. Arnow, M. H. Kalos, Michael A. Lee and K. E. Schmidt, J. Chem. Phys., **77**, 5562 (1982).
16. J. W. Moskowitz, K. E. Schmidt, M. A. Lee, and M. H. Kalos, J. Chem. Phys. **77**, 349 (1982).
17. K. E. Schmidt and M. H. Kalos, in *Applications of the Monte Carlo Method in Statistical Physics*, K. Binder, ed; Topics in Current Physics, vol. 36 (Springer, Berlin, Heidelberg, New York, 1984).

18. P. A. Whitlock and R. M. Panoff, *Proceedings of the High Energy Excitations in Condensed Matter Conference*, Los Alamos (1984).

19. J. Carlson and M. H. Kalos, Phys. Rev. **C32**, 1735 (1985).

20. J. Carlson, R. M. Panoff, K. E. Schmidt, P. A. Whitlock, and M. H. Kalos, Phys. Rev. Lett. **55**, 2367(C) (1985).

21. J. W. Moskowitz and K. E. Schmidt, J. Chem. Phys. **85**, 2868 (1986).

22. K. E. Schmidt and J. W. Moskowitz, J. Stat. Phys. **43**, 1027 (1986).

23. R. M. Panoff, in *Condensed Matter Theories, Vol. II*, P. Vashista, R. Kalia, and R. Bishop, eds. (Plenum, New York, 1987).

24. P. A. Whitlock and R. M. Panoff, Can. J. Phys. **65**, 1409 (1987).

25. K. E. Schmidt, in *Models and Methods in Few-Body Physics*, Lecture Notes in Physics, Vol. 273, (Springer, Berlin, 1987), p. 363.

26. P. A. Whitlock, G. V. Chester and M. H. Kalos, Phys. Rev. **B38**, 2418 (1988).

27. R. M. Panoff and J. Carlson, Phys. Rev. Lett. **62** 1130 (1989).

28. R. M. Panoff and P. A. Whitlock, in *Momentum Distributions*, ed. by R. Silver, (Plenum, New York, 1989).

29. R. A. Aziz, V. P. S. Nain, J. S. Cerley, W. L. Taylor, and G. T. McConville, J. Chem. Phys. **70**, 4330 (1979).

30. E. Krotscheck, R. A. Smith, J. W. Clark, and R. M. Panoff, Phys. Rev. **B24**, 6383 (1981).

31. V. R. Pandharipande and H. A. Bethe, Phys. Rev. **C7**, 1312 (1972).

32. N. Metropolis, A. W. Rosenbluth, M. N. Rosenbluth, A. M. Teller, and E. Teller, J. Chem. Phys. **21**, 1087 (1953).

33. E. Manousakis, V. R. Pandharipande, and Q. N. Usmani, Phys. Rev. **B31**, 7022 (1985).

34. D. M. Ceperley and B. J. Alder, Phys. Rev. Lett. **45**, 566 (1980).

35. P. E. Sokol, K. Sköld, D. L. Price and R. Kleb, Phys. Rev. Lett. **54**, 909 (1985).

36. H. A. Mook, Phys. Rev. Lett. **51**, 1454 (1983).

37. Ronald A. Aziz, Frederick R. W. McCourt, and Clement C. K. Wong, Mol. Phys. **61**, 1487 (1987).

BOSE CONDENSATE IN SUPERFLUID ^{4}He AND MOMENTUM DISTRIBUTIONS BY DEEP INELASTIC SCATTERING

Richard N. Silver

Theoretical Division and Los Alamos Neutron Scattering Center
MS B262 Los Alamos National Laboratory
Los Alamos, New Mexico 87545

Paul E. Sokol

Department of Physics
The Pennsylvania State University
University Park, PA 16802

INTRODUCTION

In 1938 London [1,2] offered an explanation of the observation earlier that year of superfluid behavior in liquid ^{4}He when it is cooled below a critical temperature of 2.17 °K. He argued that the superfluid transition was analogous to the Bose condensation of an (*ideal*) gas of non–interacting atoms obeying the same Bose–Einstein spin–statistics relation as ^{4}He atoms. This relation requires the many–atom wave function to be completely symmetric in the atomic coordinates, resulting in a preference for the atoms to occupy the same single–particle states. For a finite system of atoms the momenta are quantized in spacings proportional to the inverse of the system size. At high temperatures the fraction of atoms occupying any one of the momentum states also scales as the inverse of the size. However, as the temperature is reduced below a critical *Bose condensation temperature* a significant fraction of the atoms, independent of the system size, begins to occupy the zero–momentum state. The *Bose condensate fraction* of an ideal gas approaches one at zero temperature. For ^{4}He, by analogy, at high temperatures in the normal fluid the condensate fraction should be zero, but as temperatures are reduced below the superfluid transition temperature the condensate fraction should rise to a non–zero value. The effect of the strong interactions among the (*non–ideal*) ^{4}He atoms is to deplete the zero temperature condensate fraction from one in an ideal gas to a value much less than one for ^{4}He. While the analogy between superfluidity and Bose condensation is imperfect, the concept of a Bose condensate in

Recent Progress in Many-Body Theories, Volume 2
Edited by Y. Avishai
Plenum Press, New York, 1990

the superfluid phase has survived. A variety of increasingly sophisticated many–body calculations have predicted a condensate fraction of about 10 % at zero temperature in superfluid ^{4}He at SVP. Because of the importance of superfluidity and the related phenomenon of superconductivity to condensed matter physics, this simple prediction has motivated a more than twenty year effort involving up to one hundred scientists to measure the Bose condensate fraction in ^{4}He.

More generally, the goal has been to measure the *momentum distribution* [3], *n(p)*, which is the probability for an atom to occupy a single–particle state of momentum *p*. The most promising technique has been neutron scattering experiments at up to 100's of meV energies, such that the neutron energy is much larger than the characteristic energies for the collective behavior of the system such as the phonon–roton spectrum. If the energy transferred by the neutron is also much larger than the potential energies between atoms, then (hopefully!) the *impulse approximation* (IA) may be invoked in which the scattering from the many–atom system is described as the sum of independent single–atom scatterings. This predicts a simple relation between the momentum distribution and the measured scattering cross section. Analogous experiments are of interest in all sub–fields of physics, because the momentum distribution is an experimentally accessible characterization of the many–particle wave function. This includes x–ray Compton scattering at 10's of KeV energies to measure electron momentum distributions, quasi–elastic electron scattering from nuclei at GeV energies to measure nucleon momentum distributions, and electron scattering from nucleons at 100's of GeV energies to measure quark momentum distributions. Such experiments we term *deep inelastic scattering*. Compton scattering experiments for electronic momentum distributions have been very successful. However, experiments on atomic momentum distributions of interest in condensed matter physics, and on nucleon momentum distributions of interest in nuclear physics, have been plagued by uncertainties about the validity of the impulse approximation for the values of energy and momentum transfer which can be realistically achieved.

Thus, there are several reasons for the high interest in the recent experimental and theoretical progress in understanding deep inelastic neutron scattering from liquid ^{4}He: it tests the fundamental London hypothesis of a connection between superfluidity and Bose condensation; it provides a quantitative test of *ab–initio* calculational methods for all systems with strong correlations which are the focus of current quantum many–body research; and it establishes the range of validity of deep inelastic scattering as a method for measuring momentum distributions. In the following sections we introduce the concepts of impulse approximation in more detail, we describe recent progress in the theory for final state corrections to the impulse approximation, we present quantitative predictions for neutron scattering experiments, we compare with recent high energy pulsed neutron source experiments on liquid ^{4}He by P. Sokol and colleagues as well as other attempts to extract the Bose condensate fraction from the neutron scattering data, and we discuss the implications of this progress for future momentum distribution experiments in other systems such as liquid ^{3}He and quasi–elastic electron nucleus scattering.

THE IMPULSE APPROXIMATION AND THE BOSE CONDENSATE

For the sake of clarity, in this section we explicitly consider only neutron scattering from many–atom systems, although the concepts we shall discuss are common to many other sub-fields of physics.

In the impulse approximation, the neutron scattering law may be written as the sum of collisions from individual atoms which have a probability of having an initial momentum, *p*, given by the momentum distribution, $n(p)$. The energy and momentum transferred by the neutron are assumed to be high enough that the collective dynamics and interactions among He atoms only show up in initial state momentum distribution, $n(p)$, and not in the final state. Then, the scattering law may be written:

$$S(Q,\omega) \approx S_{IA}(Q,\omega) \equiv \frac{1}{\varrho}\int \frac{d^3\vec{p}_i}{(2\pi\hbar)^3}\, n(p_i)\, \delta\left(\hbar\omega - E(p_f) + E(p_i)\right) \quad . \qquad (1)$$

Here, $\vec{p}_i$ is the initial momentum of an atom, $\vec{p}_f = \vec{p}_i + \hbar\vec{Q}$ is the final momentum of the recoiling atom after the neutron has transferred momentum $\hbar Q$ and energy $\hbar\omega$, $E(p) = \hbar^2p^2/2M$ is the kinetic energy, and ϱ is the atomic density. The delta function expresses the energy conservation of the scattering process. $S(Q,\omega)$ is normalized so that the integral of it over $\hbar\omega$ equals one. Equation (1) provides the sought for simple relation between the momentum distribution and the scattering law. The impulse approximation predicts that the scattering law peaks at the recoil energy from a particle at rest, $\hbar\omega = \hbar^2Q^2/2M$. The width of the peak is $\hbar Q \Delta p/M$, where Δp is the width of the momentum distribution, $n(p)$. An experimental test to establish that deep inelastic scattering conditions, but (as we shall dicuss) not necessarily impulse approximation conditions, have been reached is the observation of this peak position varying as Q^2 and peak width varying as Q.

In 1966 Hohenberg and Platzman [4] suggested that the Bose condensate fraction in ^{4}He could be measured by high momentum transfer neutron scattering experiments. For an isotropic system in the presence of a Bose condensate, the momentum distribution may be written

$$n(\vec{p}) = n_o\varrho(2\pi\hbar)^3\delta^{(3)}(\vec{p}) + n^*(p) \quad , \qquad (2)$$

where n_o is the Bose condensate fraction, and $n^*(p)$ is a smooth function. Combining Eqs. (1) and (2) yields

$$S_{IA}(Q,\omega) = n_o\delta\left(\hbar\omega - \hbar^2Q^2/2M\right) + S^*_{IA}(Q,\omega) \quad , \qquad (3)$$

where $S_{IA}{}^*(Q,\omega)$ is the contribution to the scattering law from $n^*(p)$. Thus, in the impulse approximation the scattering law is expected to have a sharp delta function peak with weight n_o centered at the recoil energy, which sits atop a broader peak due to the smooth part of the momentum distribution. The central goal of most deep inelastic neutron scattering experiments on liquid ^{4}He has been the observation of this sharp peak in the superfluid at temperatures less than the superfluid transition temperature, $T_\lambda = 2.17$ °K, and the absence of such a peak in the normal fluid at higher temperatures.

It is common to rewrite the scattering law as a *Compton profile*, after analogous x–ray scattering experiments on electron momentum distributions carried out by A. H. Compton and J. W. M. DuMond in the 1920's. For neutron scattering, it is given by

$$J(Y,Q) \equiv \frac{QS(Q,\omega)}{M} \quad , \qquad (4)$$

where

$$Y \equiv \frac{M}{Q}\left(\frac{\omega}{\hbar} - \frac{Q^2}{2M}\right) \quad . \tag{5}$$

Specializing to the impulse approximation, one obtains

$$J_{IA}(Y) = n_o\delta(Y) + \frac{1}{4\pi^2\hbar^2 Q}\int_{\hbar|Y|}^{\infty} dp\ p\ n^*(p) \quad . \tag{6}$$

Energy conservation in Eq. (1) requires Y to equal the longitudinal wave vector of the atom, or the component of initial momentum which is parallel to the momentum transfer. In the impulse approximation, the Compton profile is centered at and symmetric about $Y = 0$. It is also independent of Q, which is termed *Y–scaling*. This is equivalent to the earlier statement that the peak in $S(Q,\omega)$ has a position varying linearly in Q^2 and a width varying linearly in Q. Figure 1 [6] shows measured Compton profiles for superfluid ^{4}He at Q's of 7, 12 [7] , and 24 A^{-1} [8]. Although the data are instrument broadened wtih a full–width–half–maximum (FWHM) of 0.6 A^{-1}, the fact that these curves for different Q's approximately lie on top of one another is strong evidence for Y–scaling, especially at higher Q's. However, Y–scaling is a necessary, but not a sufficient condition, for the validity of the impulse approximation. G. West [5] was the first to point out that Y–scaling can be true even if the impulse approximation is false. In a following section on final state effects we present an explicit counterexample to the hypothesis that the observation of Y–scaling confirms the impulse approximation.

Quantitative predictions for deep inelastic experiments on ^{4}He can be obtained from the impulse approximation by inputing theoretical momentum distributions, $n(p)$. The solid line in Fig. 2 shows momentum distributions calculated by Greens' Function Monte Carlo (GFMC) [9] at $T = 0$ °K, which predicts an n_o of 9.24 %. The dashed line shows the momentum distribution calculated by Path Integral Monte Carlo (PIMC) [10] at $T = 3.3$ °K, in which n_o is predicted to be zero. Figure 3 shows the corresponding predictions for Compton profiles. The solid line is the GFMC–IA prediction for the superfluid, which has a delta function peak at $Y = 0$ of weight n_o. The dashed line is the PIMC–IA prediction for the normal fluid, which is a smooth function.

The experimental data, such as Fig. 1, show a sharpening around $Y=0$ as the temperature is reduced into the superfluid. However, there is no direct evidence for a sharp delta function peak in the superfluid at any Q. The issue has been how to infer the value of n_o from the limited data available which inevitably include instrumental broadening, statistical and background uncertainties, and corrections to the impulse approximation at the finite values of Q's which can be achieved. There have been many reports [11] of determinations of n_o, which have involved a succession of improvements in understanding spectrometers, improvements in the data analysis procedure, and increases in the values of Q to better approach the conditions for the validity of the impulse approximation.

The most extensive data analysis of reactor experiments was carried out in 1982 [12] for data in the range 4 $A^{-1} < Q < 7$ A^{-1}. At lower Q's (< 3 A^{-1}) the neutron scattering law is dominated by the collective behavior such as the phonon and roton spectrum and at higher Q's the neutron flux from reactors becomes prohibitively small for high resolution experiments. These data begin to approach the impulse approximation predictions, i.e. the peaks are centered at the recoil energy of a free particle and the widths are proportional to Q. However, there are signifi-

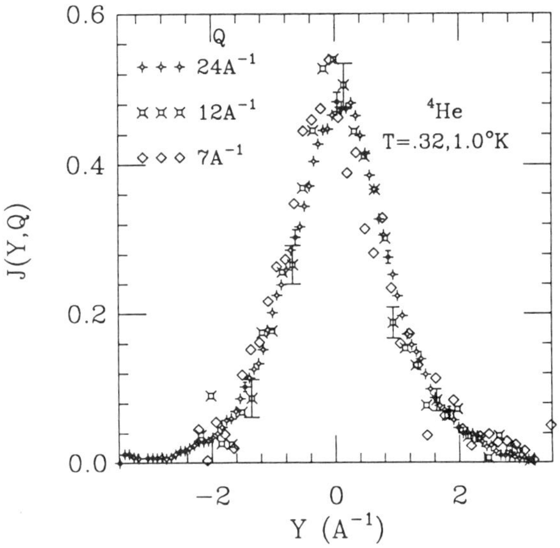

Fig. 1 Inelastic neutron scattering from liquid ^{4}He at momentum transfers of 7 A^{-1}, 12 A^{-1} (at $T = 1.0$ °K) and 24 A^{-1} (at $T = 0.35$ °K). The results, plotted as $J(Y)$, all fall on approximately the same curve, illustrating the Y–scaling behavior. The instrumental resolution is about 0.6 A^{-1}.

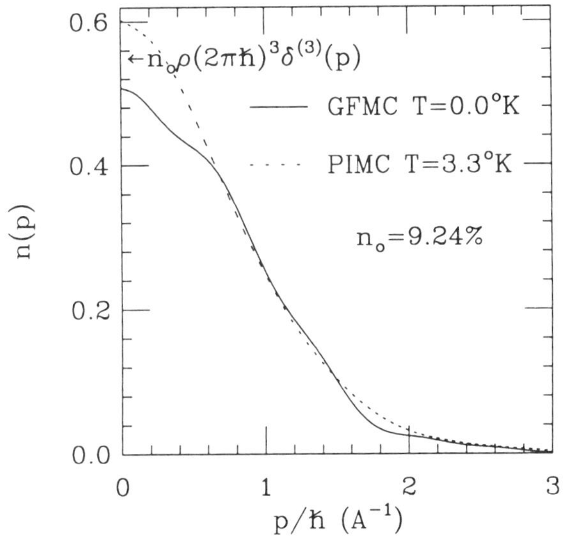

Fig. 2 Theoretical momentum distributions for liquid ^{4}He. The solid line is the Greens' Function Monte Carlo (GFMC) prediction for the superfluid at $T = 0$ °K, which has a delta function at p = 0 with a 9.2 % Bose condensate fraction. The dashed line is the Path Integral Monte Carlo (PIMC) prediction for the normal fluid at $T = 3.33$ °K.

cant deviations from the impulse approximation in the form of asymmetry of the peak shape, as shown in the $Q = 7\ A^{-1}$ data in Fig. 1, and oscillations [13] of the peak width and center with Q. To correct for these deviations, which can be quite significant at lower Q's, the data have been symmetrized about $Y = 0$ and averaged over several Q's. This gives a resultant scattering which is consistent with the basic predictions of the impulse approximation (peaks that are symmetric and centered at the recoil energy) and where the effects of the corrections to the impulse approximation are minimized. The resultant data is then converted from $S(Q, \omega)$ to $n(p)$, which involves a differentiation and a division by p according to Eq. (6).

Attempts to directly compare the experimental momentum distribution, obtained using the procedure above, with theoretical predictions always exhibited significant disagreement. Thus, attention turned to other methods to extract information on the momentum distributions and, especially, the condensate fraction, n_o. The most popular procedure has been modeling of the momentum distribution with n_o as an adjustable parameter.

The model used for the momentum distribution in the superfluid will strongly influence the value of n_o inferred. In fact, several previous studies have inferred a vanishingly small value for the condensate based on a particular model. In the 1982 analysis, only the momentum distribution at small values of p was needed. The model for the uncondensed component of the momentum distribution (i.e. everything but the condensate itself) was:

$$n^*(p) = n_o\left\{a/p^2 + b/p\right\}\ \theta(p_c - p)\ +\ (1 - n_o)\ n(p, T > T_\lambda) \quad . \tag{7}$$

The first term in this model represents singular behavior that is induced in $n^*(p)$ by the appearance of the condensate [14]. This is due to the coupling of long wavelength phonons to fluctuations in the condensate. The second term simply represents a scaled down version of the normal liquid behavior. The $n(p, T > T_\lambda)$ is fixed by measurements at high temperature where n_o is zero. Based on this model, the 1982 analysis (and subsequent analysis using the same procedure) extracted values of n_o around 10 %, in surprisingly good agreement with theory.

Unfortunately, this apparent agreement between theory and experiment was fortuitous. The small–p singular behavior above, which played a very important role in the 1982 analysis, was the result of an incorrect combination of the small–p and large–p limiting behaviors. Griffin pointed out this error [15]. Using a more correct form for the small–p singular behavior, he obtained values of n_o of 4–5 %, about half the theoretical predictions. More recently, Sokol, Silver and Clark [3] have pointed out that information on the small–p singular behavior is extremely difficult to obtain by deep inelastic neutron scattering measurements, a point to which we will return later.

FINAL STATE EFFECTS

What has been left out of all of the data analysis procedures discussed above has been a detailed understanding of the corrections to the impulse approximation, to which we now turn. These were first addressed in the original paper of Hohenberg and Platzman [4]. The additional physical effect which must be included is the scattering of the recoiling atom from neighboring atoms in the condensed phase, resulting in collisional lifetime broadening. A finite collisional lifetime, τ, results in an uncertainty, $\hbar/\tau$, of the energy of the final state of the He atom after being struck by a neutron, and so this broadening is termed *final state effects* (FSE). This can be important for He even at high Q's, because the potential energy between He atoms is steeply re-

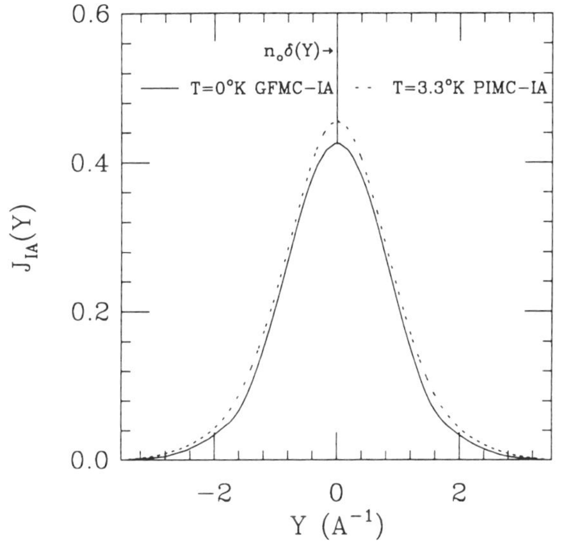

Fig. 3 Compton profiles for ^4He calculated in the impulse approximation, $J_{IA}(Y)$. The solid line is the GFMC prediction for the superfluid at $T = 0$ °K, which has a delta function at $Y = 0$ with weight equal to the Bose condensate fraction, n_o. The dashed line is the PIMC prediction for the normal fluid at $T = 3.33$ °K.

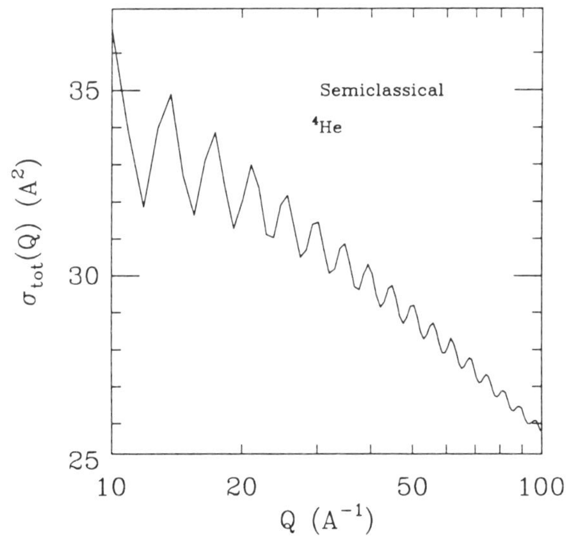

Fig. 4 Semiclassical ^4He–^4He total cross section for momentum, $\hbar Q$.

pulsive at short distances and never negligible compared to the kinetic energy of the recoiling atom, $\hbar^2Q^2/2M$.

We now describe this in a fashion which, while not following the original derivation of Hohenberg and Platzman, will naturally lead to our current understanding of final state effects. Assuming that $|\vec{p}_i| \ll \hbar Q$, the rate at which the recoiling atom scatters from neighboring atoms is given by

$$\frac{1}{\tau} \approx \varrho\ \sigma_{tot}(Q)\ \frac{\hbar Q}{M} \quad , \tag{8}$$

where $\sigma_{tot}(Q)$ is the He–He total scattering cross section for incident momentum $\hbar Q$. The resulting uncertainty in the energy of the recoiling atom is expressed by adding to $E(p_f)$ in Eq. (1) a self energy $\Sigma = -i\hbar/2\tau$. We rewrite the energy conservation delta function in Eq. (1) in terms of its integral representation. Then, the effect of collisions is to transform it as follows:

$$\delta(\Delta E) = \frac{1}{2\pi\hbar}\int_{-\infty}^{\infty} dt\ \exp\left(\frac{it\Delta E}{\hbar}\right) \underset{FSE}{\Rightarrow} \frac{1}{2\pi\hbar}\int_{-\infty}^{\infty} dt\ \exp\left(\frac{it\Delta E - i|t|\Sigma}{\hbar}\right) \quad , \tag{9}$$

where $\Delta E = \hbar\omega - E(p_f) + E(p_i)$ and t can be interpreted as the time after the neutron collision. Then by straightforward manipulations the Compton profile becomes

$$J(Y,Q) = \int_{-\infty}^{\infty} dY'\ R_{FS}(Y-Y',Q)\ J_{IA}(Y') \quad . \tag{10}$$

The broadening function due to final state effects is given by a Lorentzian form,

$$R_{FS}(Y,Q) = \frac{1}{\pi}\frac{\Gamma}{Y^2+\Gamma^2} \quad , \tag{11}$$

where $\Gamma = \varrho\sigma_{tot}(Q)/2$ is the collision rate per unit distance traveled by the recoiling atom.

Let us first consider those properties of this theory which we believe remain correct in the more modern theories. First, the full–width–half–maximum (FWHM) of the broadening is given by

$$\Delta Y_{FWHM} \approx \varrho\sigma_{tot}(Q) \quad . \tag{12}$$

(Actually, the original prediction of Hohenberg and Platzman was a factor of two larger than this, and the parameter Y had not yet been suggested as a natural variable for the problem.) In the limit that the He–He potential at short distances can be approximated by a hard sphere, $\sigma_{tot}(Q)$ is independent of Q and equals $2\pi r_o^2$. Here r_o is the hard sphere radius and the factor of 2 is due to forward diffractive scattering. In this limit, the broadening is independent of Q and the Compton profile would satisfy Y–scaling even though the impulse approximation would be invalid. Also, the collision rate per unit distance traveled by the recoiling atom would be $\varrho\pi r_o^2$, which is the value given by classical mechanics. In reality, the He–He potential is steeply repulsive, but not hard sphere, at short distances resulting in a $\sigma_{tot}(Q)$ which decreases approximately logarithmically with increasing Q, as shown in Fig. 4. The small *glory oscillations* with Q are a quantum mechanical forward–backward scattering interference effect due to Bose statistics [16]. Then the final state broadening would decrease only logarithmically with increasing Q, resulting in a very slow approach to the impulse approximation for real neutron scattering experiments on He.

However, sum rule arguments show that the Lorentzian lineshape predicted by this theory cannot possibly be correct. The second moment sum rule on the neutron scattering law can be rewritten in terms of the Compton profile as

$$\int_{-\infty}^{\infty} J(Y,Q)\; Y^2\; dY = \frac{2M}{3} < K.E. > + O(Q^{-2}) \quad . \tag{13}$$

The impulse approximation satifies Eq. (13) without the $O(Q^{-2})$ term. If the final state broadening takes the convolution form in Eq. (10), then one must conclude

$$\int_{-\infty}^{\infty} R_{FS}(Y)\; Y^2\; dY = O(Q^{-2}) \quad . \tag{14}$$

That is, final state effects cannot alter the second moment of the Compton profile from the impulse approximation value in the high Q limit. However, the Lorentzian prediction, Eq. (11), has an infinite second moment. Evidently, final state effects if they exist must be much smaller than given by Eq. (11).

Gersch and Rodriguez [17,18] were the first to address this problem in 1973. They showed how the second moment sum rule could be satisfied with a non–Lorentzian lineshape for the final state broadening. They identified the physical origin of a non–Lorentzian shape as due to a dependence of the He–He scattering rate on the distance traveled by the recoiling atom. This distance dependence was due to the real space correlations which exist among atoms in liquid He. As we shall show, they calculated a final state broadening in qualitative agreement with current neutron scattering measurements. Unfortunately, this work was ahead of its time and did not receive the attention it deserved. One possible source of difficulty was that the derivation used a many–body theory involving time ordered cumulant expansions which were unfamiliar and have not been further developed. Another is that the numerical predictions were buried in an experimental paper which incorrectly obtained a 2 % Bose condensate fraction. Also many of the inputs to the calculation were not yet available and had to be severely approximated, such as approximating the radial distribution function $g(r)$ by a step function.

Since then there have been more than twenty papers on the theory for final state effects in He, and there has been a comparable level of inconclusive activity on the analogous problem in quasielastic electron nucleus scattering in nuclear physics. Many theories continued to obtain quasi–Lorentzian lineshapes which decreased as $O(\ln Q)$ with increasing Q, while other popular theories claimed that the leading correction to the impulse approximation was asymmetric in Y and decreased as $O(Q^{-1})$. Following the Hohenberg Platzman theory several experimental papers [7,19] at low Q (< 12 A^{-1}) have identified apparent oscillations in Q of the width of $J(Y,Q)$ with the glory oscillations in $\sigma_{tot}(Q)$. However, the widths observed are much narrower than the Hohenberg Platzman theory [20]. The premature claims to have observed a 10 % Bose condensate in superfluid ^4He in apparent agreement with many–body theory may have contributed to a complacency in the scientific community, i.e. there was a false sense that final state effects were unimportant and the quest to measure the Bose condensate fraction had been achieved [12]. The popular view of the subject up to 1987 has been reviewed by Svensson and Glyde and by Sears and Svensson [11]. In reality, a distinct delta–function peak in the scattering law had never been directly observed, and the most credible values of the Bose condensate fraction obtained by ignoring final state effects were 4–5 % [15] in serious disagreement with many–body theory.

In 1988 Silver [21,22] published an independent derivation of final state effects which confirmed and built on the original work of Gersch and Rodriguez. We first present a heuristic derivation of the main results of this theory following our derivation of the Hohenberg & Platzman Lorentzian broadening theory presented earlier. There are two elements of new physics: first, the collision rate depends on recoil distance because of correlations in the positions of atoms in the condensed phase; and second, at high Q the De Broglie wavelength of the recoiling atom is short compared to interatomic distances so that a semiclassical description of motion of the recoiling atom is adequate. The distance along the classical trajectory is given by $x \equiv \hbar Qt/M$ where, again, t is the time after the neutron collision. The probability density of finding two atoms a distance x apart is given by $\varrho\, g(x)$, where $g(r)$ is the radial distribution function of the liquid. The collision rate at distance x is given by

$$\frac{1}{\tau(x)} \approx \varrho g(x)\sigma_{tot}(Q)\frac{\hbar Q}{M} \quad . \tag{15}$$

The self energy consequently depends on recoil distance, $\Sigma(x) = -i\hbar/2\tau(x)$.

This essential physics is illustrated in Fig. 5, which shows the He–He potential and the radial distribution distribution function for liquid ^{4}He. The potential is steeply repulsive at short distances and has a weak Van der Waals attraction at larger distances. The radial distribution function shows that the atoms in the condensed phase sit in the attractive part of the potential well far from the steeply repulsive core of the potential. After being struck by a neutron, the recoiling atom must travel for some distance before it begins to collide with the steeply repulsive cores of the potentials from neighboring atoms. Therefore, $1/\tau(x)$ is zero for small x and approaches the value given by Eq. (8) only at large x. So the final state broadening should be much smaller than the Lorentzian broadening prediction.

Then the energy conservation delta function in Eq. (1) becomes

$$\delta(\Delta E) \Rightarrow \frac{1}{2\pi\hbar}\int_{-\infty}^{\infty} dt\, \exp\left(\frac{it\Delta E - i|t|\Sigma(|t|\hbar Q/M)}{\hbar}\right) \quad . \tag{16}$$

The final state broadening retains the convolution form, Eq. (10), but now

$$R_{FS}(Y,Q) \approx \frac{1}{2\pi}\int_{-\infty}^{\infty} dx\, \exp\left[iYx - |x|g(|x|)\Gamma\right] \quad ; \quad \Gamma \equiv \frac{\varrho\sigma_{tot}(Q)}{2} \quad . \tag{17}$$

Let us examine the properties of this final state broadening function. It reduces to the Lorentzian broadening result, Eq. (11), in the limit of a structureless fluid, $g(r) \rightarrow 1.0$. The FWHM continues to be approximately given by Eq. (12). Again, a Y–scaling correction to the impulse approximation is obtained in the limit of a hard sphere potential. However, the lineshape is non–Lorentzian and negative at large $/Y/$, and it satisfies the second moment sum rule because $g(0) = 0.0$ in He.

Y–scaling now acquires a geometrical interpretation in the limit of hard sphere interactions between He atoms. Y loses its impulse approximation interpretation as the component of initial momentum parallel to the momentum transfer. Instead, Y becomes the canonically conjugate variable to the distance, x, along the classical trajectory of the recoiling atom. The final state broadening function is the Fourier transform of the probability of no collisions as a function of recoil distance. In the hard sphere limit, the final state effects depend only on the relative positions and sizes of atoms, so that the (potentially 10^{23}) extra variables in the problem drop out. For real potentials, Y–scaling is obtained to the extent that $\sigma_{tot}(Q)$ varies slowly with Q.

A formal derivation of a somewhat more correct result has been discussed in detail elsewhere [21,22], so that here we only sketch the approach. The goal is to correctly describe the *asymptotic limit* of the final state broadening, which is defined by the approximations of very high Q and hard sphere short distance interactions between He atoms. Real experiments are sufficiently close to these conditions that the extension to finite Q and to the real He–He interactions should not introduce serious error. The results of many–body calculations (e.g. GFMC, PIMC, variational) of ground state properties, such as $n(p)$ and $g(r)$, should be used as inputs without recalculating them. Therefore, a projection superoperator method was chosen which enables a perturbation expansion for the dynamical response about the strongly interacting ground state. The final results, Eqs. (18–20), have been subsequently derived by Rinat and Butler [23] using a multiple scattering formalism which provides a more familiar derivation for some readers.

The projection approach relies on an exact relation between the neutron scattering law and a projection superoperator for single particle–hole excitations out of the true ground state. A Dyson equation can be derived by a Mori–Zwanzig–type [24] perturbative expansion of the Liouville equation for the time evolution of the single particle–hole excitations. All atom–atom scatterings are resummed to all orders in the interaction potential in terms of two–body t–matrices, and all interactions between three or more atoms are ignored. The "effective interaction" is then a product of a two–body t–matrix times the two–body density matrix in the ground state, which effectively screens the strongly repulsive short distance behavior resulting in a small parameter expansion. All terms which do not survive in the asymptotic limit are also ignored. Since Q is large, a semi–classical on–shell approximation [25] can be used for the two–body t–matrix. The two–body density matrix is approximated as a factorizable product of $n(p)$ and $g(r)$ in a way which satisfies the sum rules. The resulting Dyson equation may then be solved analytically.

A Feynman diagram representation of the resulting Dyson equation is shown in Fig. 6. The neutron scattering law describes the propagation of a particle–hole excitation above the ground state, which removes a particle of wave vector k and creates a particle of wave vector $k+Q$. Arrows denote the direction of the flow of momentum. A right arrow denotes a particle line and the left arrow a hole line. Only the particle lines can carry high momentum on the order of Q, whereas the hole lines must carry low momenta characteristic of the ground state wave function. The hatched area denotes the exact result including all scatterings of particles and holes. In the impulse approximation, the particle and hole are assumed to propagate without scattering. The final state effects come from the scattering of the high momentum particle creating new particle–hole excitations. The approximation indicated keeps only a single additional particle–hole excitation. The shaded box represents the two–particle density matrix describing the correlations between the two holes in the ground state, which is related by sum rules to the radial distribution function, $g(r)$. The dashed line is the two particle t–matrix which describes the scattering. The fact that the hatched area appears in both the exact result and the final state effects diagram im-

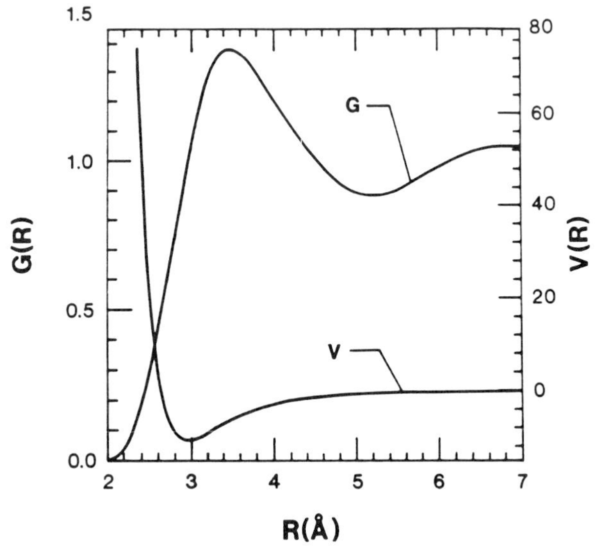

Fig. 5 Radial distribution function, *g(r)*, for liquid ^{4}He as measured by neutron diffraction, and the He–He potential, *V(r)*.

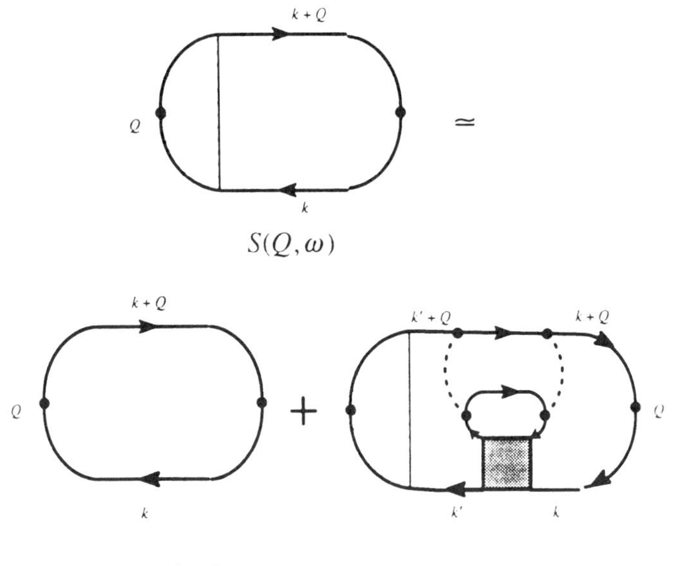

Fig. 6 Feynman diagram representation of the Dyson equation for deep inelastic neutron scattering. The left arrow denotes a hole line, the right arrow a particle line. The dashed line represents the He–He t–matrix. The box represents the two–particle density matrix in the ground state. The apperance of the hatched area on both sides of the equation indicates that the equation must be solved self–consistently.

plies that the scattering processes which transform a $(k+Q,k)$ excitation to a $(k'+Q,k')$ excitation must be calculated self–consistently.

The predicted final state broadening obtained again takes the form of an integral over a classical trajectory,

$$R_{FS}(Y,Q) = \frac{1}{2\pi} \int_{-\infty}^{\infty} dx \, \exp\left[iYx - \int_0^{|x|} \Gamma(x')dx' \right] \quad , \tag{18}$$

which should be compared to Eq. (17). The collision rate per unit distance traveled is given by

$$\Gamma(x) = -\pi\varrho \int_0^{\infty} db^2 \, f_b \, g(\sqrt{x^2+b^2}) \quad , \tag{19}$$

where b is the impact parameter for the collision. The quantity f_b can be written in terms of the scattering phase shifts, $\delta\,(b)$, as

$$f_b = e^{2i\delta(b)} - 1 + e^{2i\delta(b) - i\pi Qb/2} \quad . \tag{20}$$

The $\delta(b)$ are calculated in the Jeffreys–Wentzel–Kramers–Brillouin (JWKB) approximation, which is valid at high Q no matter how steeply repulsive the potential. We note that $\Gamma(\infty) = \varrho\ \sigma_{tot}(Q)/2$. The third term in Eq. (20) results in the glory oscillations of $\sigma_{tot}(Q)$. More generally, Eqs. (18–20) may be interpreted as the result of a WKB classical trajectory calculation of the final state broadening in which the quantity $\Sigma(x) \equiv -i\hbar^2 Q\Gamma(x)/2M$ serves as an *optical potential.*

PREDICTIONS FOR EXPERIMENT [26]

Figure 7 shows the final state broadening calculated for $Q = 30$ A^{-1} from Eqs. (18–20) using as input the experimental $g(r)$ obtained by neutron diffraction. One can see that the FWHM approximately obeys Eq. (12), but the broadening function goes negative at large $/Y/$ in order to satisfy the second moment sum rule. The final state broadening is not a narrow function, but rather it shifts $J(Y,Q)$ intensity among different Y regions. Figure 8 shows the integrand of the second moment sum rule over a larger range of Y. While the area under the curve is zero as required, the broadening function oscillates between positive and negative values at large $/Y/$, the scale of which is primarily controlled by the structure in $g(r)$ shown in Fig. 5.

Figure 9 shows the effect of the final state broadening on the measureable Compton profile for the normal fluid at $T = 3.3$ °K and $Q = 30$ A^{-1}, using the Path Integral Monte Carlo (PIMC) momentum distribution [10] as input to the calculation. The dashed line is the impulse approximation prediction. The solid line (FS) is the prediction after including final state broadening. Final state effects are very small for the normal fluid. The calculated momentum distribution for the normal fluid is almost Gaussian, and the second moment sum rule requires that final state effects cannot alter the Gaussian width of the Compton profile. In contrast, the pluses in Fig. 9 shows the prediction of the Lorentzian broadening theory (LZ), Eq. (11).

Figure 10 shows the same comparison for the superfluid at $T = 0.0$ °K, using as input the Green's Function Monte Carlo [9] momentum distribution. The dashed line is the impulse ap-

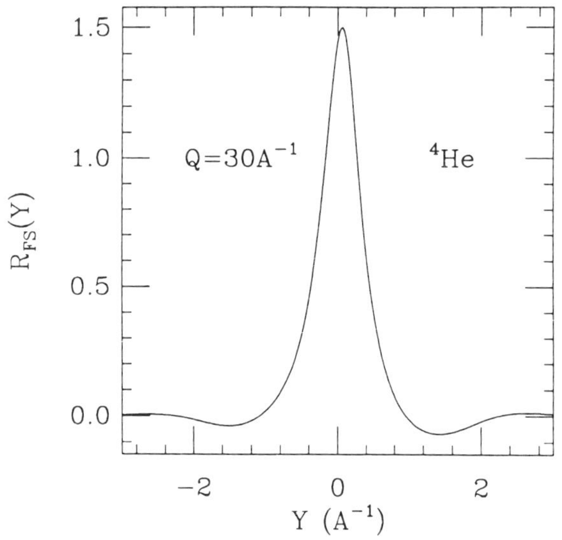

Fig. 7 Final state broadening function for ^{4}He, $R_{FS}(Y)$, predicted by Eqs. (18–20) at $Q = 30$ A^{-1}.

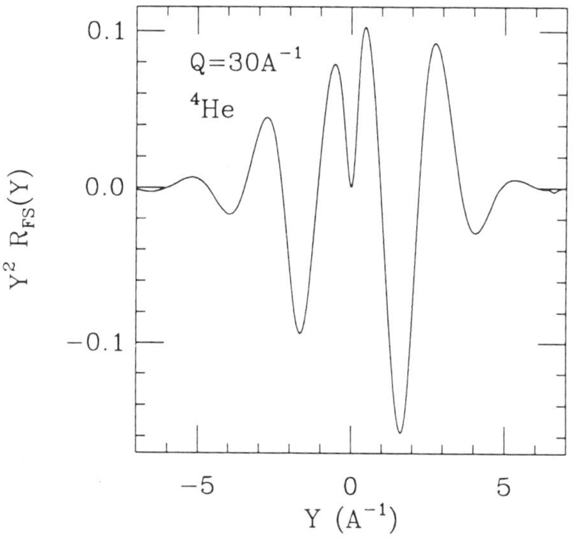

Fig. 8 A plot of $Y^2R_{FS}(Y)$ vs. Y for ^{4}He at $Q = 30$ A^{-1}. The area under this curve must be zero in order to satisfy the second moment ("kinetic energy") sum rule on the neutron scattering law.

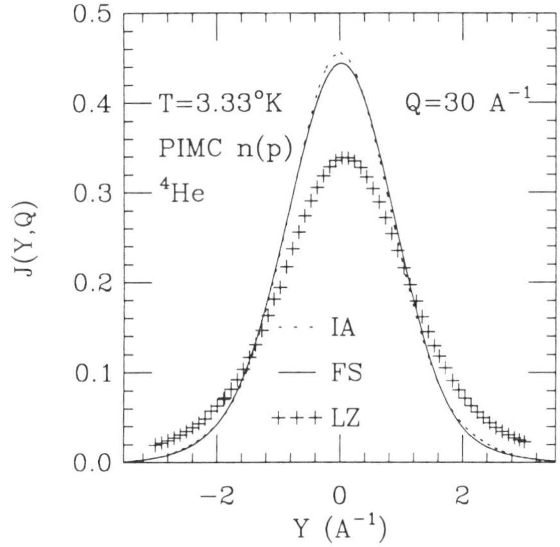

Fig. 9 Compton profiles, *J(Y,Q)*, predicted for normal liquid ^{4}He at $T = 3.33$ °K and $Q = 30$ A^{-1} using the PIMC momentum distribution. The dashed line is the prediction of the impulse approximation (IA). The solid line is the prediction of the final state effects theory (FS), Eqs. (18–20). The pluses represent the prediction of the Lorentzian broadening theory (LZ), Eq. (11).

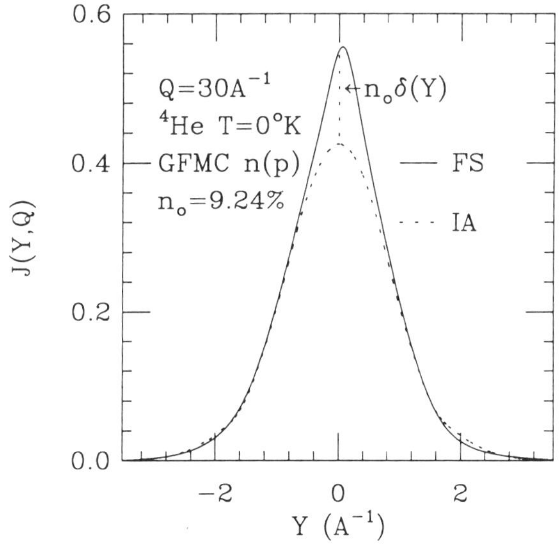

Fig. 10 Compton profiles predicted for superfluid ^{4}He at $T = 0$ °K and $Q = 30$ A^{-1} using the GFMC momentum distribution. The dashed line is the prediction of the impulse approximation (IA), and the solid line is the prediction of the final state effects theory (FS), Eqs. (18–20).

proximation prediction, where the vertical line at $Y = 0$ indicates the delta function due to the Bose condensate of weight 9.24 %. The solid line is the prediction after including final state broadening. Final state effects are much larger for the superfluid. The FWHM is scarcely changed in accordance with the second moment sum rule. But the Compton profile predicted for the superfluid is distinctly more sharply around $Y = 0$ than for the normal fluid. The Bose condensate no longer results a distinct peak sitting atop a broader background.

Figure 11 shows another way to view the results for the superfluid. The dashed line is the GFMC momentum distribution shown earlier. The solid line is the "apparent momentum distribution" which might be inferred from the final state broadened Compton profile shown in Fig. 10, obtained by analyzing the data using the impulse approximation expression, Eq. (6). This looks very similar to some earlier results on momentum distributions obtained from reactor data [11].

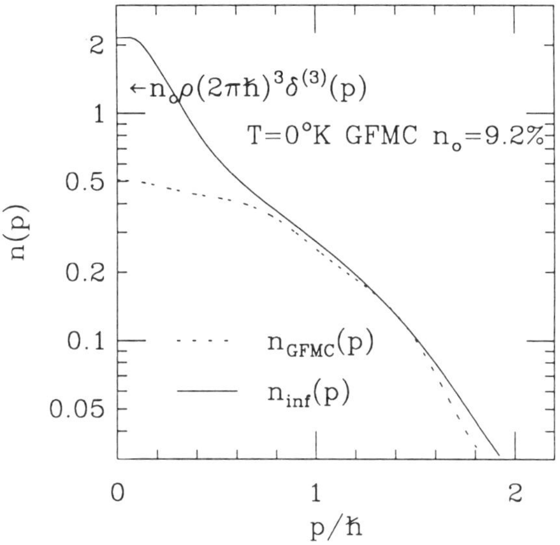

Fig. 11 The solid line is the apparent momentum distribution which would be obtained by analyzing final state broadened data, such as the solid line in Fig. 10, as if the impulse approximation was correct according to Eq. (6). The dashed line is the GFMC momentum distribution.

However, some words of caution are in order regarding comparing the results of Eqs (18–20) directly to experiment.

First, the derivation has left out a large number of terms which vary as $O(Q^{-1})$ with increasing Q. While these are not important in the asymptotic limit, they may be of importance for the Q's achievable in real experiments. These include the initial self energy of the atom in the many–body state before the neutron collision, the off–shell behavior of the t–matrix, corrections to the semiclassical approximation, etc. For example, the semiclassical methods begin to perceptibly fail at $Q < 10$ A^{-1}. Eqs. (18–20) may be fairly accurate for the high Q's achievable at pulsed neutron sources. They may require significant corrections in order to describe reactor experiments at $Q < 10$ A^{-1}. The asymmetry in the lineshapes observed at low Q's is not contained within Eqs (18–20). The apparent width oscillations observed at low Q's are not predicted by Eqs. (18–20) because the $g(r)$ in Eq. (19) suppresses the small b contribution from the third term in Eq. (20) responsible for glory oscillations of the He–He total cross section.

Second, the convolution form of the final state broadening depends on the approximation used for the two–body density matrix as a factorizable product of $n(p)$ and $g(r)$. Ristig and Clark [27] have pointed out that this choice does not satisfy a number of other known properties of the two–body density matrix. The convolution form, Eq. (10), does not hold when a more correct two–body density matrix is used as input. Clark and Ristig have recently calculated two–body density matrices for ^{4}He using a Jastrow ansatz, and they have deduced the general structural form for the two–body density matrix for any wave function. These should be used in a more accurate calculation of the scattering law. Even if the quantitative changes turn out to be small, such a calculation is conceptually important in view of the classical trajectory interpretation of the Y–variable discussed earlier.

Nevertheless, the semi–quantitative predictions of Eqs. (18–20) should be valid for $Q > 10$ A^{-1}, and the Dyson equation represented by Fig. 6 should form the basis for more precise calculations of final state effects for momentum distribution experiments by deep inelastic scattering.

EXPERIMENTAL RESULTS

A new set of deep inelastic experiments [8,28,29] on liquid ^{4}He have been performed at the Intense Pulsed Neutron Source at the Argonne National Laboratory by a team led by P. Sokol and which included T. R. Sosnick, W. M. Snow, and K. Herwig. The experiments were performed on the PHOENIX chopper spectrometer, which was designed as a dedicated instrument for deep inelastic scattering experiments. Compared with the earlier low Q reactor experiments, the new experiments used the very high epithermal neutron flux of the pulsed source to reach much higher Q's, and they also had a much better characterization of the instrument resolution function which is essential to accurate lineshape studies. Compared with earlier measurements at comparable or higher Q's, the new experiments are the first to have sufficient instrumental resolution to reliably extract the lineshape.

The data analysis avoided the model fitting procedure with n_o as a free parameter which has been attempted with reactor data taken up to 1982. Instead, experiment and *ab initio* theory were directly compared without any adjustable parameters. The theory is obtained by combining the new accurate many–body calculations of momentum distributions published in 1984–87 and the theory for final state effects [21,22] first presented in 1987. It is essential to recognize that the data are further broadened by the instrument resolution function, which should be convoluted with the theoretical predictions for the Compton profile. The spectrometer resolution function is calculated by a Monte Carlo simulation. At $Q = 23$ A^{-1}, the FWHM of the instrumental broadening is comparable to that of the final state broadening predicted by Eqs. (18–20). There was also an absolute intensity calibration of the data within 5 %. Details of the instrument and how the raw data were processed to extract the Compton profiles are discussed in ref. [28].

The data [8] in Fig. 12 were taken at $Q = 23$ A^{-1} in the normal fluid at $T = 3.5$ °K. The data in Figure 13 were taken at $Q = 23$ A^{-1} in the superfluid at $T = 0.35$ °K. One can see that the data are somewhat more sharply peaked around $Y = 0$ in the superfluid, as expected if a Bose condensate were to form. The dashed line in Fig. 11 is the PIMC–IA prediction of Fig. 8, convoluted with the instrumental resolution function. The solid line is obtained after further convoluting with final state effects. One can see that the normal fluid data are in excellent agreement with both the im-

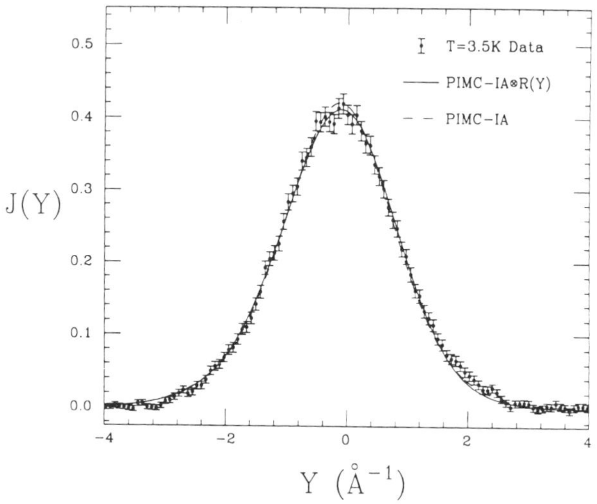

Fig. 12 The data is the measured Compton profile in normal liquid ^{4}He at $T = 3.5$ °K and $Q = 23$ A^{-1}. The dashed line is the prediction of the PIMC momentum distribution and the impulse approximation, broadened by the instrumental resolution function. The solid line is the prediction after further broadening by the final state effects function shown in Fig. 7.

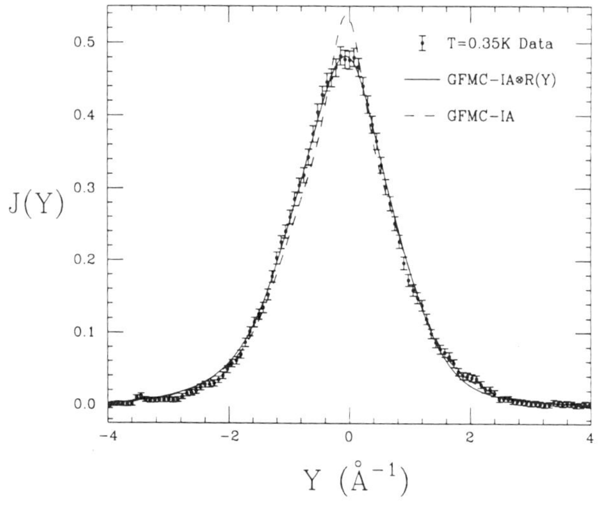

Fig. 13 The data is the measured Compton profile for superfluid liquid ^{4}He at $T = 0.35$ °K and $Q = 23$ A^{-1}. The dashed line is the prediction of the GFMC momentum distribution and the impulse approximation, broadened by the instrumental resolution function. The solid line is the prediction after further broadening by the final state effects function shown in Fig. 7.

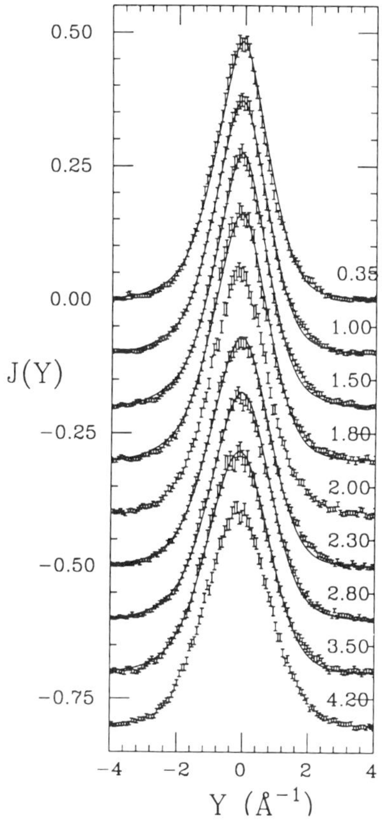

Fig. 14 Observed scattering at temperatures of 0.35, 1.0, 1.5, 1.8, 2.0, 2.3, 2.8, 3.5 and 4.2 °K. The solid lines are the theoretical predictions with instrumental resolution and final state effects included. GFMC calculations are used for comparison with the 0.35 °K results. PIMC calculations are used for the remainder of the temperatures. No calculations are available for comparison with the 2.0 and 4.2 °K measurements.

pulse approximation and the theory for final state effects given by Eqs. (18–20). However, these data disagree with Lorentzian broadening curve shown in Fig. 9 and given by Eq. (11). For the superfluid data in Fig. 12, the dashed line is again the GFMC–IA prediction of Fig. 10 again convoluted with the instrumental broadening, and the solid line is the result obtained after further broadening with final state effects. The superfluid data disagree with the impulse approximation, but they are in excellent agreement with the GFMC–IA theory broadened by final state effects.

Thus, *ab initio* theory and experiment are in excellent agreement for both the normal fluid and the superfluid with no adjustable parameters, and they are consistent with a Bose condensate fraction in the superfluid of 9.24% at $T = 0$ °K! Experiment and ab initio theory are in fact consistent for all temperatures measured, as shown in Fig. 14.

One may wonder about the importance of the condensate induced small–p singularities which played a significant role in the data analysis of the reactor experiments. Clearly the GFMC momentum distribution shown in Fig. 2 does not show these singularities, presumably because of the finite system size which can be simulated in a feasible Quantum Monte Carlo calculation. However, a variational Hypernetted Chain calculation [30] which correctly includes these singularities is also in excellent agreement with the superfluid data, and it predicts almost the same Compton profile as GFMC. The reason the singularities have a relatively small effect is that in Eq. (6) $n^*(p)$ is multiplied by p. This suppresses the contribution of the p^{-1} term in Eq. (7). However, a singular term varying as, say, p^{-4} would still result in strongly singular behavior in $J_{IA}(Y)$. Thus, the Bose condensate which is infinitely singular still shows up strongly in $J_{IA}(Y)$, even though the predicted [14] small–p singular terms do not. More generally, the inference of the low p part of $n^*(p)$ from the Compton profile data is an extremely ill–posed problem [3,28,31] in the presence of statistical error.

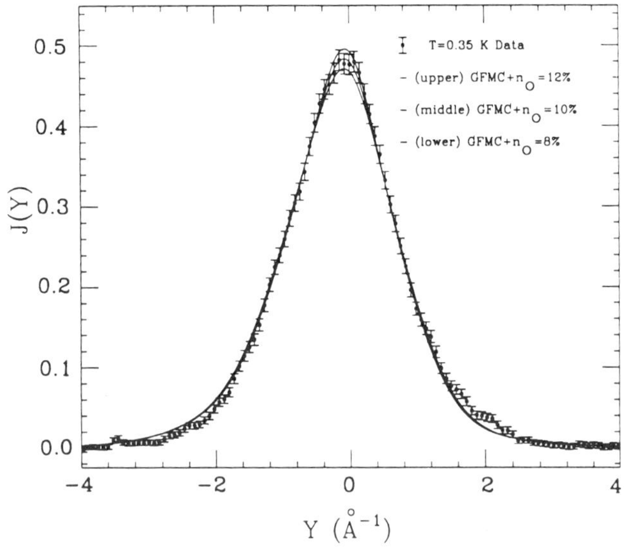

Fig. 15 Sensitivity of the observed scattering to the magnitude of the condensate fraction at $T = 0.35$ °K. GFMC calculations have been used for the uncondensed component, and a narrow Gaussian to represent the condensate. The best agreement is obtained for $n_o = 10$ %. The two limiting values, $n_o = 8$ and 12 %, are the lower and upper lines respectively.

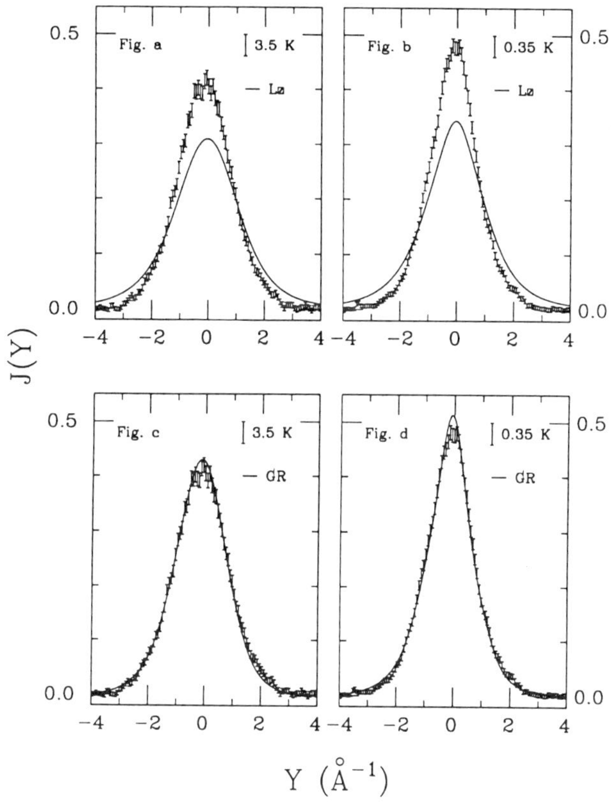

Fig. 16 Comparison of the normal and superfluid data at $Q = 23$ A $^{-1}$ with other theories for final state effects: the lines in a) & b) are the predictions of the 1966 Lorentzian broadening theory (LZ), Eq. (11); the lines in c) & d) are the predictions of the 1973 theory of Gersch and Rodriguez (GR).

One can ask how sensitive the experiment is to the actual value of the Bose condensate fraction. Fig. 15 shows the effect of changing the condensate fraction to 8 and 12 %, from the best fit value of 10 %, assuming GFMC for the non–condensed atoms and the final state broadening theory. Application of the model fitting procedure [12] to these new high Q data would yield a Bose condensate around 9 %, provided that the low–p singularities are treated correctly [15] and that corrections due to final state effects are added to the procedure. A more complete discussion is presented in ref. [28].

Assuming that the momentum distribution calculations are correct, one can use these data to test the other theories for final state broadening [32]. Figure 16 (a & b) shows the comparison of the data with the Lorentzian broadening prediction [33] obtained by taking the limit of a structureless fluid, i.e. $g(r) \rightarrow 1.0$ in Eqs. (18–20). The disagreement is strong for both the normal and the superfluid. Figure 16 (c & d) shows the comparison with the original theory of Gersch and Rodriguez [17,18,34]. The final state broadening is qualitatively correct, but it is quantitatively too small compared to the data. However, considering the necessary crudeness of the approximations made (eikonal approximation for the t–matrix, step function approximation for $g(r)$, etc.) at the time, the agreement between the Gersch and Rodriguez 1973 theory and the latest 1989 experiment is impressive.

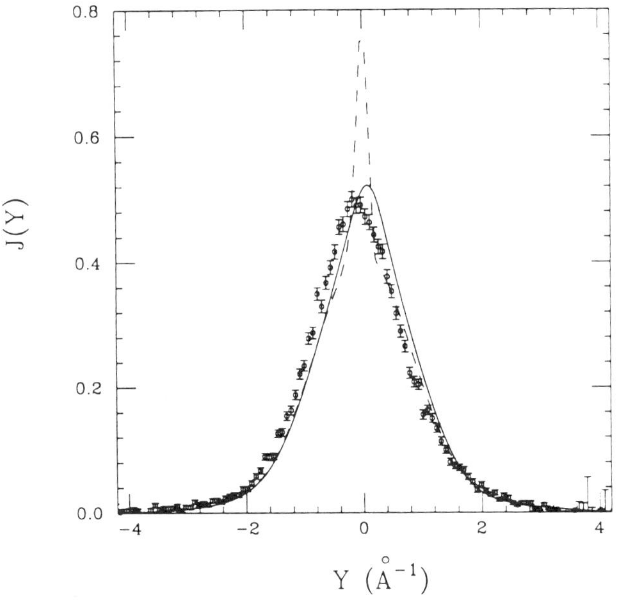

Fig. 17 Preliminary data for the superfluid at Q = 13 A^{-1} and T = 0.7 °K. The dashed line is the prediction of the impulse approximation using the GFMC momentum distribution and broadened by the instrumental resolution function. The solid line is the prediction after further broadening by the final state effects predicted by Eqs. (18–20). See the text.

Recently new experiments [29] have been performed at lower Q's for which the instrumental resolution is much smaller than the final state broadening. Figure 17 shows preliminary data for the superfluid at $T = 0.7$ °K and Q = 13 A^{-1}. Again the dashed line is the GFMC–IA prediction at $T = 0.0$ °K convoluted with the (much narrower) instrumental resolution function at this value of Q. The instrument broadened contribution of the Bose condensate is much sharper than the data. The solid line shows the prediction after including final state broadening. Now, the peak shape and height are in excellent agreement with experiment.

However, there is a shift in the data to negative Y compared with theory by about one channel, approximately 0.07 A^{-1}. The origin of this discrepancy is unclear. A simple argument [35] suggests that this shift is inconsistent with f– and second moment sum rules on $S(Q, \omega)$ by more than nine standard deviations. A similar shift of the same magnitude is seen at $Q = 18$ A^{-1} and at a variety of temperatures, and indeed the original data at $Q = 23$ A^{-1} were also shifted by the same amount. If the Y–scale were shifted by the small amount required to satisfy the f–sum rule, the agreement between the final state broadened GFMC–IA theory and experiment would again be excellent, presenting an even more convincing case for the final state effects theory in view of the superior instrumental resolution at these lower Q's. However, the correct calibration of the Y–scale should ultimately be resolved by the experimentalists.

IMPLICATIONS FOR FUTURE MOMENTUM DISTRIBUTION STUDIES

Despite the few uncertainties discussed in the previous sections, one must conclude that *ab initio* many–body theory, the theory of final state effects, and neutron scattering experiment have all converged for deep inelastic neutron scattering on liquid ^{4}He. Although a sharp peak in the scattering law due to a Bose condensate appears to be unobservable by deep inelastic scattering, experiment and theory are consistent with a Bose condensate fraction in the superfluid of 9.2 % at zero temperature. In this section, we discuss the implications of this achievement for future deep inelastic scattering studies.

First, there remains much to do on liquid ^{4}He. An analysis of the existing extensive data at very low Q ($\lesssim 10$ A^{-1}) may require significant additonal inputs to the final state effects theory for the many $O(Q^{-1})$ effects which have been left out in the asymptotic limit of high Q and hard sphere interactions. In the currently accessible Q range (< 30 A^{-1}), higher accuracy and lower background measurements may begin to reveal the two–body density matrix effects which have been discussed by Clark and Ristig [27]. These include possible deviations from the convolution form of the final state broadening. Most importantly, higher Q and higher resolution experiments on ^{4}He are needed to further test and refine the theory of final state effects. Figure 18 shows the predictions in the region near $Y = 0$ for superfluid ^{4}He for much larger Q's. Even at the experimentally unfeasible value of $Q = 270$ A^{-1} no distinct condensate peak is predicted in $J(Y,Q)$. These predictions depend on the He–He potential at short distances [36] which has been calculated, but has not been adequately measured, by atom–atom scattering experiments. However, since higher Q experiments require a percentage energy resolution which varies as $O(Q^{-1})$, they may be extremely difficult to perform. There is also great interest in experiments aimed at measuring the Bose condensate fraction for ^{4}He in other physical conditions such as in 2–dimensional films, in disordered media, ^{3}He/^{4}He mixtures, etc. Presumably the final state effects theory discussed here can be adapted with minimal changes to describe these situations.

The atomic system of most interest to study next will be normal liquid ^{3}He, because of the controversy about the existence of a sharp Fermi surface discontinuity in the momentum distribution, as shown in Fig. 19. A non–interacting (*ideal*) gas of Fermions would have a discontinuity of 1.0, several many–body calculations for real ^{3}He [37,38] predict a discontinuity of approximately 0.3, and novel alternative wave functions have been proposed [39] which predict a Fermi surface discontinuity is 0.0. In the impulse approximation the effect of a Fermi surface disconti-

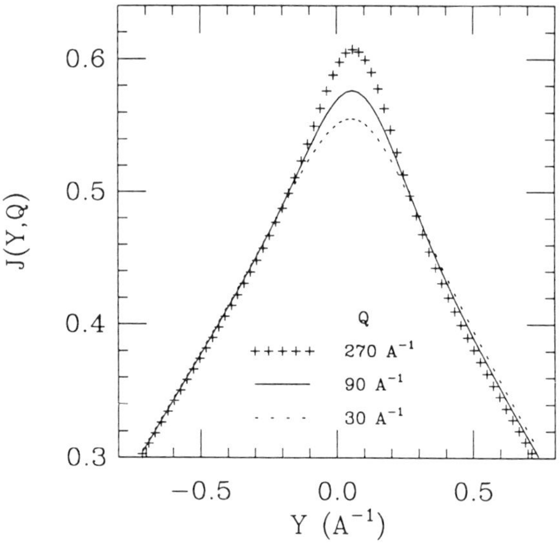

Fig. 18 Compton profiles with increasing Q for superfluid ^{4}He at $T = 0$ °K predicted by the final state effects theory. Only the region near $Y = 0$ A^{-1} relevant to the Bose condensate is shown. The condensate does not result in a distinct peak at any experimentally feasible Q.

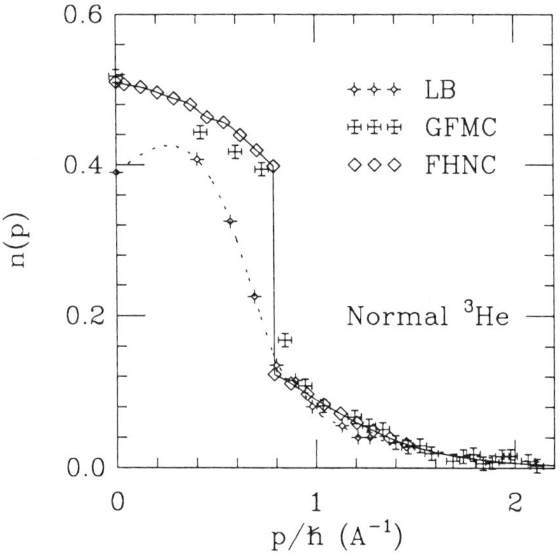

Fig. 19 Theoretical momentum distributions for normal liquid ^{3}He. The diamonds represent the Fermi Hypernetted Chain (FHNC) prediction with Fermi surface discontinuity at $p_F = 0.789$ A^{-1}. The crosses represent the Greens' Function Monte Carlo (GFMC) result. The small crosses represent the prediction of the BCS pairing theory of Lhuillier and Bouchaud (LB) which lacks a Fermi surface discontinuity.

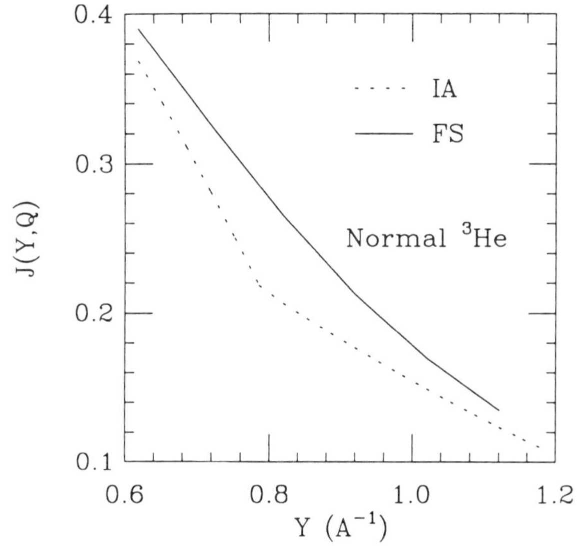

Fig. 20 Part of the Compton profile for normal liquid ^{3}He in the region near $Y = p_F/\hbar$. The dashed line is the impulse approximation (IA) prediction using the FHNC momentum distribution, showing a change in slope at $Y = p_F/\hbar$. The solid line (FS) is the prediction after broadening by final state effects, in which the discontinuity in slope is smoothed out.

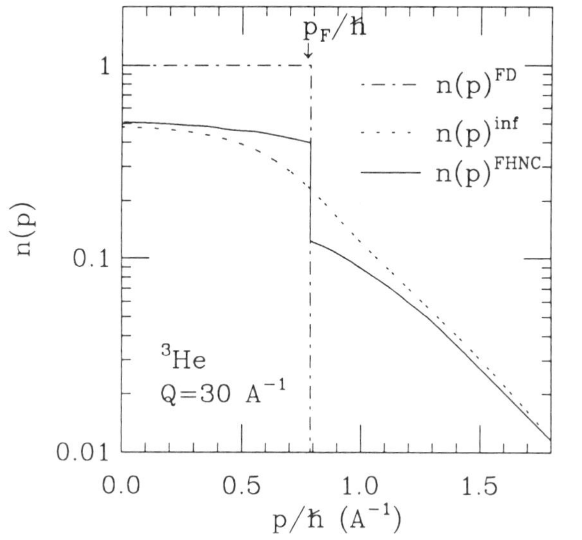

Fig. 21 Comparison of momentum distributions for ^{3}He. The curve labeled $n(p)^{FD}$ is the Fermi–Dirac distribution for a non–interacting gas at the same density, with discontinuity of 1.0 at the Fermi surface. The curve labeled $n(p)^{FHNC}$ is the Fermi Hypernetted Chain (FHNC) prediction with discontinuity of 0.3. The dashed line, $n(p)^{inf}$, is the apparent momentum distribution which would be obtained by analyzing final state broadened data as if the impulse approximation were correct.

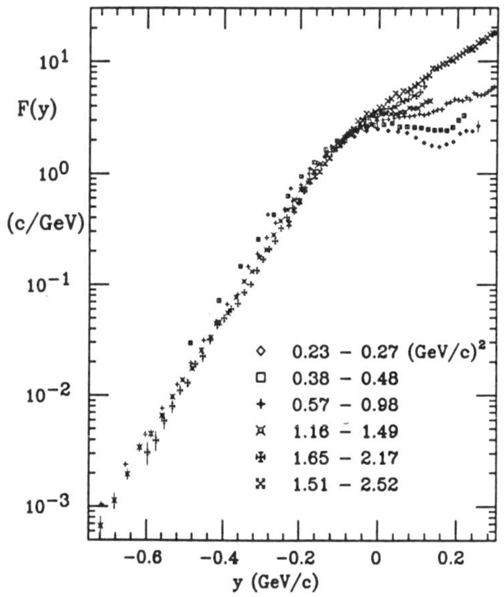

Fig. 22 Quasielastic electron nucleus scattering (QENS) from ^{12}C at several different relativistic 4–momentum transfers. For $-Y > 0.1$ GeV/c the results all fall on the same curve, illustrating the *Y*–scaling behavior. *Y*–scaling breaks down for larger values of *Y* due to the excitation of internal degrees of freedom of the nucleon.

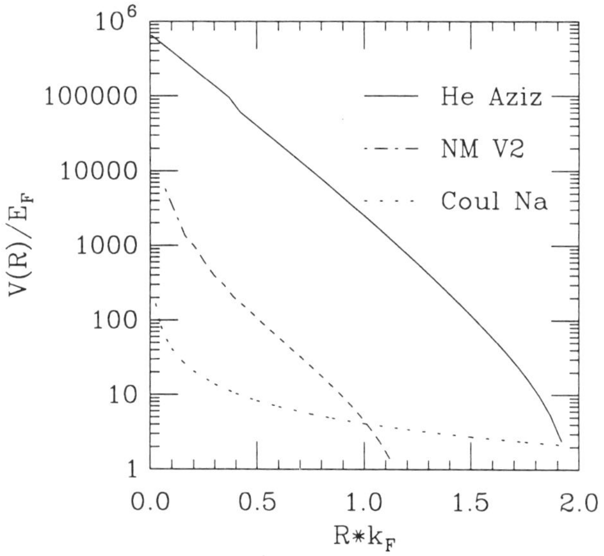

Fig. 23 Comparison of repulsive cores of the interaction potentials for atomic, electronic and nuclear systems. The potentials are scaled by the Fermi energy, E_F, and the distances are multiplied by the Fermi wave vector, k_F. The solid line is the ^{3}He Aziz potential, the long dashed line is the nuclear matter v_2 potential, and the short dashed line is the Coulomb potential for Na.

nuity on the Compton profile is a change in slope at Y equal to the Fermi wave vector, $k_F = 0.789$ A^{-1}. The predictions for the Compton profile for Y near k_F are shown in Fig. 20. With the inclusion of final state broadening, even this change in slope is predicted to be smoothed over. In Fig. 21 the apparent momentum distribution for ^3He is displayed in an analogous fashion to Fig. 11 for ^4He. The apparent Fermi surface obtained by analyzing the final state broadened Compton profile assuming the impulse approximation relation, Eq. (6), is unobservable due to final state effects. Nevertheless, as was the case for the "unobservable" Bose condensate peak in ^4He, one should still be able to test the *ab initio* many body calculations for ^3He by deep inelastic scattering experiments. Such experiments are a formidable undertaking because of the additional complication of the very strong neutron absorption in ^3He.

Finally, we turn to quasielastic electron nucleus scattering [40,41]. In this case, the term "quasi–elastic" refers to the absence of change in the rest mass of the nucleon in the scattering process, even though the energy and momenta transferred are large compared to the energies characterizing collective behavior. The term "deep–inelastic" is reserved for experiments which probe the quark substructure of the nucleon. Despite this unfortunate semantic confusion, the physics is the same as for neutron scattering from He: i.e. the energies and momenta transferred by the electron are much larger than the binding energies of nucleons inside nuclei. Figure 22 shows a Y–scaling plot for electron scattering from the ^{12}C nucleus, where the relativistic 4–momentum range for each data set is shown and the Y–variable has been generalized to relativistic kinematics. This is meant to be analogous to Fig. 1 for deep inelastic neutron scattering from ^4He. Figure 22 shows that the Y–scaling is observed for negative Y, but that *Y–scaling is broken* for positive Y because of the possibility of exciting internal degrees of freedom such as the Δ (1238 MeV) resonance. The critical question is whether the impulse approximation can be applied to extract the momentum distribution from these data, and in particular whether the quasi–exponential dependence of the Compton profile on Y observed over four orders of magnitude reflects a quasi–exponential dependence of the nucleon $n(p)$ in nuclei.

To understand the possible importance of final state effects in this problem, we consider a comparative plot of the nucleon–nucleon potential [42], the He–He potential [36], and the Coulomb potential for Na shown in Fig. 23. The potentials are scaled by the Fermi energy and the distances are scaled by the Fermi momentum. The Coulomb potential is the softest at short distances, which is the reason that final state effects are relatively unimportant for x–ray Compton scattering experiments on electron momentum distributions. The nucleon–nucleon potential is much more steeply repulsive at short distances, and the He–He potential is several orders of magnitude harder than that. Thus, we expect that the impulse approximation will be approached more quickly with increasing Q for nuclear physics than for He. However, kinematics requires that Q and ω should be low enough to avoid excitation of internal degrees of freedom of the nucleon. Comparatively, while neutron scattering on ^3He can reach Q/k_F of more than fifty, quasielastic electron nucleus scattering is restricted to Q/k_F less than ten. The combination of these two effects leads to the expectation that final state effects should be as important for quasielastic electron nucleus scattering as they are in deep inelastic neutron scattering from He.

Detailed calculations for quasielastic electron nucleus scattering have not yet been carried out [6]. They suffer from uncertainties regarding the appropriate choice of nuclear potential. They potentially require inclusion of all the $O(Q^{-1})$ effects which were left out of the asymptotic theory for neutron scattering from He. And the relativistic effects are a further essential complication. Nevertheless, we expect that the leading term of the theory to be represented by the same Dyson equation shown in Fig. 6.

ACKNOWLEDGEMENTS

Richard N. Silver is supported by the U. S. Department of Energy under Office of Basic Energy Sciences / Division of Materials Science (OBES/DMS) support of the Manuel Lujan Jr. Neutron Scattering Center (LANSCE). Paul E. Sokol is supported by NSF Grant DMR–8704288 and OBES/DMS support of the Intense Pulsed Neutron Source at Argonne National Laboratory. We thank our colleagues whose work we report here, in particular T. R. Sosnick, W. M. Snow, and K. Herwig. We also thank J. W. Clark for earlier collaborations on much of this work.

REFERENCES

[1] F. London, Nature **141**, 643 (1938).

[2] For a tutorial discussion of the relation beween Bose condensation and superfluidity see D. R. Tilley, J. Tilley, *Superfluidity and Superconductivity*–2nd edition, Adam Hilger Ltd., 1986.

[3] For a complete review of the current state of the art, see *Momentum Distributions*, R. N. Silver, P. E. Sokol, eds., Plenum Press, 1989. For an introductory survey of momentum distribution studies across all of physics, see the article by P. E. Sokol, R. N. Silver, J. W. Clark, p. 1–38.

[4] P. C. Hohenberg, P. M. Platzman, Phys. Rev. **152**, 198 (1966); see also A. Miller, D. Pines, P. Nozieres, Phys. Rev. **127**, 1452 (1962).

[5] G. B. West, Physics Reports **18C**, 263 (1975); see also G. B. West, in ref. [3], p. 95–110.

[6] J. W. Clark, R. N. Silver, *Proceedings of the Third International Conference on Nuclear Reaction Mechanisms*, Varenna, Italy, June 13–18, 1988, E. Gadioli, ed. (Ricerca Scientifica ed Educazione Permanente, Universita degli Studi di Milano), Supplemento 66, p. 531–540 (1988).

[7] W. G. Stirling, E. F. Talbot, B. Tanatar, H. R. Glyde, J. Low Temperature Physics, **73**, 33 (1988)

[8] T. R. Sosnick, W. M. Snow, P. E. Sokol, R. N. Silver, Europhysics Letters **9**, 707 (1989).

[9] P. Whitlock, R. M. Panoff, Can. J. Phys. **65**, 1409 (1987).

[10] D. M. Ceperley, E. L. Pollock, Can. J. Phys. **65**, 1416 (1987); see also D. M. Ceperley, in ref. [3], p. 71–80.

[11] For reviews, see H. R. Glyde, E. C. Svensson, in Methods of Experimental Physics, V. **23B**, D. L. Price, K. Skold, eds., Academic Press, 1987, p. 303–404; E. C. Svensson, V. F. Sears, Physica **137B**, 126–140 (1986).

[12] V. F. Sears, E. C. Svensson, P. Martel, A. D. B. Woods, Phys. Rev. Lett. **49**, 279 (1982).

[13] P. Martel, E. C. Svensson, A. D. B. Woods, V. F. Sears, R. A. Cowley, J. Low Temp. Phys. **23**, 285 (1976).

[14] J. Gavoret, P. Nozieres, Ann. Phys. (N.Y.) **28**, 349 (1964); P. C. Hohenberg, P. C. Martin, Ann. Phys. (N. Y.) **34**, 291 (1965).

[15] A. Griffin, Phys. Rev. B**32**, 3289 (1985).

[16] R. K. B. Helbing, J. Chem. Phys. **50**, 493 (1969).

[17] H. A. Gersch, L. J. Rodriguez, Phys. Rev. A**8**, 905 (1973).

[18] L. J. Rodriguez, H. A. Gersch, H. A. Mook, Phys. Rev. A**9**, 2085 (1974).

[19] P. Martel, E. C. Svensson, A. D. B. Woods, V. F. Sears, R. A. Cowley, J. Low Temp. Phys. **23**, 285 (1986).

[20] See also H. R. Glyde, W. G. Stirling, in ref. [3], p. 111–122.

[21] R. N. Silver in *Condensed Matter Theories* V. 3, J. S. Arponen, R. F. Bishop, M. Manninen, eds., p. 131–142, Plenum Press, 1988.

[22] R. N. Silver, Phys. Rev. B**37**, 3794 (1988); ibid B**38**, 2283 (1988). The latter paper contains a rather complete list of theoretical papers on final state effects in deep inelastic neutron scattering.
[23] A. Rinat, M. Butler, to be published.
[24] R. W. Zwanzig, Physica (Utrecht) **30**, 1109 (1964); see also P. N. Argyres, J. L. Sigel, Phys. Rev. Letts. **31**, 1397 (1973); Phys. Rev. B**9**, 3197 (1974).
[25] R. K. B. Helbing, J. Chem. Phys. **50**, 493 (1969).
[26] R. N. Silver, Phys. Rev. B**39**, 4022 (1989).
[27] M. L. Ristig, J. W. Clark, in ref. [3], p. 365–370; M. L. Ristig, J. W. Clark, Phys. Rev. B**40**, 4355 (1989).
[28] P. E. Sokol, T. R. Sosnick, W. M. Snow, in ref. [3], p. 139–158.
[29] K. Herwig, W. M. Snow, P. E. Sokol, to be published.
[30] E. Manousakis, V. R. Pandharipande, Q. N. Usmani, Phys. Rev. B**31**, 7022 (1985); E. Manousakis, V. R. Pandharipande, ibid., 7029; see also, E. Manousakis, in ref. [3], 81–94.
[31] D. S. Sivia, R. N. Silver, in ref. [3], p. 377–380.
[32] P. E. Sokol, R. N. Silver, T. R. Sosnick, W. M. Snow, in ref. [3], 385–392; W. M. Snow, T. R. Sosnick, P. E. Sokol, R. N. Silver, to be published.
[33] This is essentially the same as P. M. Platzman, N. Tzoar, Phys. Rev. B**30**, 6397 (1984).
[34] We thank Prof. Gersch for coming out of retirement to perform these calculations.
[35] R. N. Silver, to be published.
[36] R. Feltgen, H. Kirst, K. A. Koehler, F. Torello, J. Chem. Phys. **26**, 2360 (1982).
[37] P. Whitlock, R. M. Panoff, Can. J. Phys. **65**, 1409 (1987); see also R. M. Panoff, P. A. Whitlock, in ref. [3], p. 59–70.
[38] E. Manousakis, S. Fantoni, V. R. Pandharipande, Phys. Rev. B**28**,3370 (1983).
[39] J. P. Bouchaud, C. Lhuillier, Europhys. Lett. **3**, 1273 (1987); J. P. Bouchaud, C. Lhuillier, Z. Phys. B**75**, 283 (1989).
[40] For an experimental review, see I. Sick, in ref. [3], p. 175–186; D. Day, in ref. [3], p. 319–332.
[41] For a theoretical discussion, see O. Benhar, A. Fabrocini, S. Fantoni, in ref. [3], p. 187–202.
[42] V. R. Pandharipande, R. B. Wiringa, B. D. Day, Phys. Lett. **57**B, 205 (1975).

BALLISTIC ELECTRONIC TRANSPORT IN SEMICONDUCTOR STRUCTURES

M. Pepper

Cavendish Laboratory
Madingley Road
Cambridge, CB3 0HE
U.K.

A brief discussion is presented on aspects of ballistic transport in small structures exploiting the high mobility electron gas in GaAs–AlGaAs heterojunctions. In this regime of transport the conductance is quantised and has value $\frac{2e^2 i}{h}$ where i is the number of one dimensional subbands. Lifting the spin degeneracy with magnetic field, either transverse or parallel, results in removal of the factor of 2 in the quantised conductance. It is shown that a variety of quantum interference phenomena can be observed with such structures, including Aharonov-Bohm oscillations in singly connected structures.

INTRODUCTION

Recent developments in the technology of the growth of semiconductors allow the formation of very high quality GaAs–AlGaAs heterojunctions. This results in a two dimensional electron gas with values of mobility at 4.2K well in excess of 10^6 cm^2 volt^{-1} sec^{-1} at values of carrier concentration of order 10^{11} cm^{-2} to 5.10^{11} cm^{-2}, corresponding to a scattering mean free path > 10^{-4} cm. This is considerably greater than the size of samples which can be fabricated by high resolution lithography.

Fabrication of a MOS Silicon type structure with a controlling gate allows control of carrier concentration. However, as first demonstrated by Thornton et al,[1] application of a negative voltage to two Schottky gates on top of a GaAs-AlGaAs heterojunction separated by about 10^{-4} cms results in the 2D electron gas being "electrostatically squeezed" to form a one dimensional (1D) line, so allowing a comprehensive investigation of transport phenomena in this dimension.[2] If the length of the electron gas defined by the split gates is sufficiently small, then the ballistic range can be entered in which the scattering mean free path exceeds the channel length. As first demonstrated by Wharam et al[3] and van Wees et

Recent Progress in Many-Body Theories, Volume 2
Edited by Y. Avishai
Plenum Press, New York, 1990

al[4] this leads to a new regime of behaviour in which the resistance is quantised. This will now be discussed in further detail.

QUANTISATION OF THE BALLISTIC RESISTANCE

A simple, physical model, of the quantisation can be obtained in the following way. The net current , I, flowing through a 1D channel can be written as

$$J = \int_0^{E_F} \frac{eN(E)v}{2}\, dE - \int_{eV}^{E_F} \frac{eN(E)v}{2}\, dE \qquad (1)$$

where V is the applied voltage, e is the electronic charge, v is the group velocity at energy E, E_F is the Fermi energy and the factor of 1/2 arises from the need to consider only that half of the electron density which flows along each direction in the channel.

The density of states, including spin degeneracy N(E) is $\frac{2}{h}\left(\frac{2m}{E}\right)^{1/2} = \frac{4}{hv}$, giving a value of J,

$$J = \frac{2e^2}{h} V. \qquad (2)$$

Thus the conductance σ is quantised having a value of $\frac{2e^2}{h}$, arising from a cancellation of the group velocity. If there are i 1D subbands due to the absence of intersubband scattering, the total channel conductance becomes $\frac{2e^2 i}{h}$. As the width of the channel, or the Fermi energy, changes so in turn i changes by one and the conductance changes by $\frac{2e^2}{h}$. In the split gate technique, changing the negative gate voltage results in a jump between quantised values, as first found by Wharam et al[3] and van Wees et al[4], an example of which is shown in figure 1. The theory of one dimensional quantisation was first considered by Landauer[5], Imry[6] and Buttiker[7], as pointed out by these authors it is necessary to include the transmission ratio T, giving

$$\sigma = \frac{2Te^2 i}{h}. \qquad (3)$$

Lifting the spin degeneracy by a magnetic field results in the conductance change decreasing from $\frac{2e^2}{h}$ to $\frac{e^2}{h}$. This is most readily shown by a magnetic field which is in the plane of the 2D electron gas, figure 2, as a transverse field will produce a transition to the Quantum Hall Effect. As seen in figure 2, the lifting of the spin degeneracy is most effective at high quantum number, i. This has been explained by a coupling between the spin and the 1D quantised wavefunction.[8]

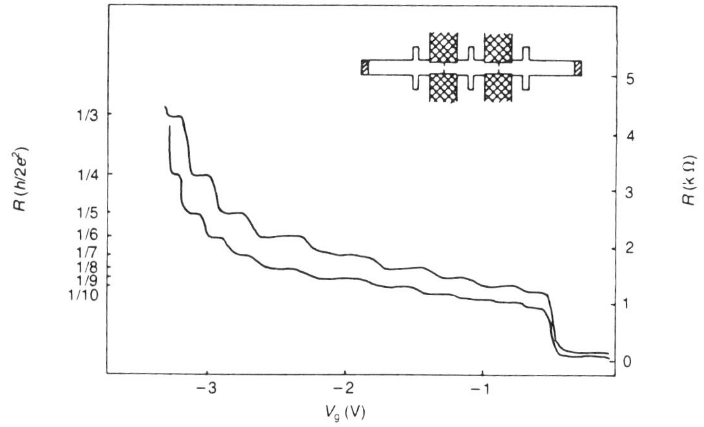

FIGURE 1

Here we illustrate the quantised plateaux which appear as a function of the "squeezing" voltage V_g at 0.1K. The results are shown for two different values of carrier concentration, requiring different voltages to produce a particular plateau, and the resistance is shown in both kΩ and units of $h/2e^2$. The inset illustrates narrow pairs across a Hall bar, the shaded region indicates the large area gate metallisation. Only one narrow pair is accessed for a particular experiment, current passes through the Hall probes; from reference 3.

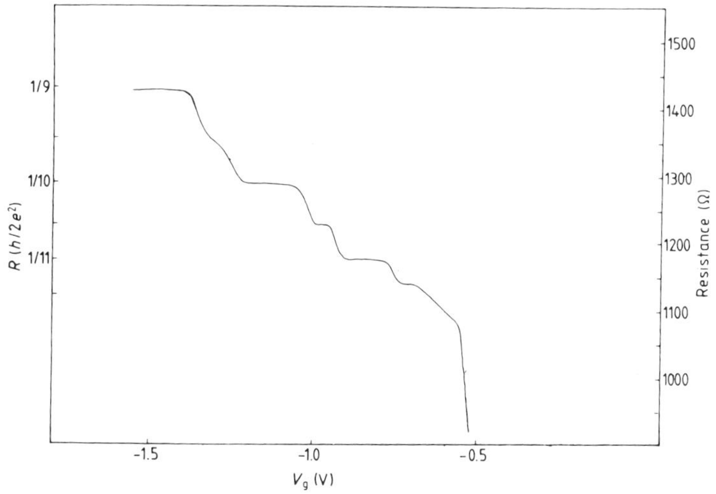

FIGURE 2

Similar results as in figure 1 but in a parallel magnetic field of 13.6 Tesla spin split plateaux are clearly apparent at half integer values of $\frac{1}{i}$, at $\frac{1}{9.5}$, $\frac{1}{10.5}$ and $\frac{1}{11.5}$ the temperature of measurement is 0.1K.

As is clear from the preceding derivation of the quantised conductance, the relation breaks down when the applied voltage is sufficiently great that $eV \geq E_F$. From equation (3) it is seen that now the current carried by the subband becomes a constant, independent of V, at a value $\frac{2e^2}{h} E_F$.[9.10] This saturation will not be seen if many subbands are occupied. However an increase in resistance is observed and is illustrated in figures 3 and 4. Calculations on the transmission of ballistic electrons[9] show that the factor T decreases as V increases leading to the possibility of differential negative resistance (DNR). Measurements of the current voltage characteristic show that the increase in resistance is accompanied by considerable noise which may indicate instabilities due to the onset of DNR, figure 4.

QUANTUM INTERFERENCE

Smith et al[11] first showed that deviations from the quantised ballistic resistance could be seen clearly by inserting a quantum box between two ballistic resistors. Destructive interference in the zero dimensional box results in resistance peaks interposed between minima which are near quantised values. This technique has been recently extended into the Fabry-Pérot interferometry of electron waves in which a 1D resistor is placed in a cavity defined by reflector gates[12].

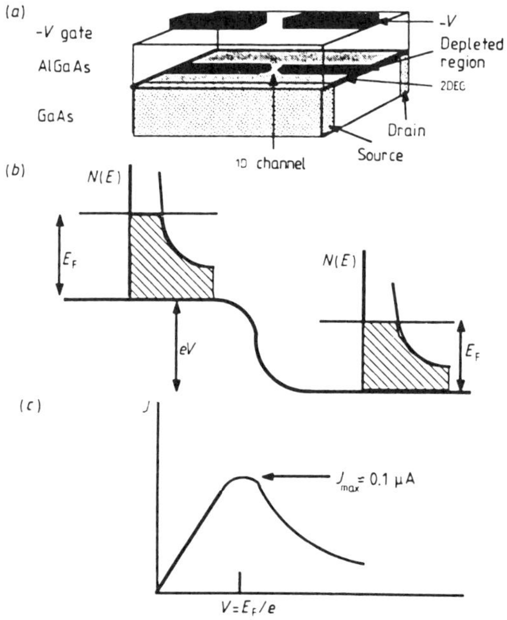

FIGURE 3

A diagram from Kelly et al[9] illustrating (a) the action of the split gate in narrowing the channel, (b) the action of a high applied voltage on the density of states N(E) and preventing electrons diffusing in one direction by lowering the Fermi energy and (c) the suggested current-voltage curve showing differential negative resistance.

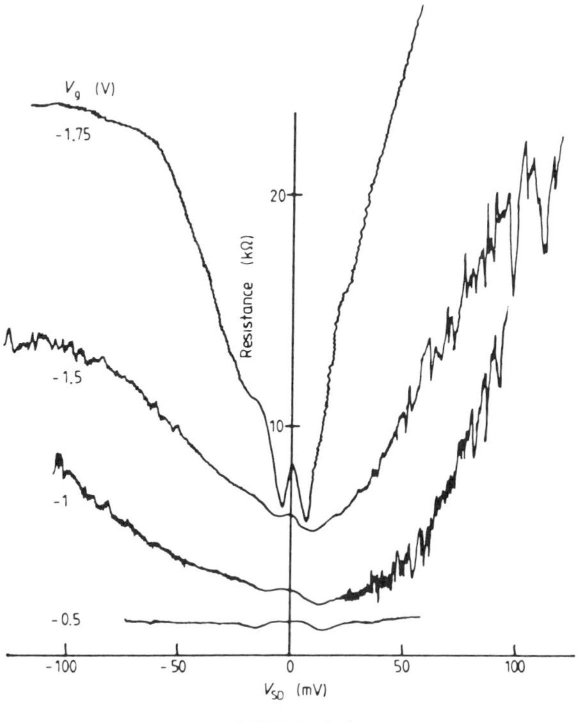

FIGURE 4

Resistance plotted against applied source-drain voltage, V_{SD}, for different values of gate voltage V, from Brown et al[10]. The value of E_F is ≈ 15 meV, and the temperature of measurement 4.2K.

Interference effects can be seen in the presence of a magnetic field. Sivan and Imry first suggested that Aharonov-Bohm (A-B) oscillations would be present in a singly connected (non-ring) geometry sample,[13], by virtue of interference between electrons which were in edge states. Electrons flow through the sample in both bulk Landau levels and edge states, which correspond to Landau levels at the sample edge where the energy increases with decreasing separation from the edge. When the Fermi energy is between the bulk Landau levels, the edge states determine the conductivity. As they are localised within a cyclotron radius of the edge, a ring geometry is effectively formed through which flux flows so giving rise to an Aharonov-Bohm effect. If the entrance and exit regions are sufficiently narrow then wavefunction overlap occurs and a consequent interference pattern can be observed. Confirmation of this rests upon observation of A-B oscillations, the period of which is a function of the area as observed by Wharam et al,[14] van Wees et al[15] and Brown et al[16] The work of Wharam et al showed that as the width of the channel was decreased by electrostatic squeezing so the length increased by the same amount. If the depletion width is d, due to the electrostatic squeezing voltage, then the width of the channel is D – 2d where D is the separation of the split gates. The same reasoning suggests a length L + 2d where L is the length of the gates. Confirmation of this simple model gives confidence in understanding the mechanism of "electrostatic squeezing".

The relation between gate voltage and width has been investigated, a change in both carrier concentration and width, W, prevent the simple assumption that $k_F = \frac{i\pi}{W}$, with a consequent extraction of the width from the gate voltage required to depopulate the ith subband, where k_F is the Fermi wavevector for a wide channel. However analysis of magnetic depopulation produced by a transverse magnetic field which forces levels through the Fermi energy can be used to obtain information on the confining potential and width of channel.

In the experiments investigating edge current interference, the interference period is due to the change in flux through the sample changing by one quantum. However in the case of a well defined quantum box an additional periodicity arises due to a resonance in the scattering at entrance and exit which is necessary to produce the wavefunction overlap and hence the interference. This results in a further modulation of resistance of $\sin^2\phi$ type which has been explained by Yosephin and Kaveh[17]. Figure 5 illustrates the two period of quantum interference superimposed on plateaux of quantised resistance. The plateaux are related simply to the Quantum Hall Effect and arise from the reflection of edge states at the entrance to the quntum box.

QUANTISED BALLISTIC RESISTORS IN PARALLEL CONFIGURATION

Recently the nature of transport in parallel ballistic resistors has been investigated[18] A particular feature of this work was that it was possible to determine if the resistors were

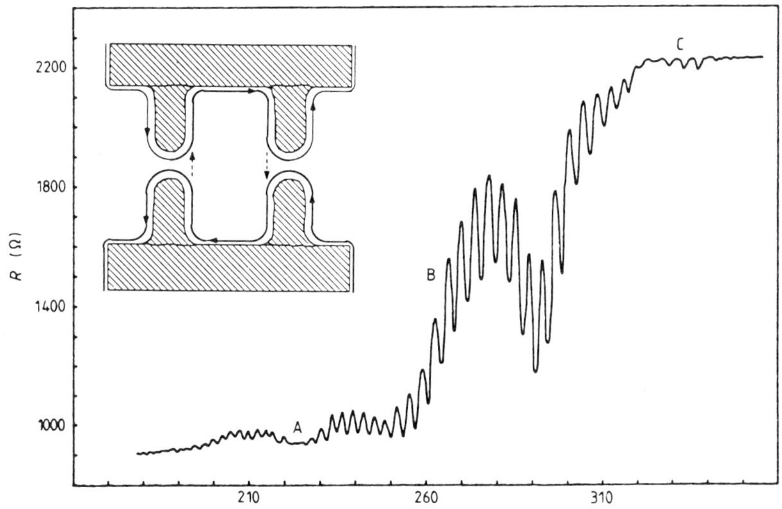

FIGURE 5

Resistance R is plotted against the voltage (in milli-volts) applied to the metal pattern forming the quantum box structure (defined in the inset) at 0.1K and in a field of 2 Tesla. This figure from reference 16 shows the depletion pattern in the inset with the electron flow indicated by arrows. The dotted line illustrates tunnelling across the exit and entry regions necessary to form an interference loop. In the absence of the loop plateaux are observed without the interference pattern superimposed, the high frequency oscillations are Aharonov-Bohm and the low frequency arises from the wavelength dependence. Region A corresponds to a particular edge current scattered around the box and interfering with the incoming wave; B occurs as the same edge current in A is progressively excluded and C is the weak effect due to this edge current tunnelling into the box.

within a phase coherence length by the observation of an A-B effect due to current circulating between the two. When the resistors were within this length it was found that they locked together, both depopulating at the same gate voltage and producing a change in conductance $\frac{4e^2}{h}$, figure 6. On the other hand when the resistors were not within a phase coherence length, as indicated by the absence of A-B oscillations, the locking mechanism was not present. In this case the combined resistance characteristics were essentially additive, clear plateaux were not seen as only one showed plateau device for a particular gate voltage whereas at this voltage the other was not in the plateau condition. At present the origin of the locking mechanism is not clear, but it may arise from a wavefunction hybridisation which occurs as a result of the form of the 1D density of states. Thus, if a small number of electrons, δn, transfers from one resistor to another then the increase in Fermi energy, δE_F, which occurs is given by $\delta E_F = \frac{\delta n}{N(E_F)}$. As $E_F \rightarrow 0$ so $N(E_F) \rightarrow \infty$ and $\delta E_F \rightarrow 0$, consequently the locking mechanism could result in a net lowering of energy. Calculations on this effect would be very relevant to the development of experiments.

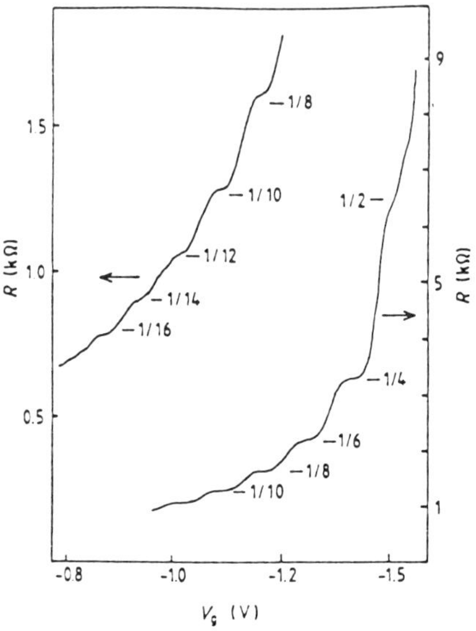

FIGURE 6

This figure, from reference 13, illustrates the locking of ballistic resistors in parallel. The plateau indices are in units of $\left(\frac{h}{2e^2}\right)$, hence a single resistor would exhibit a sequence of indices $\frac{1}{16}, \frac{1}{15}, \frac{1}{14}, \frac{1}{13}$ . . . . .; the temperature of measurement is 0.1K.

SUMMARY

A very short description of electron transport in small semiconductor structures has been given. The combination of advanced technology in the form of high resolution electron beam lithography and Molecular Beam Epitaxy has allowed entry into a new physical regime of ballistic transport. The phenomena found to date include quantisation and a range of quantum interference effects. Experience suggests that new information will emerge as the technology develops and an additional range of samples becomes available for experimentation.

ACKNOWLEDGEMENTS

This work was performed in collaboration with many colleagues including H Ahmed, R J Brown, C J B Ford, J E F Frost, D G Hasko, G A C Jones, M J Kelly, R Newbury, D C Peacock, D Ritchie, C G Smith, T J Thornton, D A Wharam.

Support was provided by the Science and Engineering Research Council of the United Kingdom, in part, by the European Research Office of the U.S. Army and, also in part, by the EEC.

REFERENCES

1. Thornton, T.J., Pepper, M., Ahmed, H., Andrews, D., Davies, G J., Phys. Rev. Lett., 56; 1198 (1986)

2. Berggren, K-F., Thornton, Pepper, M., Ahmed, H., Andrews, D., Davies, G. J., Phys. Rev. Lett. 57; 1769 (1986)

3. Wharam, D.A., Thornton, T.J., Newbury, R., Pepper, M., Ahmed, H., Frost, J.E.F., Hasko, D.G., Peacock, D.C., Ritchie, D.A., Jones, G.A.C., J. Phys. C Solid State Physics 21; L209 (1988)

4. van Wees, B.J., van Houton, H., Beenakker, C.W.J., Williamson, J.G., Kouwenhoven, L.P., van der Marel, D., Foxon, C.T., Phys. Rev. Lett. 60; 848 (1988)

5. Landauer, R., Phys. Rev. Lett., 85A; 91 (1981)

6. Imry, Y., in "Directions in Condensed Matter Physics", ed. G. Grinstein and G. Mazenko (World Scientific, Singapore, 1986) Vol. 1, 102

7. Buttiker, M., Phys. Rev.Lett. 57; 1761 (1986)

8. Glazman, L.J., Lesovik, G.B., Khmelnitskii, D.E., and Shekhter, R J. JETP Lett. 48; 238 (1988)

9. Kelly, M.J., Brown, R.J., Smith, C.G., Wharam, D.A., Pepper, M., Ahmed, H., Hasko, D.G., Peacock, D.C., Frost, J.E.F., Newbury, R., Ritchie, D.A., Jones, G.A.C., Electronics Letters, 25; 993 (1989)

10. Brown, R.J., Kelly, M.J., Pepper, M., Ahmed, H., Hasko, D.G., Peacock, D.C., Frost, J.E.F., Ritchie, D.A., Jones, G.A.C., J.Phys: Condens. Matter 1 6285 (1989) Kelly, M.J., J. Phys; Condens. Matter 1; 7643 (1989)

11. Smith, C.G., Pepper, M., Ahmed, H., Frost, J.E.F., Hasko, D.G., Peacock, D.C., Ritchie, D.A., Jones, G.A.C., J.Phys. C21; L893 (1988)

12. Smith, C.G., Pepper, M., Newbury, R., Ahmed, H., Hasko, D.G., Peacock, D.C., Frost, J.E.F., Ritchie, D.A., Jones, G.A.C., Hill, G., J. Phys: Condens. Matter 1; 9035 (1989)

13. Sivan, U., Imry, Y., Phys. Rev. Lett. 61; 1001 (1988)

14. Wharam, D.A., Pepper, M., Newbury, R., Ahmed, H., Hasko, D.G., Frost, J.E.F., Peacock, D.C., Ritchie, D.A., Jones, G.A.C., J.Phys; Condens. Matter 1; 3369 (1989)

15. van Wees, B.J., Kouwenhoven, L.P., Harmans, C.J.P.M., Williamson, J.G., Timmering, C.E., Broekaart, M.E.F, Foxon, C.T., Harris, J.J., Phys. Rev. Lett. 62; 2523 (1989)

16. Brown, R.J., Smith, C. G., Pepper, M., Kelly, M.J., Newbury, R., Ahmed, H., Hasko, D.G., Frost, J.E.F., Peacock, D.C., Ritchie, D.A., Jones, G.A.C., J.Phys: Condens. Matter 1; 6291 (1989)

17. Yosephin, M. and Kaveh, M., J. Phys; Condens. Matter, in the press

18. Smith, C.G., Pepper, M., Newbury, R., Ahmed, H., Hasko, D.G., Peacock, D.C., Frost, J.E.F., Ritchie, D.A., Jones, G.A.C., Hill, G., J. Phys: Condens. Matter 1; 6763 (1989)

RANDOM-MATRIX-MODELS FOR FLUCTUATIONS IN MICROSCOPIC AND MESOSCOPIC SYSTEMS

Hans A. Weidenmüller

Max Planck-Institut für Kernphysik
6900 Heidelberg, West Germany

ABSTRACT

In microscopic systems, statistical fluctuations arise as a consequence of chaotic motion. In mesoscopic systems, their origin resides in disorder. In both cases the complex Hamiltonian dynamics can be simulated in terms of random-matrix models having the correct symmetry properties. From such models, mean values and variances of observables can be calculated with the help of a technique that uses a generating function written as an integral over commuting and anticommuting variables. The following examples are discussed. (i) Statistical nuclear cross-sections; (ii) Observables testing the breaking of parity (and time-reversal) symmetry in statistical nuclear reactions; (iii) Signals for chaotic motion in molecular physics; (iv) Conductance fluctuations in mesoscopic systems.

1 INTRODUCTION

In this paper I consider stochastic fluctuations in microscopic and mesoscopic systems. Here, microscopic stands for atoms, molecules, atomic nuclei—in short for quantum systems in which fluctuations arise not because of a coupling to an external heat bath, but because the dynamics of the system is inherently chaotic. The term mesoscopic denotes metallic or semiconductor devices (with linear dimensions that are typically of the order μm) for which, when investigated at sufficiently low temperature, the inelastic mean free path of the electron is larger than the sample size. In mesoscopic systems, stochastic fluctuations of the conductance are due to random-impurity elastic scattering of the electron on its way through the sample.

While the physical origin of the fluctuations (chaos or disorder) is quite different in both kinds of systems, it is my contention that in both cases fluctuating observables can be modelled successfully in terms of random matrices. The techniques of handling such models analytically have been developed to the point that mean values, variances

Recent Progress in Many-Body Theories, Volume 2
Edited by Y. Avishai
Plenum Press, New York, 1990

and autocorrelation functions can be calculated reliably. This permits the physical description of, and the deduction of useful information from, experimental data on fluctuating observables.

I present four examples to illustrate these claims. The examples are introduced in section 2, the associated random-matrix models are described in section 3, and some results obtained from the models are presented in section 4.

Lack of space and the complexity of the method do not allow me to address a very important point: The theoretical tools used to calculate mean values and higher moments of observables from random-matrix models. Central ideas were developed in refs. 1, 2; the application to scattering theory—needed in all the examples given below—was made in ref. 3. Further references are given as I proceed.

At first sight, the topic of this paper does not seem to belong to the theme "many-body theories". I wish to recall, however, that the generic classical many-body problem is non-integrable and at least partially chaotic; the associated quantum problem therefore displays stochastic fluctuations in suitably chosen observables and at suitably chosen energies.

2 FOUR EXAMPLES

2 (i) Compound-Nucleus Cross-Sections

Fig. 1 shows the total neutron cross-section on the target isotope ^{232}Th versus neutron energy. The figure shows a sequence of nearly isolated resonances, all of spin and parity $1/2^+$, with widths of about 1 eV and spacings of about 10 eV. These values must be compared with the excitation energy of about 10 MeV in the compound nucleus ^{233}Th to which the resonances belong: Counting from the ground state, we are looking at the millionth or so excited states of the ^{233}Th nucleus of fixed spin and parity. Data of this type, taken in the 1970's by Rainwater et al on about a dozen heavy nuclei, provided the main input for the statistical analysis of nuclear level-spacing distributions by Haq et al.[6]

This analysis showed in a statistically significant fashion that the nuclear level spacing distribution at neutron energy agrees with that of a random-matrix ensemble (the Gaussian orthogonal ensemble (GOE) introduced in section 3). It is now known that the level-spacing distribution of fully chaotic Newtonian systems also agrees locally with that of the GOE. Since chaos implies instability, these facts indicate that a theoretical calculation, using the nuclear Hamiltonian, of the properties (locations and widths) of the resonances shown in figure 1 is neither feasible (because of instability of the calculation against rounding or other errors which exceed 10 eV) nor desirable (because of instability, the individual resonances carry no information content). For all intents and purposes, the cross-section shown in figure 1 depends stochastically on energy. Nevertheless, it is of interest to calculate theoretically the energy-average of the cross-section, its variance, and its autocorrelation function: Comparing the results with the data, one gains information on average quantities characterizing the compound nucleus (transmission coefficients, lifetimes etc.). Moreover, the applied physics community is lively interested in such information. Finally, this type of analysis and information is also needed in the tests of fundamental symmetries described under (ii). It is not advisable

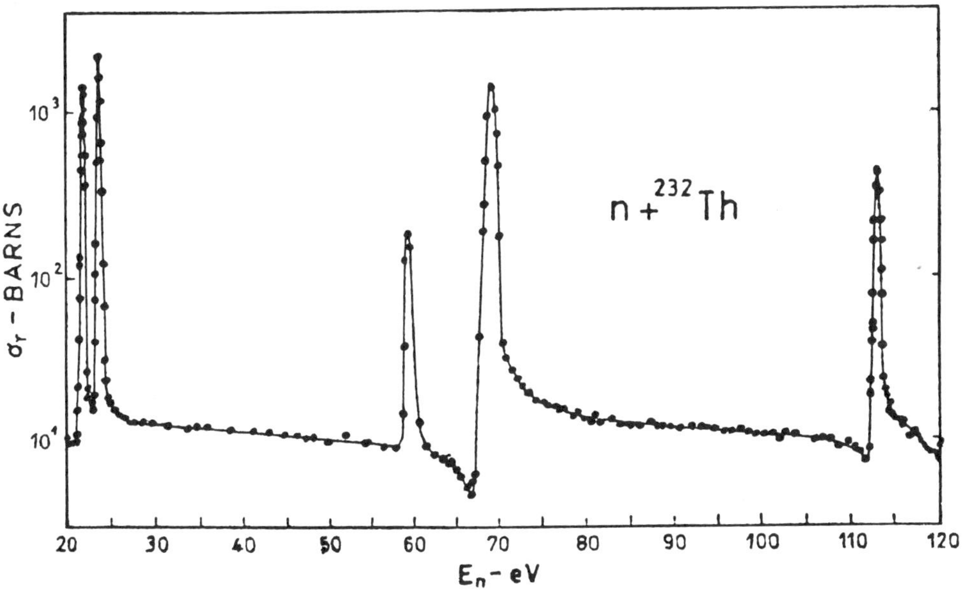

Figure 1 The total neutron cross-section on ^{232}Th versus neutron energy. (Taken from ref. 4).

to start the calculation ab initio but rather to employ a suitable random matrix model that is known to correctly reproduce the fluctuations.

The situation just described is not confined to excitation energies near neutron threshold, or to heavy nuclei: It is typical for nuclear reaction cross-sections at bombarding energies small enough for compound-nucleus processes to be important. Figure 2 shows another example. While figure 1 displays resonances for which the mean width Γ is small compared to the mean spacing d, we have $\Gamma \gg d$ in figure 2: The maxima (and minima) of the stochastic cross-section result from an accidentally constructive (and an accidentally destructive) superposition of many overlapping resonances. All the remarks made above on the case of nearly isolated resonances apply here, too. This shows that, given a random-matrix model, we need to calculate mean values, variances and autocorrelation functions of cross-sections in the entire regime from $\Gamma \ll d$ to $\Gamma \gg d$.

2 (ii) Tests of Fundamental Symmetries in Compound-Nucleus Reactions

I cite one figure to show that the compound nucleus provides a sensitive laboratory for tests of parity and time-reversal symmetry. Parity is conserved by the strong and broken by the weak interactions; parity violation in nuclei is therefore expected (and has been found) to amount typically to one part in 10^6 or 10^7. Parity violation can be tested by comparing the transmission of neutrons polarized in and opposite to their momentum direction through a nuclear target. Such an experiment, first performed at Dubna[7] on ^{138}La in the regime of isolated compound-nucleus resonances, has yielded a signal on the 10 per cent level, showing that the closeness of levels of opposite parity

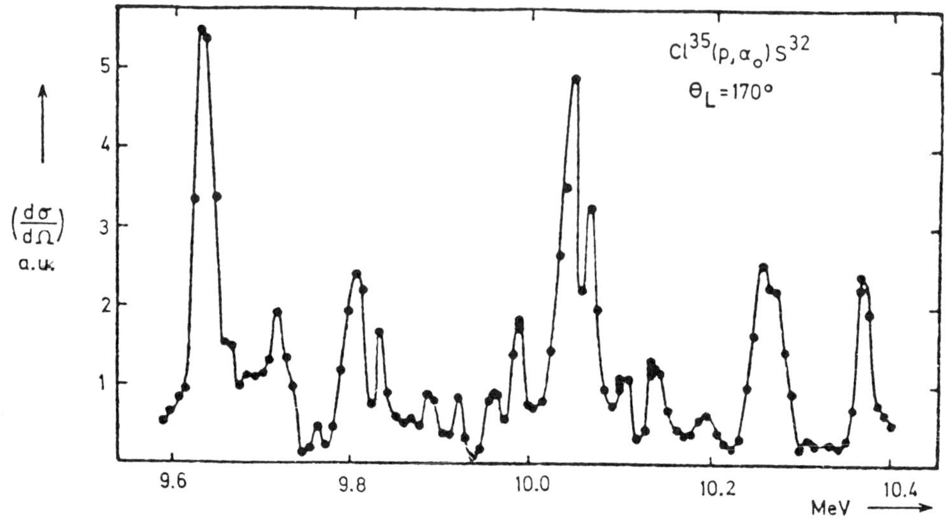

Figure 2 The cross-section for the reaction $^{35}Cl(p, \alpha_0)^{32}S$ at a lab scattering angle of 170° versus proton energy. (Taken from ref. 5).

in the compound nucleus enhances symmetry breaking by 5 or 6 orders of magnitude! This observation has spurred considerable experimental activity aimed at a systematic investigation of parity violation in compound-nucleus reactions, and carried by the hope that in the future more stringent tests of time-reversal symmetry will become possible. The analysis of the data aims at determining (or putting an upper bound on) the relative strength of the symmetry-breaking part in the nuclear Hamiltonian; in the stochastic regime, it must rely on a random-matrix model.

2 (iii) Chaotic Motion in Molecules

To understand chemical reactions and energy transport in molecules, it is important to address the question whether in a given range of excitation energies, the level spacing fluctuations do or do not signal classical chaos. A pertinent experiment[8] is schematically shown in figure 3. A singlet state at an excitation energy of about 26 000 cm^{-1} in Methylglyoxal is laser-excited; the resonance fluorescence yield from this level is observed. The s-state is embedded in a dense background of triplet-states. Changing an externally applied magnetic field, one shifts the $m = \pm 1$ components of the t-states past the s-state, thereby inducing many Landau-Zener crossings. At each crossing, the s-state wave function mixes with one of the t-state wave functions; the mixing enables the s-state to gamma decay into other modes and reduces the resonance fluorescence yield. The upper part of figure 4 displays many hundred such crossings. The resonances overlap at least partially. This makes a direct determination of the level-spacing distribution from the data impracticable. The insert in the lower part of the figure shows that the Fourier transform of the intensity-autocorrelation function displays a "correlation hole". It was suggested[8] that this hole indicates GOE statistics and thereby chaos in the classical limit. This statement needs to be verified in terms of a random-matrix calculation modelling the situation shown in figure 3. The problem is of interst not only in its own right, but also because the signal (the correlation hole) is obtained without tediously analysing the data with regard to individual resonances and their spacing distribution.

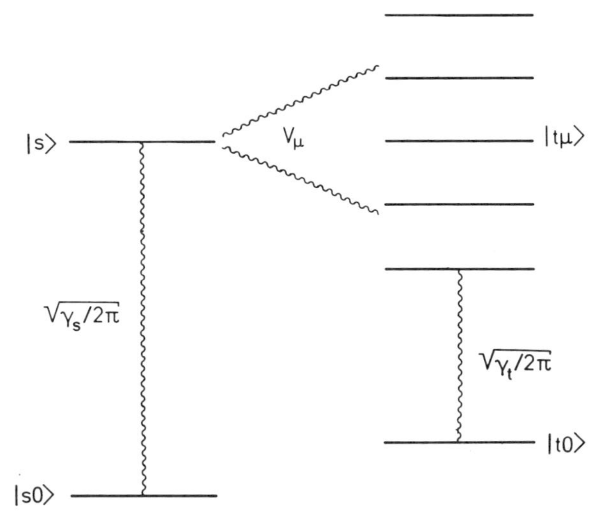

Figure 3 The Laser experiment on Methylglyoxal (schematic). (Taken from the first of ref. 18).

2 (iv) Universal Conductance Fluctuations

In the last few years, it has become possible to manufacture metallic wires of μm length, and semiconductor devices in which electrons move in a virtually two-dimensional layer. At sufficiently low temperature, the conductance in both kinds of samples is due to elastic impurity scattering of the electrons. (It is commonly assumed that the electron density is sufficiently small to render the electron-electron interaction unimportant). The conductance G is measured as function of an external magnetic field H or, in the case of MOSFETs, also as function of the Fermi energy E_F (the gate voltage). The dependence on H or on E_F is supposed to simulate the fluctuations of G over an ensemble of macroscopically identical probes. The intuitive, classical argument is that in changing H (or E_F), one changes the curvature (the velocity) of the electron which therefore experiences a different set of random scatterings on its way through the sample. With $g = Gh/e^2$ a dimensionless variable, the remarkable finding of such experiments is this: The average $\bar{g}$ of g over the fluctuations depends on the mate-

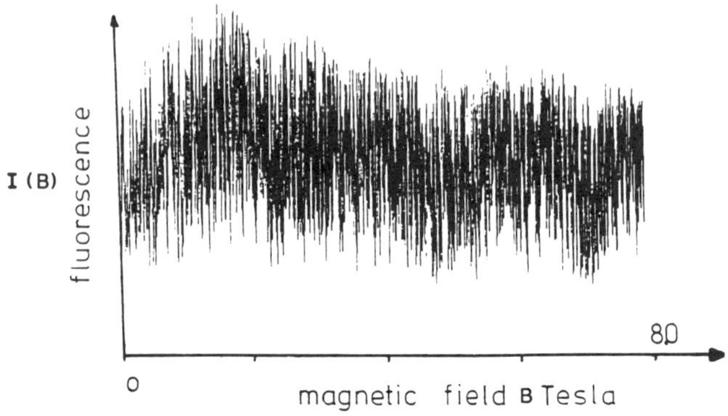

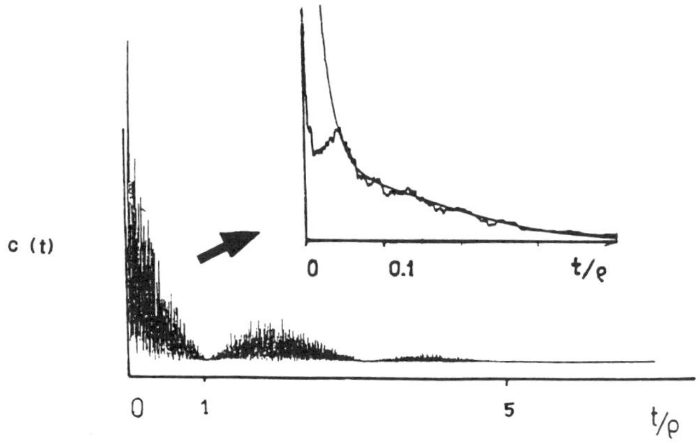

Figure 4 Upper part: Resonance fluorescence yield $I(B)$ versus applied magnetic field B. Lower part: The Fourier transform $\int_{-\infty}^{\infty} dB \cdot \int_{-\infty}^{\infty} db \exp(ibt) \cdot I(B + b/2)I(B - b/2)$ of the autocorrelation function versus t. The insert shows a correlation hole for small values of t. (Taken from ref. 8).

rial and the geometrical dimensions of the wire. However, the size of the fluctuations, quantified by $\overline{g^2} - \overline{g}^2 = \text{var}(g)$, is always of order unity, see figure 5. This phenomenon has been named "Universal Conductance Fluctuations". It is of obvious interest for applications in microelectronics. Perhaps more importantly, the fluctuations relate to the effect which disorder has on the coherence properties of wave functions. This is an important topic in several areas of physics. (See the contributions by Kaveh, Stephen and Stern to this conference). The importance of phase coherence is manifest in figure 6 which shows the fluctuations of G for a ring geometry rather than for a simple wire. The rapid oscillations are Ahavanov-Bohm interference patterns superposed on the longer-wavelength random fluctuations. I will show that G can be modelled with random matrices, and that the results, while consistent with other approaches, offer new insight into the behaviour of G.

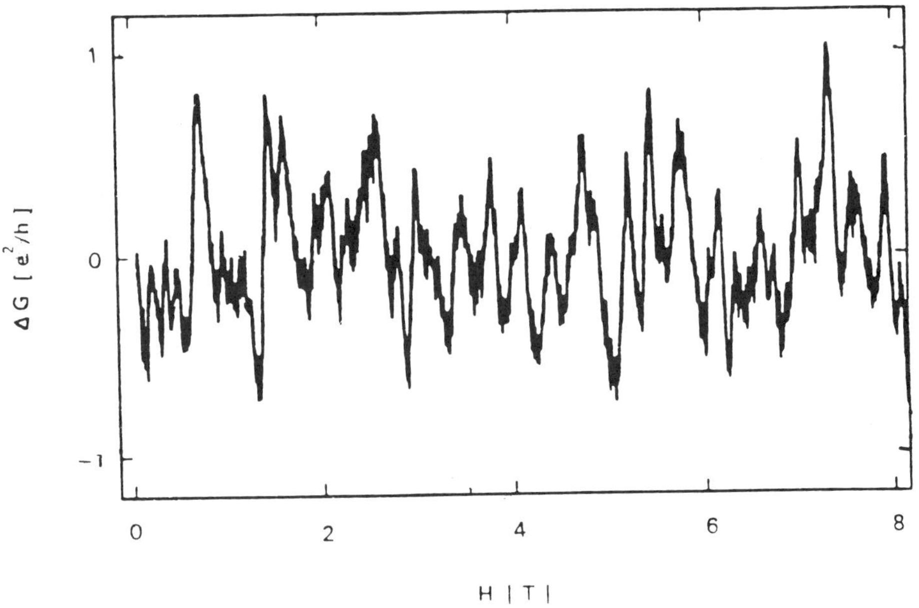

Figure 5 The fluctuations of $\Delta G = G - \overline{G}$ in units e^2/h versus an external magnetic field H in tesla. The wire was 310 nm long and 25 nm wide, the temperature 0.01 K. (Taken from ref. 9).

3 RANDOM-MATRIX MODELS

Modelling the four examples of section 2, I will for simplicity always use the GOE. An external magnetic field breaks the time-reversal symmetry of the intrinsic Hamiltonian in the cases 2 (iii) and 2 (iv); I do not pay attention to this fact and use the time-reversal symmetric GOE throughout.

The GOE is the generic ensemble for time-reversal symmetric systems. It consists of real, symmetric matrices $H_{\mu\nu}$ of dimensions N. The independent matrix elements $H_{\mu\nu}(\mu \leq \nu)$ are uncorrelated Gaussian distributed random variables with zero mean value and a second moment given by $\overline{H_{\mu\nu}^2} = (1 + \delta_{\mu\nu})\lambda^2/N$. The bar denotes the ensemble mean, and λ is a parameter of dimension energy which determines the mean level spacing. In the GOE, the fluctuations of the level spacings about their mean value are parameter-free. In the applications listed below, we always take the limit $N \to \infty$.

3 (i) Compound-Nucleus Scattering.

Let a, b, c,... denote the open channels feeding compound states μ, ν,... of fixed spin and parity. In the absence of direct reactions and under omission of elastic phase shifts (both of which can easily be included), the S-matrix elements have the form

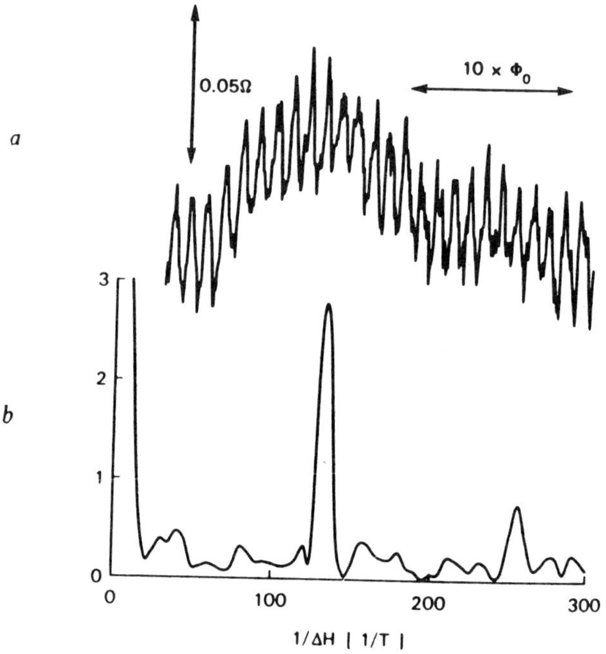

Figure 6 Fluctuations of G versus H for a ring geometry. (Taken from ref. 9).

$$S_{ab}(E) = \delta_{ab} - 2i\pi \sum_{\mu\nu}^{N} W_{a\mu}[D^{-1}]_{\mu\nu} W_{\nu b}. \tag{3.1}$$

Here, E is the energy and $W_{a\mu}$ are matrix elements coupling channels and levels. The $W_{a\mu}$ typically change very little when E ranges over a mean level spacing and are taken as constants. The inverse propagator is

$$D_{\mu\nu}(E) = E\delta_{\mu\nu} - H_{\mu\nu} + i\pi \sum_{c} W_{\mu c} W_{c\nu}. \tag{3.2}$$

To simulate stochastic fluctuations of S_{ab} and, thereby, of the cross-section (which is proportional to $|S_{ab}|^2$), we consider an ensemble of S-matrices by keeping $W_{\mu a}$ fixed and by letting $H_{\mu\nu}$ run over the GOE. The parameters in this model are λ (the GOE parameter) and the $W_{\mu a} = W_{a\mu}$ (time-reversal symmetry). Because of orthogonal invariance of the GOE, the distribution function of the S-matrix elements depends only on the dimensionless orthogonal invariants $\sum_\mu W_{a\mu} W_{b\mu}/\lambda$. The number of these latter input parameters is equal to the number of average S-matrix elements $\overline{S}_{ab}$. Given the average S-matrix, the model therefore describes the fluctuations of S in a parameter-free fashion.

The form (3.1),(3.2) of the S-matrix is generic to resonance theory; it is obtained from any of the standard nuclear reaction theories; it can be read as an N-level Breit-Wigner formula which preserves the unitarity of S. Aside from the resonance energies determined by $H_{\mu\nu}$, the denominator (3.2) also contains an imaginary width term which accounts for the coupling to the open channels.

3 (ii) Tests of Fundamental Symmetries

To be brief, I confine myself to parity violation and consider two classes of compound states, i.e. states $|+\sigma>$ with positive and states $|-\tau>$ with negative parity. Since the enormous enhancement of symmetry violation described in section 2 (ii) is known to be due to resonance mixing, I assume that in eqs. (3.1),(3.2), the channels a, b, $c, \ldots$ have well-defined parity, and that the matrix elements $W_{a\mu}$ do not break parity. Then parity violation occurs only in $H_{\mu\nu}$. When account is taken of the existence of two classes of states, $H_{\mu\nu}$ can be written as a block matrix,

$$\begin{pmatrix} \langle\sigma+|\, H\, |+\sigma'\rangle & \langle\sigma+|\, H\, |-\tau'\rangle \\ \langle\tau-|\, H\, |+\sigma'\rangle & \langle\tau-|\, H\, |-\tau'\rangle \end{pmatrix}. \tag{3.3}$$

The matrices $\langle\sigma+|\, H\, |+\sigma'\rangle$ and $\langle\tau-|\, H\, |-\tau'\rangle$ in the diagonal blocks are each replaced by a GOE; the two GOE's are assumed uncorrelated. The two matrices in the off-diagonal blocks are linked by Hermitecity. With time-reversal symmetry preserved, the matrix elements $\langle\sigma+|\, H\, |-\tau'\rangle$ are real; they are taken to be uncorrelated Gaussian distributed random variabless with zero mean value and a common variance. With all other parameters of this model fixed through the analysis of parity-conserving compound-nucleus reactions, it is the aim to relate the variance of $\langle\sigma+|\, H\, |-\tau'\rangle$ to experimental data, and to infer from the knowledge of this variance the strength of the effective parity-violating nucleon-nucleon force in the shell model. The former is achieved by considering $\langle\sigma+|\, H\, |-\tau'\rangle$ as a small perturbation; the latter, by using methods of statistical nuclear spectroscopy.[10]

3 (iii) Chaos in Molecules

We deal with resonant scattering of light. Therefore, the S-matrix again has the form (3.1),(3.2) where a now denotes the channel in which Methylglyoxal is in its ground state (so that $|\, S_{aa} - 1\,|^2$ is proportional to the resonance fluorescence yield) while b, $c, \ldots$ are channels populated by the gamma decay of the triplet states. Let the index 1 denote the s-state and the indices μ, $\nu = 2, 3, \ldots$ denote the t-states. Then, $W_{a\mu} = \delta_{\mu 1} W_{a1}$, while all the other $W_{b\nu}$'s connect only triplet states with gamma channels. The Hamiltonian has the form

$$\begin{pmatrix} \varepsilon_s + i\pi W_{a1}^2 & V_\nu^{st} \\ V_\mu^{st} & H_{\mu\nu}^T + i\pi(\cdots) \end{pmatrix} \tag{3.4}$$

The dots indicate width terms that account for the decay of the triplet states; ε_s is the s-state resonance energy, and V_μ^{st} are the singlet-triplet coupling matrix elements. The model attains stochastic features by replacing the triplet-state Hamiltonian $H_{\mu\nu}^T$ by a GOE. To describe the shift caused by a magnetic field H of the triplet states past the singlet states, one either adds a diagonal matrix of strength $\sim H$ to $H_{\mu\nu}^T$, or simply shifts ε_s.

To simulate the experiment, a somewhat more complex model must be considered: Two groups of states with magnetic quantum numbers $+1$ and -1 must be accounted for, and within each group there are several uncorrelated classes of states pertaining to different symmetry.

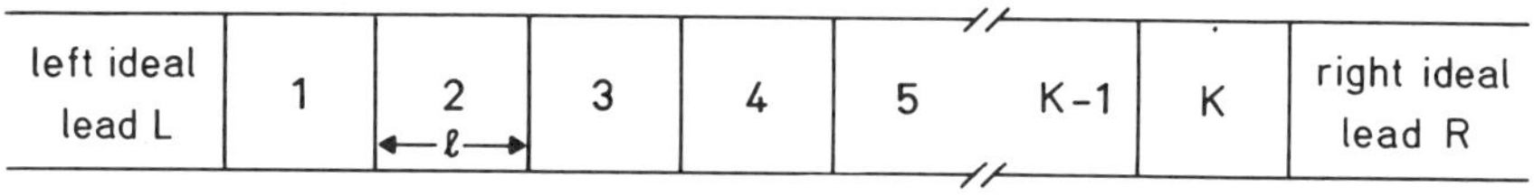

Figure 7 A random-matrix model for conductance fluctuations in a two-lead experiment (schematic). L and R stand for left and right, respectively. The disordered sample is divided into K slices of length l each.

3 (iv) Universal Conductance Fluctuations

Although experiments on mesoscopic wires usually employ a four-lead geometry, I consider for simplicity here a two-lead setup: A disordered wire of length L_x connected at either end to an ideal lead. In each lead, channels a, b, $c, \ldots$ are defined by the transverse modes of the electron. The number of such channels at the Fermi energy is given by $\Lambda = L_x L_z K_F^2/\pi$ where $L_x L_z$ is the cross-section area of the wire and K_F the Fermi wave number. For typical experiments, Λ is at least several thousand. With $G = (e^2/h)g$, we use the many-channel approximation to Landauer's formula which reads $g = 2\sum_{ab} |S_{ab}^{LR}|^2$. Here, S_{ab}^{LR} is the element of the scattering matrix connecting channel a on the left with channel b on the right (figure 7); $\sum_{ab} |S_{ab}^{LR}|^2$ is obviously the total transmission through the disordered sample; the factor 2 is due to spin. To determine S_{ab}^{LR}, we consider the longitudinal motion of the electron in the leads as free. Then, the S-matrix has a structure similar to eqs. (3.1),(3.2): The Hamiltonian $H_{\mu\nu}$ can be thought of as an Anderson tight-binding Hamiltonian for the electron in the disordered sample in the absence of any coupling to external leads. To implement a random-matrix model, we use a different construction. The disordered sample is thought to be divided (see figure 7) by transverse cuts into K equal slices of length l, the elastic mean free path. (We obviously assume $L_x = Kl$ with K integer). In each slice, the Hamiltonian for that slice is replaced by a GOE; GOE's in different slices are uncorrelated but have the same variances. States in neighbouring slices are coupled by real matrix elements that are Gaussian distributed uncorrelated random variables with zero mean and fixed variance (the same for all matrix elements, and all pairs of neighbouring slices). States in two slices separated by at least one further slice have zero coupling matrix elements. The resulting Hamiltonian is an extension of the block-form[4] (3.3) to K classes. Labelling the slices by running indices j, $l = 1, \ldots, K$ and the states in each slice by running indices μ, $\nu = 1, \ldots, N$ we can write the Hamiltonian as $H_{\mu\nu}^{jl}$. For $j = l$, this a GOE; for $j = l \pm 1$, the matrix elements are Gaussian; for $|j - l| > 1$, the matrix elements vanish.

This model for $H_{\mu\nu}$ encompasses the idea of a mean free path for random impurity scattering. It applies to probes with transverse dimensions of the order of l. It contains two parameters: The GOE parameter λ which is adjusted to the mean level density, and the variance of the block non-diagonal elements $H_{\mu\nu}^{jj+1}$. Dividing the latter variance by λ^2 we obtain a dimensionless quantity which, for $K \gg 1$, is related to the diffusion constant D.

To construct the S-matrix, it is consistent with this model to assume that an electron entering from the left (right) lead does not travel beyond the first (the last) slice before it undergoes its first random scattering. We accordingly assume that the matrix elements $W^L_{a\mu}(W^R_{a\nu})$ are zero unless $\mu(\nu)$ is a state in the first (the last) slice. Then the S-matrix has the form

$$S^{LR}_{ab} = -2i\pi \sum_{\mu\nu} W^L_{a\mu}[D^{-1}]^{1K}_{\mu\nu} W^R_{\nu b}. \tag{3.5}$$

The propagator is given by

$$D^{jl}_{\mu\nu} = E\delta_{jl}\delta_{\mu\nu} - H^{jl}_{\mu\nu} + i\pi\delta_{jl}\left[\delta_{j1}\sum_a W^L_{a\mu}W^L_{a\nu} + \delta_{jK}\sum_b W^R_{b\mu}W^R_{b\nu}\right]. \tag{3.6}$$

We work in the metallic regime $K_F l \gg 1$ so that $N \gg \Lambda \gg 1$. Orthogonal invariance in the end slices reduces the $2N \cdot \Lambda$ input parameters $W^{L,R}_{a\mu}$ to the quantities $\sum_\mu W^L_{a\mu}W^L_{b\mu}$ and $\sum_\nu W^R_{a\nu}W^R_{b\nu}$; without loss of generality it may be assumed that the latter quantities are diagonal in the channel indices; it is physically reasonable to postulate that the remaining quantities $\sum_\mu (W^L_{a\mu})^2$ and $\sum_\nu (W^R_{b\nu})^2$ are all equal. This reduces the model to a three-parameter model: It depends on the mean level spacing, on the diffusion constant, and on the strength of the coupling between the two end-slices and the adjacent lead.

4 RESULTS

I summarize results on items (i) to (iii) without going into any details, and am more explicit on point (iv).

(i) The average cross-section, proportional to $\overline{|S_{ab}|^2}$, has been worked out[11] analytically for values of $\Gamma \ll d$ (isolated resonances) to $\Gamma \gg d$ (strongly overlapping resonances). The result is given in terms of a three-dimensional integral. Limiting values and asymptotic expansions are derived in several papers.[12,13] The variance of the cross-section has also been calculated over the full range of Γ/d values.[14] The cross-section autocorrelation function has been worked out[11] for $\Gamma \gg d$, the only case of interest experimentally.

(ii) Optimal choices for observables which test fundamental symmetries have been determined.[15,16] I recently had the opportunity to review this topic and refer to my paper,[17] in which further references can be found.

(iii) The model described under 3 (iii) has been worked out by a combination of analytical and numerical methods.[18] For the actual case of Methylglyoxal, it turns out that the correlation hole in the lower part of figure 4 does indeed correlate to GOE statistics of the triplet states and thus to chaos.

(iv) The large-length behaviour ($L_x \gg l$) of both $\bar{g}$ and $\mathrm{var}(g)$ coincides[19,20] with that of earlier calculations, indicating that the random-matrix model of section 3 (iv) which maps a three-dimensional problem onto a many-channel one-dimensional model is successful. Differences arise[19,20] for smaller values of L_x/l; these are quantified below. This shows that coupling to the leads does have an influence on the conductance and its fluctuations. A particular strength of the approach lies in the ease with which one can deal with geometries that are more complex than the one shown in figure 7.

To be more specific, let me recall that the model of section 3 (iv) contains three parameters. The first of these (the variance of $H^{jj}_{\mu\nu}$) is uniquely related to the mean level spacing d in the disordered sample. We obviously have $d \sim L_x^{-1}$. The second parameter (the variance of $H^{jj+1}_{\mu\nu}$), when expressed in units of the first, relates to the diffusion constant D for electron propagation through the disordered sample. This constant D in turn appears in the diffusion equation for the probability density $P(x,t)$ to find an electron at time t at the longitudinal position x in the sample,

$$\frac{\partial}{\partial t} P(x,t) = D \frac{\partial^2 P}{\partial x^2} \quad . \tag{4.1}$$

Dimensional arguments show that the diffusion time T_{diff} through the sample is proportional to $[D/L_x^2]^{-1}$. The energy E_c associated by the uncertainly principle with T_{diff} (the Thouless energy) is given by

$$E_c = \frac{\pi^2 \hbar D}{L_x^2} \quad . \tag{4.2}$$

Working out E_c from the model one sees that E_c is (except for a factor of order unity) what nuclear physicists call the spreading width. It is the FWHM of the probability to find an eigenfunction of the unperturbed Hamiltonian in the disordered sample (defined as $H^{jj\pm1} = 0$ for all j) mixed into an eigenfunction of the full problem. The third parameter (the strength $\sum_a W^2_{a\mu}$ of the coupling to external channels) determines another energy scale, the decay width of Γ of the sample. It is given by

$$\Gamma = \frac{d}{2\pi} 2\Lambda\alpha \tag{4.3}$$

where the coefficient α obeys the inequalities $0 \le \alpha \le 1$. Eq. (4.3) is intuitively obvious: $d/(2\pi\hbar)$ is the frequency with which a time-dependent quasiperiodic wave packet in the disordered sample returns to its original position; α is the average probability with which it escapes into one of the open channels; $(d/2\pi\hbar)2\Lambda\alpha$ is the total escape rate into all channels. For good conductors (good coupling to the leads) we expect $\alpha \cong 1$. In this case, $\Gamma \gg d$: We deal with the case of strongly overlapping resonances. We can interpret $T_{\text{dec}} = \hbar/\Gamma$ as the decay time: An electron localized at time $t = 0$ within the sample will, for $t \to \infty$, attain equal probability density everywhere (in leads and disordered sample); since the leads are infinitely extended by definition, this implies that the probability for finding the electron within the disordered region goes to zero; the rate

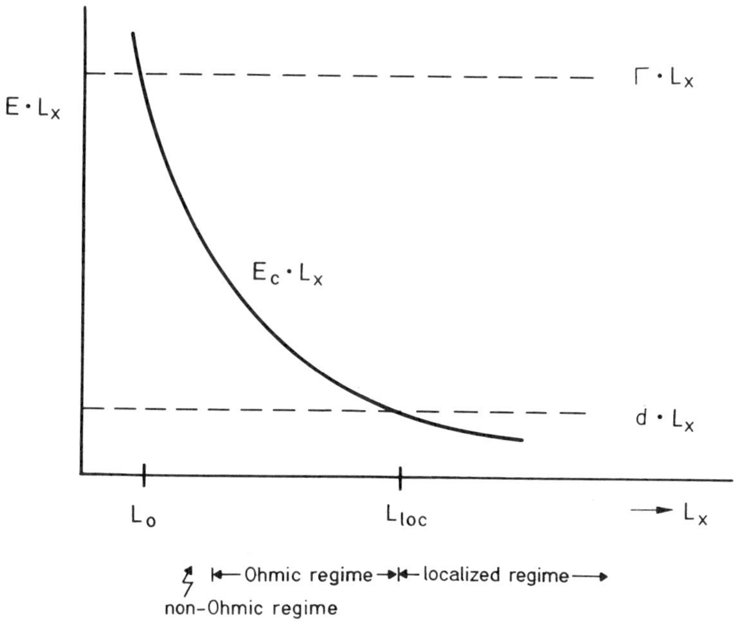

Figure 8 The three characteristic energy scales of the model, multiplied by L_x, versus L_x (schematic).

of the associated exponential decay is given by T_{dec}. The novel energy scale resulting from our approach is Γ. We note that $\Gamma \sim L_x^{-1}$. Figure 8 displays the significance of the three energy scales d, E_c and Γ on the behaviour of $\overline{g}$ which for $\alpha = 1$ is explicitly given by

$$\langle g \rangle = \frac{2\Lambda}{2 + (K-1)\frac{\pi^2}{2}\frac{\Gamma}{KE_c}} + \cdots \tag{4.4}$$

The dots indicate the term of next order in the asymptotic expansion (the weak-localization correction to $\overline{g}$). (Note that $\Gamma/(KE_c)$ is independent of length L_x). In figure 8, we plot schematically energy $\cdot L_x$ versus length L_x. The dotted straight line gives $d \cdot L_x$, the solid straight line, $\Gamma \cdot L_x$, and the dashed and dotted curve, $E_c \cdot L_x$ (which falls off with L_x^{-1}). The curves for $E_c \cdot L_x$ and $d \cdot L_x$ intersect at a length L_{loc} determined by $E_c = d$ or by the condition that spreading width and mean level spacing are equal. For $L_x > L_{\text{loc}}$, the mixing of wave functions seizes to be effective over the entire sample, the eigenfunctions fall off exponentially at either end, and so does $\overline{g}$: We have reached the localization transition with L_{loc} the localization length. Inspection of the terms indicated by dots shows that the asymptotic expansion (4.4) is valid for L_x-values less than L_{loc}. — The curves for $E_c \cdot L_x$ and $\Gamma \cdot L_x$ intersect at another critical length, L_0. Eq. (4.4) shows that for $L_0 < L_x < L_{\text{loc}}$, i.e. for $T_{\text{diff}} > T_{\text{dec}}$, $\overline{g}$ is nearly Ohmic, i.e. $\overline{g} \sim K^{-1}$. This is because the diffusion time through the sample determines the behaviour of the system. For $L_x < L_0$, on the other hand, we have $T_{\text{diff}} < T_{\text{dec}}$, so that

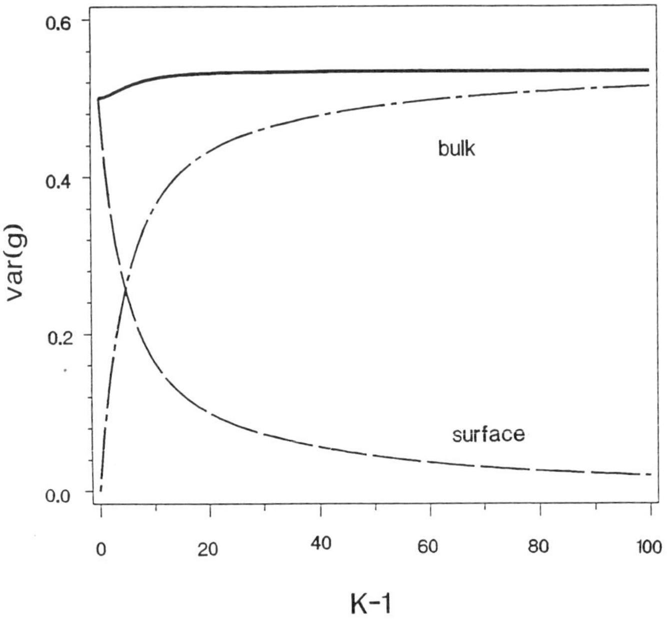

Figure 9 The dependence of var(g) on $K = L_x/l$ for $\alpha = 1$ (see text). (Taken from ref. 20).

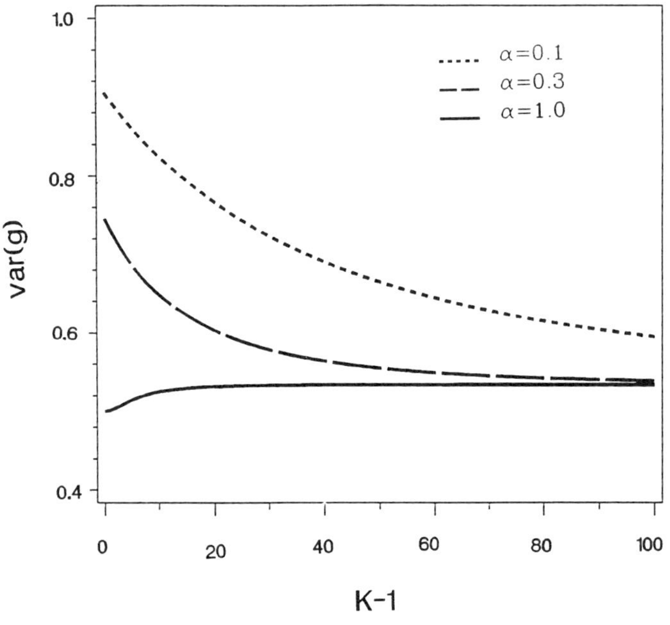

Figure 10 The dependence of var(g) on $K = L_x/L$ for three values of α (see text). (Taken from ref. 20).

internal diffusion is more rapid than decay into the open channels. The electron density fills the sample uniformly before decay sets in. Therefore, $\bar{g}$ is (nearly) independent of length L_x and (nearly) equal to half its maximum possible value 2Λ because electron emission into both leads is equally probable.

Fig. 9 shows var(g) for $\alpha = 1$ versus $K = L_x/l$. The dashed and dotted curve gives the bulk contribution to var(g). Asymptotically ($L_x \rightarrow L_{\text{loc}}$) it approaches the value 8/15 known from previous studies of the problem. For small values of L_x/l, it is dwarfed by the surface contribution (proportional to the coupling to the channels (dashed curve)). The two curves add to a nearly constant curve (full line) for var(g). This is fortuitous, as shown by figure 10 where var(g) is plotted versus $K = L_x/l$ for three values of α. We observe that var(g) is influenced by the surface terms, and that this influence extends over dozens of mean free paths.

5 CONCLUSIONS

I have shown how problems of chaos and disorder can be modelled by random matrices, and that such models can be worked out either exactly or via suitable asymptotic expansions. In this way, a quantitive understanding of the relevant features of fluctuations in diverse physical systems can be attained.

Acknowledgements

I wish to express my gratitude to those collaborators and friends who over the years have worked with me on these problems, and without whom the results described above could not have been obtained: D. Boosé, E. D. Davis, T. Guhr, H. L. Harney, U. Hartmann, S. Iida, H. Nishioka, J. J. M. Verbaarschot, S. Yoshida, M. R. Zirnbauer, and J. A. Zuk.

REFERENCES

1. L. Schäfer and F. J. Wegner, Z. Phys. B**38** (1980) 113
2. K. B. Efetov, Adv. Phys. **32** (1983) 53
3. H. A. Weidenmüller, Ann. Phys. (N.Y.) **158** (1984) 120
4. A. Bohr and B. R. Mottelson, Nuclear Structure II, W. A. Benjamin, Reading (1975)
5. T. Ericson and T. Mayer-Kuckuk, Ann.\Rev. Nucl. Sci. **16** (1966) 183
6. R. W. Haq, A. Pandey, and O. Bohigas, Phys. Rev. Lett. **48** (1982) 1086
7. V. P. Alfimenkov, Sov. Phys. Usp. **27** (1984) 797 and references therein
8. R. Jost and M. Lombardi, in Proc. Int. Conf. on Quantum Chaos and Stat. Nucl. Phys., T. M. Seligman and H. Nishioka editors, Lecture Notes in Physics Vol. 263 (1986)
9. S. Washburn and R. A. Webb, Adv. Phys. **35** (1986) 375
10. J. B. French and V. K. B. Kota, Rochester preprints UR-1116 und UR-22, 1989
11. J. J. M. Verbaarschot, H. A. Weidenmüller and M. R. Zirnbauer, Phys. Rep. **129** (1985) 367

12. J. J. M. Verbaarschot, Ann. Phys. (N.Y.) **168** (1986) 368
13. H. L. Harney and A. Z. Hüpper, Z. Phys. A. **328** (1987) 327
14. E. D. Davis and D. Boosé, Z. Phys. A. **332** (1989) 427
15. E. D. Davis and U. Hartmann, submitted to Phys. Rev. C
16. E. D. Davis, H. L. Harney and A. Müller, private communication (1989)
17. H. A. Weidenmüller, Proceedings of the International Symposium on Symmetries in Nuclei and Particles, Pasadena 1989. "Fundamental Symmetries in Nuclei and Particles", H. Henrikson and P. Vogel, editors, World Scientific, Singapore (1990) p. 30
18. T. Guhr and H. A. Weidenmüller (submitted to Chem. Physics); U. Hartmann, Doctoral Thesis, Heidelberg (1989)
19. S. Iida, H. A. Weidenmüller, and J. A. Zuk, Phys. Rev. Lett. **64** (1990) 583
20. S. Iida, H. A. Weidenmüller, and J. A. Zuk, Ann. Phys. (N.Y.) 280 (1990) 219

SUPER-EFFECTIVE-FIELD CAM THEORY OF STRONGLY CORRELATED ELECTRON AND SPIN SYSTEMS

Masuo Suzuki

Department of Physics, Faculty of Science, University of Tokyo

Hongo, Bunkyo-ku, Tokyo 113, Japan

1. INTRODUCTION

This is partially a review paper of the CAM theory[1-35] and the super-effective-field theory[36-44] with applications to many-body systems[2-44]. We also propose here new ideas concerning how to construct correlated-mean-field type approximations to many-body systems. Our coherent-anomaly method (CAM) is based on generalized mean-field approximations, but it gives an accurate estimation of fractional critical exponents as well as the critical point. Many mean-field-type approximations can be unified to study the true non-classical criticality. In this sense, the CAM theory may be called "met-approximation".

2. BASIC IDEA OF THE CAM

We consider a systematic series of approximations in which the relevant physical quantity shows classical critical behavior. It can be shown in the CAM theory[1,2] that the coefficients of the relevant classical critical singularity show an anomaly systematically as the degree of approximation increases, namely as the approximate critical point T_c obtained in each mean-field approximation approaches the true critical point T_c^*.

More explicitly, the relevant physical quantity $Q_n(x)$ obtained in the n-th approximation takes, in general, the following classical singularity

$$Q_n(x) \simeq \frac{\bar{Q}_n}{(x - x_n)^{\varphi_0}} \tag{1}$$

near the critical point x_n. (Here, the classical exponent φ_0 is unity in most cases.) Then, the critical coefficient $\bar{Q}_n$ is shown[1,2] to have the following coherent anomaly

$$\bar{Q}_n \sim \frac{1}{(x_n - x^*)^{\psi}} \tag{2}$$

Recent Progress in Many-Body Theories, Volume 2
Edited by Y. Avishai
Plenum Press, New York, 1990

near the true critical point x^*. Therefore, we may write as

$$Q_n(x) \sim \frac{1}{(x - x_n)^{\varphi_0}} \cdot \frac{1}{(x_n - x^*)^{\psi}}. \tag{3}$$

Then, the asymptotic behavior of the quantity $Q(x)$ is given by

$$Q(x) \sim \frac{1}{(x - x^*)^{\varphi_0 + \psi}} \tag{4}$$

This is derived by the envelope theory[1], the degree-of-approximation scaling theory[1] (i.e., $x - x_n \sim x_n - x^*$) and by the ordinary scaling theory of correlation functions[2].

Thus, we can estimate non-classical fractional critical exponents by analyzing the mean-field critical coefficients $\{\bar{Q}_n\}$ systematically[1–35]. This scheme seems very simple and quite universal. In fact, it yields a new approach to many-body systems, by which intrinsic fluctuation can be treated in any accuracy.

Using the CAM plot, we can compare different kinds of approximations with each other concerning their degrees of approximation and the corresponding coherent anomalies.

In this CAM scheme, it is substantial to construct an efficient systematic series of mean-field approximations which approach the true criticality in the sense of the CAM.

Owing to the discovery of the CAM, the mean-field approximation is recovering its proper dignity in the theory of phase transitions.

Thus, many people[2–44] have resumed studying mean-field or effective-field approximations seriously from a new view-point of the CAM. Even old results[16–19,33] of phase transitions based on mean-field approximations are now reviving in the CAM.

3. SYSTEMATIC CLUSTER MEAN-FIELD APPROXIMATIONS

A standard CAM canonical series[1] is a systematic series of cluster mean-field approximations[1–3] in which the self-consistency condition is imposed that the average of the order parameter at the center of each cluster should be equal to the corresponding average at the boundary which is multiplied by some appropriate modular factor due to the gauge change of the order parameter at each lattice point. For explicit construction, see the original papers[2–35].

In order to show that this construction of cluster mean-field approximations makes a CAM canonical series[1], we have to prove[2] that mean-field critical points $\{T_c\}$ approach the true one T_c^*, as the cluster size increases. This gives a challenging problem in mathematical physics for general Hamiltonians of many-body systems.

Here it should be instructive to show a typical result of the CAM. In the two-dimensional Ising model, Hu et al[4,5]. obtained $\gamma \simeq 1.749$ for the susceptibility exponent γ, using the transfer-matrix effective-field theory. This agrees very well with the exact value $\gamma = 7/4 = 1.75$. There are many other applications[2-35] of the CAM.

4. SUPER-EFFECTIVE-FIELD THEORY

In order to apply the CAM to many-body systems, we have to construct mean-field approximations for the relevant phase transition. Then, one may ask whether or not it is possible to construct mean-field or effective-field approximations even for exotic phase transitions such as spin glasses[24] and chiral orders[37].

Until now, mean-field approximations have been constructed mainly by decoupling original Hamiltonians. It is, however, impossible to perform such a procedure when the relevant order parameter is a hidden one. In such a case, we can make use of the following super-effective-field theory (SEFT) introduced by the present author[36-39]. The basic idea is to consider an effective field conjugate Λ to a possible order parameter Q and to apply such a "super-effective field" Λ to each boundary cell of the relevant cluster, for example, as shown in Fig.1.

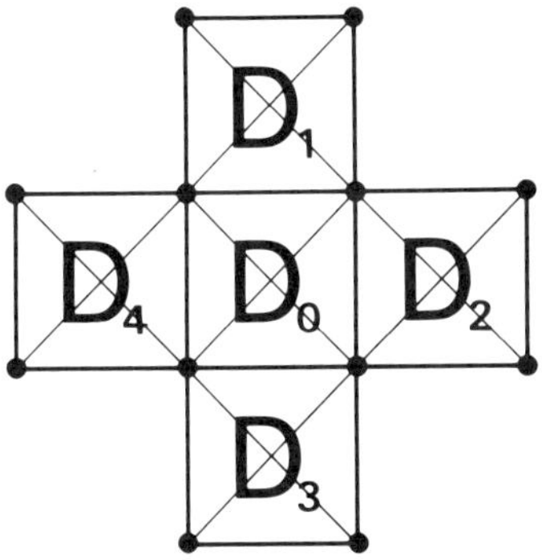

Fig.1 Typical cluster with a super effective field Λ applied to the boundary cells D_1, D_2, D_3 and D_4.

The effect of the infinite degrees of freedom outside of the cluster is included through such a super-effective field term as

$$-\Lambda \sum_{j \epsilon \partial \Omega} Q_j, \tag{5}$$

where Q_j denotes the relevant order parameter defined at the domain D_j of the boundary $\partial\Omega$ of the cluster. The super effective Hamiltonian $\mathscr{H}_{eff}$ is thus given[37] by

$$\mathscr{H}_{eff} = \mathscr{H}_{cl} - \Lambda \sum_{j\epsilon\partial\Omega} Q_j, \tag{6}$$

where $\mathscr{H}_{cl}$ denotes the original cluster Hamiltonian of the relevant system. A new point of our formulation is that at a glance the second term in (6) seems to have no explicit relation to the original Hamiltonian but that the average $< Q_0 >$ of the order parameter at the center of th cluster is induced through the effective correlation between Q_0 and Q_j which is produced by the original interaction of the system. The super effective field Λ is determined by the self-consistency condition

$$< Q_0 >= \varepsilon < Q_j > \tag{7}$$

where ε denotes a modular factor[37] due to the gauge change of the order parameter.

It should be remarked here that ordinary effective fields can, in principle, be derived by eliminating the degrees of freedom outside of the relevant cluster, but that the super effective fields are not necessarily derived in such a way as is easily seen from Fig.1. They should be considered to have been introduced ad hoc in order to describe the resultant main physical role of the remaining degrees of freedom outside of the cluster. This way of introduction of super effective fields is justified by confirming that such a series of super-effective-field approximations is canonical[1] in the sense of the CAM, namely that the corresponding approximate critical points $\{T_c\}$ approach the true one T_c^*.

Now we formulate explicitly a scheme to determine the critical point T_c. For this purpose, we expand $< Q_0 >$ and $< Q_j >$ with respect to $\lambda = \beta\Lambda = \Lambda/k_{\mathrm{B}}T$ as

$$< Q_0 >= \lambda \sum_{k\epsilon\partial\Omega} < Q_0; Q_k >_{cl} + O(\lambda^2) \tag{8}$$

and

$$< Q_j >= \lambda \sum_{k\epsilon\partial\Omega} < Q_j; Q_k >_{cl} + O(\lambda^2) \tag{9}$$

Here $< A; B >_{cl}$ denotes Kubo's canonical correlation[45] defined by

$$< A; B >_{cl} = \frac{1}{\beta}\int_0^\beta < AB(i\hbar\lambda) >_{cl} d\lambda \tag{10}$$

where

$$B(z) = e^{iz\mathscr{H}_{cl}/\hbar} B e^{-iz\mathscr{H}_{cl}/\hbar} \tag{11}$$

with the cluster Hamiltonian $\mathscr{H}_{cl}$. In the above equations (8) and (9), we have used that $< Q_j >= 0$ for $T \geq T_c$. Thus, the critical point T_c is determined from the solution of the following equation

$$\sum_{j\epsilon\partial\Omega}(< Q_0; Q_k >_{cl} -\varepsilon < Q_j; Q_k >_{cl}) = 0 \tag{12}$$

Similarly the response function $\chi_Q(T)$ defined by

$$< Q >= \chi_Q(T)H \tag{13}$$

with an external field H conjugate to the order parameter Q is given by

$$\chi_Q(T) = N\beta \frac{< Q_0; Q_0 >_{cl} \sum_{k\epsilon\partial\Omega}(t_{0j}t_{0k} - t_{jk})}{\sum_{k\epsilon\partial\Omega}(\varepsilon t_{0k} - t_{jk})} \tag{14}$$

Here $j\epsilon\partial\Omega$ and

$$t_{jk} = < Q_j; Q_k >_{cl} / < Q_0; Q_0 >_{cl} \tag{15}$$

If we make the "Kirkwood" approximation (KA)

$$< Q_j; Q_k >_{cl} \simeq < Q_j; Q_0 >_{cl} < Q_0; Q_k >_{cl} / < Q_0; Q_0 >_{cl} \tag{16}$$

then the response function is reduced[37] to

$$\chi_Q^{(KA)}(T) = N\beta \frac{< Q_0; Q_0 >_{cl}(< Q_0; Q_0 >_{cl} + \varepsilon < Q_0; Q_j >_{cl})}{< Q_0; Q_0 >_{cl} - (z-1)\varepsilon < Q_0; Q_j >_{cl}} \tag{17}$$

Here z denotes the number of nearest neighbour cells in the cluster. For example, we have z=4 in Fig.1.

In the above original (not KA) super-effective-field approximations, more correlation functions are required to obtain $\chi_Q(T)$. Anyway,in our SEFT formulation, the problem to study phase transitions in the thermodynamic limit is reduced to the estimation of Kubo's canonical correlation functions in finite systems. Therefore, to study phase transitions and critical phenomena is reduced to systematic investigations of canonical correlations in finite many-body systems, owing to the CAM theory.

Spiritually, the present SEFT may be analogous[46] to the structure of the world or universe in "Mandala".

5. CHIRAL ORDERS IN THE TWO-DIMENSIONAL ANTIFERROMAGNETIC HEISENBERG MODEL

Recently it was pointed out[47,48] that a scalar chiral order might be related to the mechanism of the high T_c superconductivity. There are two kinds of chiral orders, namely vector and scalar chiral orders in two-dimensional antiferromagnets.

The vector chiral order has already been studied by many authors[37,49–54]. It is defined by

$$Q_{ijk} = \frac{2\sqrt{3}}{3}(S_i \times S_j + S_j \times S_k + S_k \times S_i) \tag{18}$$

for a triangular cell (ijk). Quite recently Kawarabayashi and the present author[56] proved the absence of vector chiral order in the two-dimensional antiferromagnetic Heisenberg model at finite temperatures, using Bogoliubov's inequality. The criticality of the vector chiral order for the XY-model has been studied using the SEFT [37,55].

The scalar chiral order is defined by

$$X_{ijk} = S_i \cdot (S_j \times S_k) \tag{19}$$

for a triangular cell (ijk). We have applied [57] the SEFT to this problem. We have found that the SEFT of the cluster shown in Fig.1 yields no scalar chiral order at any temperature, even when there exist diagonal interactions in each square cell. It is quite interesting that even the smallest-cluster SEFT yields no chiral order in this problem. As fluctuations are apt to destroy, in general, long-range order, larger-cluster approximations will favor non-existence of the scalar chiral order in the Heisenberg model. This tentative conclusion is consistent with other works.

6. THE SEFT OF STRONGLY CORRELATED ELECTRONS

To study the Hubbard model and t-J model with holes using the SEFT will be an extremely interesting problem from a view-point of clarifying the mechanism of the high T_c superconductivity. The simplest cluster of the SEFT for these strongly correlated electron systems will be the 12-spin cluster shown in Fig.1. Investigation of the effect of one hole, two holes etc. in this cluster is now in progress.

7. PERIODIC-CLUSTER MEAN-FIELD APPROXIMATIONS

We propose here a new type of multi-effective-field[58] approximations using clusters with periodic boundary condition. Up to now, we have used clusters with open boundary condition in constructing systematic effective-field approximations. It was pointed out by the present author[58] that more multi-effective fields yield better approximations and that we can construct optimal multi-effective-field approximations[58] in small clusters.

Our new proposal is based on the observation that periodic boundary condition can be regarded as even-parity effective-fields. Thus, it is justified that we can apply odd-parity effective fields to periodic clusters.

Applications of this new scheme will be reported elsewhere in the near future.

8. EFFECTIVE CLUSTER VARIATIONAL METHOD

Here we formulate a general effective cluster variational method. Let Ψ_0 be a suitably chosen independent (or mean-field type) state in the original Hilbert space with infinite degrees of freedom. Now, our effective cluster variational state is defined by

$$\Psi_\Omega = S_{cl}\Psi_0, \tag{20}$$

where S_{cl} denotes such a cluster operator as expresses an effective correlation inside of the relevant cluster.

For example, we consider the following variational wave-function

$$\Psi_\Omega = \prod_{r_j \in \Omega} S(\{r_j\})\Psi_0, \tag{21}$$

where Ψ_0 denotes the Jastrow wave-function, and $S(\{r_j\})$ a higher correlation factor inside the relevant cluster.

By confining higher correlations into a finite cluster, we can calculate analytically any physical quantity of the system. This scheme is similar to those developed by Suzuki and Miyashita[59], by Ogawa and Kanda [60]and by Fazekas[61].

It will be instructive to present another simple example of quantum spin systems. For example, we consider the following Hamiltonian

$$\mathscr{H} = -J \sum_{<ij>} (\sigma_i^+\sigma_j^- + \sigma_i^-\sigma_j^+), \tag{22}$$

where

$$\sigma^\pm = \frac{1}{2}(\sigma^x \pm i\sigma^y) = S^x \pm iS^y, \tag{23}$$

with Pauli operators σ^x, σ^y or spin operators S^x, S^y.

It is quite difficult[62] to find exactly the ground state of the X-Y model (22) in general dimensions. Hereafter we consider the ferromagnetic case $J > 0$. The zeroth-order wave function Ψ_0 may be taken[59] as

$$\Psi_0 = (\sum_i \sigma_i^-)^{N/2} \mid \mathrm{F} >, \tag{24}$$

where $\mid \mathrm{F} >$ denotes the completely ferromagnetic state $\mid \mathrm{F} >=\mid\uparrow\uparrow\ldots\uparrow>$. Clearly[59], this state is a linear combination with equal weights of all states of half up and half down spins[59].

In the previous paper[59], we introduced the correlated trial function

$$\Psi = exp(-\alpha \sum_{ij} \sigma_i^z\sigma_j^z)\Psi_0. \tag{25}$$

It is, however, difficult to calculate exactly the expectation value $< \Psi \mid \mathscr{H} \mid \Psi >$. Then, we studied only the limiting case in which α is small. Here, we apply our effective cluster variational method to this problem, namely we consider the trial state (20) with

$$S_d = exp(-\alpha \sum_{<ij>\in\Omega} \sigma_i^z \sigma_j^z). \tag{26}$$

The difference between Ψ and Ψ_Ω is that the correlation operator is applied effectively only inside the cluster for Ψ_Ω and consequently that we can calculate analytically $< \Psi_\Omega \mid \mathscr{H} \mid \Psi_\Omega >$ for any value of α.

For simplicity, we consider the $2z$ cluster shown in Fig.2.

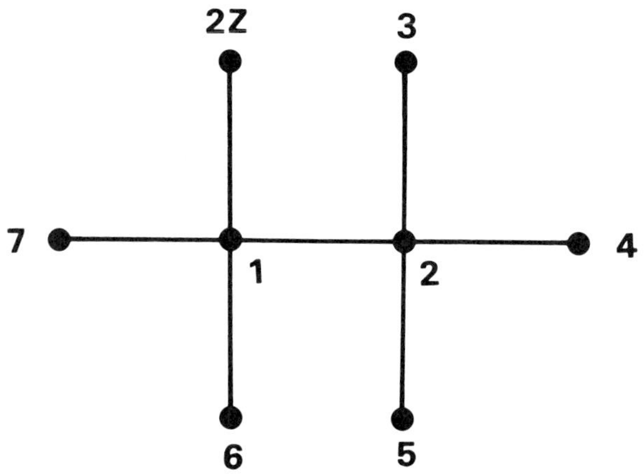

Fig.2. a typical cluster of $2z$ lattice points.

Then, we obtain

$$< \sigma_1^+ \sigma_2^- >= \frac{< \Psi_\Omega \mid \sigma_1^+ \sigma_2^- \mid \Psi_\Omega >}{< \Psi_\Omega \mid \Psi_\Omega >} = \frac{N}{4(N-1)} \frac{e^{2\alpha}}{(cosh\alpha)^{2z-1}}. \tag{27}$$

It happens that this agrees with the previous result obtained by Suzuki and Miyashita[59].

Clearly, in our cluster variational calculation, the uniformity of the system is violated. To remedy this defect, we can consider the following two methods; namely (a) to add such a self-consistent correlation at the boundary bonds as $exp(-\alpha' \sum_{bound} \sigma_i^z \sigma_j^z)$, and (b) the total energy is considered to be given by

$$E_0 = -JzN < \sigma_1^+ \sigma_2^- > . \tag{28}$$

Through this second interpretation, some correlations outside of the cluster are effectively taken into account.

Here we discuss only the second method to obtain the ground state energy of the system. The first self-consistent cluster variational method will be discussed elsewhere. Now by minimizing the total energy (28) with (27) with respect to α, we have[59]

$$\frac{E_0}{N} = -\frac{2^{2z-3}z^z(z-1)^{z-1}}{(2z-1)^{2z-1}}\left(\frac{N}{N-1}\right)J, \tag{29}$$

for

$$\alpha = \frac{1}{4}log\left(\frac{z}{z-1}\right). \tag{30}$$

This yields[59]

$$E_0/N = -1.074J \tag{31}$$

for the square lattice (z=4). This agrees very well with the numerical result $E_0^{\mathrm{BO}}/N = -1.078J$ obtained by Betts and Oitmaa[62].

In the previous paper[59] we used the compact expression (27) which is valid only up to the order of α^3 in the previous perturbational argument. In this sense, our previous argument, but the present new interpretation is more attractive.

Larger clusters will be required to obtain more accurate values of the ground state energy. It should be, however, remarked that as far as our trial function is confined into (20) with (26), the exact ground state energy can not be evaluated even in the limit $\Omega \rightarrow \bowtie$. This is a quite general remark for the variational method using wave functions confined in some subspace of the Hilbert space.

9. ADDITIVE CLUSTER VARIATIONAL METHOD

The variational methods presented in the previous section do not necessarily satisfy the variational principle. In particular, the second effective cluster variational method does not give a uniform wave function.

Now, we propose here the following "additive" cluster variational method :

$$\Psi = \sum_j S(\Omega_j)\Psi_0. \tag{32}$$

Here, $S(\Omega_j)$ denotes a cluster operator centered at the site j such as (26). Clearly this trial wave function is uniform and correlated inside each cluster.

Some explicit applications of this new "AC variational method" to many-body systems will be given elsewhere.

10. NEW SCHEME OF THE CAM ANALYSIS

In Section 2, we have briefly presented the basic idea of the CAM. It should be noted that the evalution of the coherent anomaly exponent ψ depends[1–35] on the choice of parameters such as T or $K = J/k_{\mathrm{B}}T$.

Here we propose a new scheme of the CAM analysis. Up to now, the critical point x^* and exponent ψ have been both fitting parameters to be determined. In our new scheme, we determine the value ψ using the least square fitting for a specific value of x^*. Then, such ψ depends on x^*, namely $\psi = \psi_1(x^*)$. If we use another variable y instead of x, then we obtain another ψ as $\psi = \bar{\psi}_2(y^*)$. This function can be expressed as a function of the variable x^* as

$$\psi = \bar{\psi}_2(y^*(x^*)) = \psi_2(x^*). \tag{33}$$

If the two functions $\psi_1(x^*)$ and $\psi_2(x^*)$ cross, then the crossing point x^* defined by

$$\psi_1(x^*) = \psi_2(x^*) \tag{34}$$

many be considered to be our desired critical point and the corresponding critical exponent $\psi^* = \psi_1(x^*)$ will be our desired critical exponent. As canditates of two variables, the temperature T and the inverse temperature K will be useful.

Explicit applications will be given elsewhere.

11. SUMMARY AND DISCUSSION

In the present paper, we have explained briefly the CAM and SEFT theories of phase transitions and critical phenomena. The combination of these two theories yields the powerful SEFT CAM theory of strongly correlated electron and spin systems.

We have also proposed two new methods of many-body problems, namely the periodic-cluster mean-field approximations and effective cluster variational method.

It is expected that our new schemes explained in the present paper will be useful in solving exotic many-body problems such as the high T_c superconductors.

ACKNOWLEDGEMENTS

The author would like to thank his research group, particularly T. Kawarabayashi for his collaboration on the SEFT of the chiral orders.

This presentation is partially financed by the Ministry of Education, Science and Culture.

REFERENCES

1. M. Suzuki, J. Phys. Soc. Jpn. 55 (1986) 4205.
2. M. Suzuki, M. Katori and X. Hu, J. Phys. Soc. Jpn. 56 (1987) 3092.
3. M. Katori and M. Suzuki, J. Phys. Soc. Jpn. 25 (1987) 3113.
4. X. Hu, M. Katori and M. Suzuki, J. Phys. Soc. Jpn. 56 (1987) 3865.
5. X. Hu and M. Suzuki, J. Phys. Soc. Jpn. 57 (1988) 791.

6. M. Katori and M. Suzuki, J. Phys. Soc. Jpn. 57 (1988) 807.

7. M. Suzuki, in Qauntum Field Theory, ed F. Mancini (Elsevier Science,1986) p.505.

8. M. Suzuki, Prog. Theor. Phys. Suppl. 87 (1986) 1.

9. M. Suzuki, Phys. Lett. A116 (1986) 375.

10. M. Suzuki, J. Stat. Phys. 49 (1987) 977.

11. M. Katori and M. Suzuki, in Progress in Statistical Mechanics, ed C.K. Hu (World Scientific, 1988) p.273.

12. M. Suzuki, in Dynamics of Ordering Processes in Condensed Matter ed. S. Komura and H. Furukawa (Plenum, 1988) p.23.

13. M. Suzuki, Sci. Form. 3 (1988) 43.

14. M. Suzuki, J. Phys. Soc. Jpn. 56 (1987) 4221.

15. M. Suzuki, J. Phys. Soc. Jpn. 57 (1988) 1.

16. M. Katori and M. Suzuki, J. Phys. Soc. Jpn. 57 (1988) 3753.

17. J.L. Monroe, Phys. Lett. A131 (1988) 427.

18. T. Oguchi and H. Kitatani, J. Phys. Soc. Jpn. 57 (1988) 3973; 58 (1989) 3033.

19. N. Ito and M. Suzuki, Int. J. Mod. Phys. B2 (1988) 1.

20. M. Suzuki, J. Phys. Soc. Jpn. 58 (1989) 3642.

21. M. Katori, J. Phys. Soc. Jpn. 57 (1988) 4114.

22. S. Fujiki, in Proc. 2nd. YKIS on Cooperative Dynamics in Complex Physical Systems ed. H. Takayama (Springer,1988) p.179.

23. X. Hu and M. Suzuki, Physica A150 (1988) 310.

24. M. Suzuki, Phys. Letters 127A (1988) 410.

25. M. Takayasu and H. Takayasu, Phys. Lett. A128 (1988) 45.

26. H. Takayasu, M. Takayasu and T. Nakamura, Phys. Lett. A132 (1988) 429.

27. X. Hu and M. Suzuki,submitted to J. Stat. Phys.

28. N. Konno and M. Katori, submitted to J. Phys. Soc. Jpn.

29. X. Hu and M. Suzuki, submitted to J. Phys. A.

30. N. Kawashima, M. Katori, C. Tsallis and M. Suzuki, submitted to Int. J. Mod. Phys.

31. Y. Kinoshita, N. Kawashima and M. Suzuki, in preparation.

32. N. Ito and M. Suzuki, submitted to J. Stat. Phys.

33. K. Wada and T. Uchida, submitted to J. Phys. Soc. Jpn.

34. Y. Hirata, preprint.

35. K. Nakano and H. Mano, in preparation.

36. M. Suzuki, J. Phys. Soc. Jpn. 57 (1988) 683.

37. M. Suzuki, J. Phys. Soc. Jpn. 57 (1988) 2310.

38. M. Suzuki, J. Stat. Phys. 53 (1988) 483.

39. M. Suzuki, in Ordering and Organization in Ionic Solutions, ed. N. Ise. and I. Sogami (World Scientific, 1988) p.635.

40. M. Suzuki, in Proc. 2nd. YKIS on Cooperative Dynamics in Complex Physical Systems ed. H. Takayama (Springer,1988) p.9.

41. M. Suzuki, J. de Phys. (Paris) Colloque 8 (1988) 1519.

42. N. Kawashima and M. Suzuki,J. Phys. Soc. Jpn. 58 (1989) 3123.

43. N. Kawashima, N. Ito,N. Hatano and M. Suzuki, in preparation.

44. M. Suzuki and T. Kawarabayashi , in preparation.

45. R. Kubo, J. Phys. Soc. Jpn. 12 (1957) 570.

46. M. Suzuki, Elekiteru (in Japanese) No.32 (1989) 19.

47. W.G. Wen, F. Wilczek and A. Zee, Phys. Rev. B39 (1989) 11413.

48. Yi-H Chen, F. Wilczek, E. Witten and B.I. Halperin, Int. J. Mod. Phys. B3 (1989) 1001.

49. J. Villain, J. Phys. C10: 1717 and 4793 (1977); G. Forgacs, Phys. Rev. B22(1980) 4473.

50. S. Teitel and C. Jayaprakash, Phys. Rev. B27 (1983) 598.

51. S. Miyashita and H. Shiba, J. Phys. Soc. Jpn. 53 (1984) 1145.

52. D.H. Lee, J.D. Joannopoulos, J.W. Negele and D.P. Landau, Phys. Rev. Lett. 52 (1984) 433; Phys. Rev. B33 (1986) 450.

53. B. Berge, H.T. Diep, A. Ghazali and P. Lallemand, Phys. Rev. B34 (1986) 3177.

54. R.G. Caflisch, Phys. Rev. 34B: 3185 (1986) and preprint. See also D.H. Lee, R.G. Caflisch, J.D. Joannopoulos and F.Y. Wu, Phys. Rev. 29B: (1984) 2680.

55. T. Kawarabayashi and M. Suzuki, submitted to Phys. Lett. A.

56. N.D. Mermin and H. Wagner, Phys. Rev. Lett. 17 (1966) 1133.
P.C. Hohenberg, Phys. Rev. 158 (1967) 383.

57. T. Kawarabayashi and M. Suzuki, in preparation.

58. M. Suzuki, in New Trends in Magnetism (World Scientific, 1988).

59. M. Suzuki and K. Miyashita, Can. J. Phys. 56 (1978) 902.

60. T. Ogawa and K. Kanda, Z. Physik B30 (1978) 355.

61. F. Fazekas, Physica Scripta, T29 (1989) 125.

62. D.D. Betts and J. Oitmaa, Phys. Lett. A62 (1977) 277.
J. Oitmaa and D.D. Betts, Can. J. Phys. 56 (1978) 897.

UNIFIED DESCRIPTION OF STRONGLY AND WEAKLY CORRELATED ELECTRONS

Klaus W. Becker and Peter Fulde

Max-Planck-Institut für Festkörperforschung
7000 Stuttgart 80, Federal Republic of Germany

Abstract

A method is developed which allows for a treatment of strongly and weakly correlated electronic systems. It is based on a generalized projection method and is applied here to ground state calculations. In the case of weak correlations standard quantum chemical methods like $CEPA$ or Local Ansatz are easily reproduced. Two examples are given to demonstrate that the method can be also applied to strongly correlated systems. One concerns the ground state energy of a two-dimensional Heisenberg antiferromagnet. This energy can be calculated by hand with an accuracy comparable to that of good Monte Carlo calculations.The other deals with the one of $(CuO_2)_x$ planes which are described by a model Hamiltonian.

1 Introduction

A proper treatment of strong electron correlations has been a long standing problem. Its importance has become particularly apparent after the discovery of the new class of high temperature superconducting materials by Bednorz and Müller[1]. Even before, the physics of heavy fermion systems has demonstrated the necessity of dealing with strongly correlated electrons. Despite of great efforts in the past one has only been able to deal with relatively simple model Hamiltonians. Even then approximations have to be made which one might call uncontrolled. The situation is therefore distinct from that of solids with weakly correlated electrons, for which e.g. ground state calculations can be done with quantum chemical accuracy. With this it is meant that starting from a given basis set of Gauss type orbitals ($GTO's$) correlations e.g. in the ground state of silicon [2] can be treated with the same accuracy as in small molecules such as SiH_4. In order to understand the difference in quality of our understanding of strongly and weakly correlated systems, assume that

$$H = H_0 + H_1 \tag{1}$$

where the effect of H_1 is small as compared with that of H_0. The usual way of treating the effect of H_1 on a physical system is by decomposing

Recent Progress in Many-Body Theories, Volume 2
Edited by Y. Avishai
Plenum Press, New York, 1990

$$U(\lambda) = e^{-\lambda H} \qquad (2\text{ a})$$

$$= e^{-\lambda H_0} T_\tau e^{-\int_0^\lambda d\tau\, H_1(\tau)} \qquad (2\text{ b})$$

and expanding in powers of $H_1(\tau)^3$. When $\lambda = it$ then U is the time evolution operator, while when $\lambda = (k_B T)^{-1}$ it is the statistical operator. Furthermore, T_τ is the time ordering operator and $H_1(\tau) = exp[\tau H_0] H_1 exp[-\tau H_0]$. Expanding in powers of $H_1(\tau)$ and taking the trace results in expectation values of the form

$$\int_0^\lambda d\tau_1 ... \int_0^\lambda d\tau_n \quad < T_\tau \; H_1(\tau_1)...H_1(\tau_n) > \qquad (3)$$

Here $< .... >$ denotes an average with respect to the ground state $| \phi_0 >$ of H_0 when $T = 0$, and a thermodynamic average with respect to H_0 when $T \neq 0$. When Wick's theorem holds the evaluation of expectation values of the above form can be linked to Feynman diagrams, and follows well established lines. Often Wick's theorem is also used in order to derive a linked cluster theorem. The problem with strongly correlated electron systems is that Wick's theorem does not hold there. This is so, because H_0 contains the strong two-particle Coulomb interactions and cannot be diagonalized in terms of operators, which fulfil simple fermionic anticommutation relations. For example, H_0 may contain in that case spin- or Hubbard operators[4]. Therefore the opinion has been prevailing, that weakly and strongly correlated systems require very different methods for treating them. The aim of the present work is to show that this needs not to be so. In order to avoid the problem of Wick's theorem it seems advantageous to avoid the introduction of time dependent operators $H_1(\tau)$. This can be done[5] by starting from the following decomposition of $U(\lambda)$, instead of from Eq. (2b)

$$U(\lambda) = e^{-\lambda(H_1 + L_0)} e^{-\lambda H_0} \qquad (4)$$

where the Liouville operator is a superoperator acting on other operators C according to

$$L_0 C = [H_0, C]. \qquad (5)$$

Starting from Eq. (4) an approximation scheme can be built up which may be used to treat weakly and strongly correlated electron systems within one and the same frame work.

2 Projection method

The Hamiltonian H of a system is decomposed according to Eq. (1) and it is assumed that H_0 has been diagonalized. The ground state $| \phi_0 >$ is assumed to be nondegenerate for simplicity, but the theory can be also formulated for the case of degeneracy. By starting from Eq. (4) it was shown in Refs.[5,6] that the shift of the ground state energy δE_0 due to the presence of H_1 can be expressed in the form

$$\delta E_0 = \lim_{z \to 0} < \phi_0 \mid H_1 + H_1 \frac{1}{z - H_1 + L_0} H_1 \mid \phi_0 >^c \tag{6}$$

where c implies taking the cumulant of this expectation value. For a definition of cumulants we refer to Kubo's seminal paper[7]. Both, H_1 and L_0 are subject to forming cumulants. The notation can be simplified by the introduction of the following measure in Liouville space (i.e. operator space)

$$(A \mid B) = < \phi_0 \mid A^+ B \mid \phi_0 >^c . \tag{7}$$

Furthermore it is advantageous to introduce the operator

$$\mid \Omega(z)) = \mid 1 + \frac{1}{z - L_0 - H_1} H_1). \tag{8}$$

Then Eq. (6) can be written in condensed form as

$$\delta E_0 = \lim_{z \to 0} (H_1 \mid \Omega(z)). \tag{9}$$

It should be pointed out that the $c-$ number 1 in Eq. (8) is not subject to cumulant formation.

Eq. (9) is a central result. First of all, one notices that no time dependent operators do appear. Therefore the question of Wick's theorem does not occur. Since nothing has been said about the subdivision (1) of the Hamiltonian H_1, Eq. (9) is applicable to weakly as well as strongly correlated electronic systems. Furthermore, it is noticed that the form of Eq.(6) for δE_0 is very suitable for applying the projection technique of Zwanzig and Mori[8] for its evaluation. We want to propose the following approximation scheme for solving Eq. (9). The assumption is made that the Liouville space $\Re$ can be divided into a relevant part $\Re_0$ and an irrelevant part $\Re_1$ and that $\Re_1$ may be neglected[9]. Let $\Re_0$ be spanned by a set $\{A_\nu\}$ of operators. Without loss of generality it may be assumed that the $\mid A_\nu)$ are orthogonal, i.e.

$$(A_\nu \mid A_\mu) = \chi_\mu \delta_{\mu\nu}. \tag{10}$$

The projector

$$P = \sum_\nu \mid A_\nu) \frac{1}{\chi_\nu} (A_\nu \mid \tag{11}$$

projects onto $\Re_0$. Neglecting the influence of $\Re_1$ implies the following approximation for δE_0

$$\delta E_0 = \lim_{z \to 0} (H_1 \mid 1 + \frac{1}{z - P(L_0 + H_1)P} P H_1) \tag{12}$$

By enlarging the dimension of $\Re_0$, i.e. by including more and more operators $\mid A_\nu)$ in the set $\{A_\nu\}$, the quality of the calculated value δE_0 can be improved until it satisfies the desired accuracy. Eq. (12) may serve as the basic equation for calculating ground state energies of strongly as well as weakly correlated systems. There are two ways of selecting the $\{A_\nu\}$. One is by constructing them through repeated application of H_1. Another way of selection is through physical intuition. One includes in $\{A_\nu\}$ all those variables, which are considered as being important on physical grounds.

3 Applications: weakly correlated systems

It is shown that two well known quantum chemical approximation schemes, i.e. the Coupled Electron Pair Approximation ($CEPA$) and the Local Ansatz (LA) are easily obtained from Eq. (12).

When the electron correlations are weak one chooses $H_0 = H_{SCF}$, i.e. the self-consistent field part of the Hamiltonian, and $H_1 = H_{res}$ which are the residual interactions. The energy shift δE_0 equals then the correlation energy E_{corr}. Let the $c_\mu^+(c_\mu)$ and $c_i^+(c_i)$ denote creation (destruction) operators for electrons in orthogonalized, SCF spin-orbitals. Greek indices refer to orbitals which are occupied in the ground state and Latin indices to orbitals which are unoccupied. From them transition operators

$$\omega_{\mu\nu}^{ij} = c_i^+ c_j^+ c_\nu c_\mu \quad ; \quad i > j \ and \ \mu > \nu \tag{13}$$

are defined. The $CEPA-0$ is obtained by choosing for the $\{A_\nu\}$ the set $\{\omega_{\mu\nu}^{ij}\}$. Thus, the variables are obtained by the action of H_{res} on the ground state $\mid \phi_0 >$.

In that case it follows from Eq. (12) that

$$\begin{aligned} E_{corr} &= -\sum (H_1 \mid A_\nu) L_{\nu\mu}^{-1} \ (A_\mu \mid H_1) \\ L_{\nu\mu} &= (A_\nu \mid H A_\mu). \end{aligned} \tag{14}$$

Alternatively one may write

$$E_{corr} = \sum_{ij} \ (H_1 \mid \omega_\mu^{ij}) \alpha_{\mu\nu}^{ij} \tag{15}$$

where the $\alpha_{\mu\nu}^{ij}$ fulfil the equation

$$(\omega_{\mu\nu}^{ij} \mid H_1) \ + \ \sum_{ab\rho\tau} (\omega_{\mu\nu}^{ij} \mid H \omega_{\rho\tau}^{ab}) \alpha_{\rho\tau}^{ab} \ = \ 0. \tag{16}$$

For more details see Ref. [10]. Similarly one can derive the equations for $CEPA-2$. In that case the set of $\{A_\nu\}$ is extended by including in addition the operators $\mid \omega_{\mu\nu}^{ab} \omega_{\mu\nu}^{ij})$ in it. When evaluating E_{corr} with this set of $\{A_\nu\}$ an additional simplifying approximation is made as explained e.g. in Refs.[10,11]. Note, that here Eq. (14) must be replaced by a more general form following from Eq. (12), because $(H_1 \mid \omega_{\mu\nu}^{ab} \omega_{\mu\nu}^{ij})$=0.

The Local Ansatz (LA) is similar to $CEPA-0$, and is an example in which the $\{A_\nu\}$ are selected by physical arguments. The characteristic feature of the LA is the introduction of local operators $b_{i\sigma}^+(b_{i\sigma})$. They create (destroy) electrons with spin σ in local orbitals (states) $g_i(\underline{r})$ which are usually limited to an atomic site. This construction has been described at length in Ref.[12]. From the b operators, occupation number operators $n_{i\sigma} = b_{i\sigma}^+ b_{i\sigma}$ and spin operators $\underline{s}_i = (1/2) \sum_{\alpha\beta} \ b_{i\alpha}^+ \ \underline{\sigma}_{\alpha\beta} b_{i\beta}$ are formed. Furthermore, the following operators are constructed

$$O_\nu = \begin{cases} n_{i\uparrow} n_{i\downarrow} \\ n_i n_j \\ \underline{s}_i \underline{s}_j \end{cases} \tag{17}$$

where for simplicity only one index ν has been attached to the O operators. The $\{A_\nu\}$ are identified with the $\{\delta O_\nu\}$, whereby the difference between δO_ν and O_ν is that the zero- and one-particle excitations have been subtracted from O_ν. Since for all δO_ν the relation $(H_1 \mid \delta O_\nu) \neq 0$ holds, also Eq. (14) holds. The above two examples demonstrate that conventional methods for treating weakly correlated electrons can be rederived from Eq. (12).

4 Applications: strongly correlated systems.

The advantage of Eq. (12) is that it can be also applied to strongly correlated electron systems. Two examples are given. One concerns the ground state energy of a 2 dimensional (2 D) Heisenberg antiferromagnet (AF)[6] and the other deals with a two-band Hubbard Hamiltonian at half-filling[5].

In the case of a $2D - AF$ the Hamiltonian is

$$\begin{aligned} H_0 &= -J \sum_{<ij>} S_i^z S_j^z \\ H_1 &= -\frac{J}{2} \sum_{<ij>} (S_i^+ S_j^- + S_i^- S_j^+) \\ &= -\frac{J}{2} \sum_{<ij>} A_{ij} \end{aligned} \tag{18}$$

The ground state of H_0 is the Néel state. The space $\Re_0$ is successively enlarged by including in $\{A_\nu\}$ the operators $\mid A_{ij}), \mid A_{ij} A_{k\ell})$ and $\mid A_{ij} A_{k\ell} A_{mn})$, respectively. The ground state energy E_0 is improved accordingly, i.e.

$$-\frac{E_0}{2NJ} = \frac{1}{3} \ ; \ 0.33287 \ ; \ 0.33372 \tag{19}$$

where N is the number of sites. The last result agrees well with that of Monte-Carlo calculations[13].

A two-band Hubbard Hamiltonian at half-filling is often assumed to describe the planes of $(CuO_2)_m$ in La_2CuO_4. The Cu $3d_{x^2y^2}$ orbital hybridizes with the O $2p_{x(y)}$ with a hopping matrix element t and the difference in orbital energies is $\Delta = \epsilon_p - \epsilon_d$. It is assumed that only those orbitals are containing the holes. The Coulomb interaction between two d holes on a Cu site is set equal to infinite while that of two p holes on an oxygen site U_p is assumed to be large, but finite. In the limit of zero $Cu-O$ hybridization each Cu site contains one hole. The degeneracy of the ground state is split by an antiferromagnetic Heisenberg interaction due to superexchange between the Cu site. The coupling constant is $J = -4t^4/(U_p + \Delta)\Delta^2$. The energy shift δE due to $t \neq 0$ can be calculated according to Eq. (6) by making use of the projection technique. One finds the following result

$$\delta E_0 = N(\delta\epsilon_0 - \cfrac{4x^2\Delta}{1 - \cfrac{\delta\epsilon_0}{4x^2\Delta} + \cfrac{5x^2}{1 + \cfrac{5.93x^2}{1+\delta}}}) \tag{20}$$

Hereby $x = t/\Delta$, $\delta\epsilon_0$ is the shift due to the Heisenberg interaction and N is the number of sites. The continued fraction has a radius of convergence of $x \to \infty$. Here it is terminated by the undetermined quantity δ. Eq. (20) contains all terms to order $(t/\Delta)^6$ exactly. While for $t = 0$ the hole number at an oxygen site is $< n^P >= 0$, one finds for $t/\Delta = 0.3$ a charge transfer of $< n^P >= 0.13$.

REFERENCES

1. J. G. Bednorz and K. A. Müller, Z. Phys. B 64, 189 (1986)
2. M. V. Ganduglia Pirovano, G. Stollhoff, P. Fulde and K. P. Bohnen, Phys. Rev. B 39, 5156 (1989)
3. R. P. Feynman "Statistical Mechanics", W. A. Benjamin, Inc. (Reading, Mass.) 1972
4. see e.g. in "Strong Correlation and Superconductivity" ed. by H. Fukuyama, S. Maekowa, and A. P. Malozemoff, Springer Series in Solid-State Sciences, Vol. 89, 1989
5. K.W. Becker and P. Fulde, Z. Phys. B 72, 423(1988)
6. K. W. Becker, H. Won and P. Fulde, Z. Phys. B 75, 335(1989)
7. R. Kubo, J. Phys. Soc. Japan 17, 1100 (1962)
8. see e.g. D. Forster "Hydrodynamic Fluctuations, broken Symmetry and Correlation Functions" W. A. Benjamin, Inc. (Reading, Mass.) 1975. The book contains also a complete list of references, relating to the original work of R. Zwanzig and H. Mori.
9. Compare e.g. with P.O. Lõwdin, Int. J. Quantum Chem. 21, 69(1982)
10. K.W. Becker and P. Fulde, J. Chem. Phys. (in print)
11. W. Meyer in "Methods of Electronic Structure Theory" Vol. 3 of Modern Theoret. Chemistry, ed. H.F. Schaefer III, Plenum Press (New York) 1977.
12. G. Stollhoff and P. Fulde, J. Chem. Phys. 73, 4568 (1980)
13. J. D. Reger and A. P. Young, Phys. Rev. B 37, 5978 (1988)

FERMI HYPER-NETTED CHAIN THEORY ON A LATTICE: THE HUBBARD MODEL

X. Q. Wang[a], S. Fantoni[b], E. Tosatti[a,c], L. Yu[c,d] and X. Q. G. Wang[a,1]

[a] *International School for Advanced Studies, 34014 Trieste, Italy*
[b] *Department of Physics and INFN, University of Lecce, 73100 Lecce, Italy*
[c] *International Centre for Theoretical Physics, 34100 Trieste, Italy*
[d] *Institute of Theoretical Physics, Academia Sinica, 100080 Beijing, China*

Abstract

We review a new lattice version of Fermi Hyper-Netted Chain method for the study of strongly interacting electrons. The ordinary paramagnetic and the spin density wave functions have been correlated with Jastrow-type and $e-d$ correlations, and the corresponding FHNC equations for the pair distribution function, the one body density matrix and the staggered magnetization are discussed. Results for the $1D$ chain and $2D$ square lattice models are presented and compared with the available results obtained within Quantum Monte Carlo, variational Monte Carlo and exact diagonalization of a 4×4 Hubbard cluster. Particularly interesting are the strong effects of $e-d$ correlations on E/Nt and on the momentum distribution as well as antiferromagnetic behavior away from half filling found in our FHNC calculations in agreement with other studies.

1 Introduction

The quantitative understanding of correlation effects in strongly interacting system is a challenging issue in Many-Body physics. In most cases of interest such effects can hardly be handled by means of conventional perturbation theories *built upon independent particle basis functions.* Most notable are stochastic methods, such as Green Function Monte Carlo[1], and theories using correlated basis functions (CBF)[2] which have allowed for very accurate *ab initio* calculations in systems like liquid helium, high density neutron matter and nuclear matter[3].

In solid state physics much effort has been devoted to study the behavior of electrons in narrow band systems, their localization property and the strong correlation effects. The Hubbard hamiltonian[4,5] given by

$$H=-t\sum_{<i,j>,\sigma}c^{\dagger}_{i\sigma}c_{j\sigma}+U\sum_{i}n_{i\uparrow}n_{i\downarrow}, \tag{1}$$

[1]In the future papers, X. Q. G Wang always represents Xiaoqun Wang or Wang Xiaoqun as distinct from X. Q. Wang, representing Xiaoqian Wang, a different person.

Recent Progress in Many-Body Theories, Volume 2
Edited by Y. Avishai
Plenum Press, New York, 1990

where i, j are sites of a D-dimensional lattice, t is the hopping energy, and U is the on-site repulsion, is a well-known prototype model which incorporates the main features of such systems, namely itinerant magnetism and metal-insulator transition[6]. More recently, interest in the Hubbard model has been revived by Anderson's suggestion[7] that non-Fermi liquid behavior away from $\frac{1}{2}$ filling might be at the origin of high T_c superconductivity.

In $D = 1$ dimension it has been exactly solved by Lieb and Wu[8], and recently developed methods of numerical solution[9,10] as well as variational approaches[11-13] have been widely applied to the cases with $D = 1,2,3$. Similar attention has also been devoted to related spin-$\frac{1}{2}$ antiferromagnetic Heisenberg and quantum nonlinear σ models[14]. In spite of the enormous proportions acquired by the literature, the deep physics of this model is still open to discussion.

The variational approaches have been very useful to understand the role played by certain type of correlations, like for instance the on-site (Gutzwiller) correlations and those between empty and doubly occupied sites ($e - d$ correlations)[13,15]. The variational ground state is taken of the form

$$|\Psi_0> = G|\Phi_0>, \tag{2}$$

where $|\Phi_0>$ is a reference state which can be expressed as a Slater determinant of single particle states and G is a correlations operator building into $|\Phi_0>$ those correlation which lower the energy upperbound.

Typical reference states considered so far are the ordinary paramagnetic state $|\Phi_P>$, the BCS state $|\Phi_{BCS}>$ and the spin density wave state $|\Phi_{SDW}>$. In contrast to $|\Phi_P>$ which has a Fermi surface, both $|\Phi_{BCS}>$ and $|\Phi_{SDW}>$ do not show Fermi liquid character and have long range order.

The correlation operator which has been mostly used has the Jastrow form:

$$G_J = \prod_{ij} f_J(\mathbf{r}_{ij}) \tag{3}$$

which, in the case of on-site correlations $f(r_{ij})$, has been commonly denoted as Gutzwiller projection operator. In fact, finite range two-body correlations do not lead[13] to a significant lowering of the energy. However, the optimization of the function $f(r_{ij})$ is expected to be important for a realistic evaluation of quantities of interest like the momentum distribution and the single particle spectrum[16]. Other type of correlations as the spin-dependent correlations or the four-body correlations simulating the $e - d$ correlation have been found to play an important role.

Recently[13], the Fermi Hyper-Netted Chain (FHNC) scheme has been implemented for the Hubbard discrete lattice model. The results for the $D = 1$ case are in very good agreement with the available variational Monte Carlo estimates. The FHNC scheme has a few advantages over the Monte Carlo method: (i) it can be done at the outset for the thermodynamic limit, *avoiding* the problems concerning the finite size effects which can be serious in $D = 2$, 3 dimensions; (ii) it is very flexible as far as the possible choice of correlation operator is concerned: for instance, finite range (and also long range) correlations do not constitute a serious problem; similarly three-, four-body correlations or spin-dependent correlations can be handled. Actually, this has already been done in nuclear matter[17], liquid helium[18] and, more recently, in $1D$ Hubbard model[13], (iii) the solution of hamiltonians which are more realistic than (1) for describing the copper-oxide materials, like the two-band model hamiltonian[19] are certainly at the reach of the method; (iv) it allows for applying the full *machinery* of CBF theory[20,21] to go beyond the variational estimates, which may be *crucial* to determine the phase disgram of the system as a function of the filling factor and the coupling constant U/t.

On the other hand, the FHNC diagrammatic expansion at present cannot be expressed in a fully closed form, which makes this approach intrinsically approximate. Higher levels of approximations (FHNC/n), although formally doable, require increasing amount of analytical work. By contrast, the *accuracy* of stochastic methods is in practice limited only by the computer capabilities.

It is gratifying that already at the lowest order approximation, the FHNC/0, a reasonable agreement with the available Monte Carlo results is obtained for the $1D$ Hubbard model *for* $U/t \leq 8$.

In this contribution we discuss the FHNC scheme to calculate the expectation value of the hamiltonian (1) with trial wave function (2) with G of the Jastrow form G_J and $\Phi_0 = \Phi_{SDW}$. The SDW state is a band insulator at any filling with the Fermi surface being destroyed by single particle Bragg scattering. It has an antiferromagnetic long range order (AFLRO) that is maintained in the correlated Ψ_{SDW}. It has recently been reconsidered by Schrieffer *et al*[22] in a study of high T_c superconductivity. In fact, neutron[23] and Raman[24] scattering experiments on superconducting materials have shown that finite range antiferromagnetic correlations exist in superconducting phase and that AFLRO occurs in the phase diagram near the superconducting phase. Therefore, it is interesting to calculate at T=0 the staggered magnetization of the Hubbard hamiltonian for a Jastrow correlated model as a function of U/t and the filling factor ν. The proposed method also allows for evaluating the momentum distribution, the charge and spin structure functions and the staggered magnetization of the model. Moreover, the same scheme can be used to calculate the diagonal matrix elements of (1) between correlated basis SDW states given by

$$\Psi_{SDW}(p_1, ..., p_n; h_1, ..., h_n) = G_J \Phi_{SDW}(p_1, ..., p_n; h_1, ..., h_n) \tag{4}$$

where $\Phi_{SDW}(p_1, ..., p_n; h_1, ..., h_n)$ is the Slater determinant of SDW single particle states having the n-particle states p_1, ..., p_n and the n hole-states h_1, ..., h_n. This calculation provides the variational estimate of the single particle spectrum which is a key ingredient for future CBF perturbative calculations.

The FHNC scheme to be used with the Jastrow correlated SDW model is discussed in section 2. The last section is devoted to a discussion of the results obtained within the FHNC approach and those concerning the magnetic properties of the Jastrow correlated models.

2 Correlated spin-density wave theory

The reference state Φ_{SDW} is a Slater determinant of single particle wave functions given by[22]

$$\phi^{<}_{\mathbf{k}\sigma}(r_i) = (u(\mathbf{k})e^{i\mathbf{k}\cdot\mathbf{r}_i} + \hat{\sigma}_z(i)v(\mathbf{k})e^{i(\mathbf{k}+\mathbf{Q})\cdot\mathbf{r}_i})\xi_\sigma(i) \tag{5}$$

for $\mathbf{k}$ imbedded in the Fermi sea S_F corresponding to the half-filling case, characterized by $\cos k_x a$ $(+\cos k_y a) \geq 0$ in $1D$ ($2D$ square lattice(SQL)), and

$$\phi^{>}_{\mathbf{k}\sigma}(r_i) = (u(\mathbf{k})e^{i\mathbf{k}\cdot\mathbf{r}_i} - \hat{\sigma}_z(i)v(\mathbf{k})e^{i(\mathbf{k}+\mathbf{Q})\cdot\mathbf{r}_i})\xi_\sigma(i) \tag{6}$$

for $\cos k_x a$ $(+\cos k_y a) \leq 0$. ξ_σ is the spin eigenfunction, and $\mathbf{Q}$ is the wavevector characterizing the extra-periodicity due to the antiferromagnetic ordering. In principle,

$\mathbf{Q}$ might depend on the filling factor and should be kept as a variational parameter. Here we consider a single two-sublattice antiferromagnetic structure, which implies $\mathbf{Q}$ is defined by the condition $e^{i\mathbf{Q}\cdot\mathbf{r}_i} = -1$ for all translations t which transform a sublattice into the other. Thus $\mathbf{Q}$ is $\frac{\pi}{a}$ for $1D$ and $(\frac{\pi}{a},\frac{\pi}{a})$ for $2D$-SQL.

The orthonormalization conditions $u^2(k) + v^2(k) = 1$, $u(|\mathbf{k} \pm\mathbf{Q}|) = v(k)$ and $v(|\mathbf{k} \pm \mathbf{Q}|) = u(k)$ constrain the functions $u(k)$ and $v(k)$ which can be taken of the form

$$u(k) = (\frac{1}{2}(1 + \frac{\epsilon(k)}{E_0(k)}))^{\frac{1}{2}}, \tag{7}$$

$$E_0(k) = (\epsilon^2(k) + \Delta^2)^{\frac{1}{2}}, \tag{8}$$

where $\epsilon(k) = -2t \cos k_x a$ for $1D$ and $\epsilon(k) = -2t\ (\cos k_x a + \cos k_y a)$ for $2D$-SQL. The uncorrelated single particle spectrum is given by $E_0(k)$ for $k < k_F(\nu = \frac{1}{2})$ and $-E_0(k)$ for $k > k_F(\nu = \frac{1}{2})$. The gap parameter Δ at $k = k_F(\nu = \frac{1}{2})$ is given by[29]

$$\Delta = Um/2, \tag{9}$$

where m is the spin polarization per particle in the z-direction. In the present correlated model Δ is taken as a variational parameter and the relation (9) is no longer valid. The ordinary paramagnetic correlated ground state is a special case of this function, with $\Delta = 0$.

The expectation value of the hamiltonian is given by[13]

$$E_0 =< H > /N = -Zt\rho n(\mathbf{r}_a) + U\rho^2 g(0)/2, \tag{10}$$

where N are the lattice points separated by a, $\rho = A/N = 2\nu$ is the density, Z is the coordinate and $\mathbf{r}_a = a$ is nearest neighbor lattice vector. The functions $g(\mathbf{r}_{ij})$ and $n(\mathbf{r}_{ij})$ are the pair distribution function and the one-body density matrix, respectively. In the uncorrelated model ($f = 1$) one has $g(0) = 1 - \frac{1}{2}\ l_u^2(0) - \frac{1}{2}\ l_v^2(0)$ and $n(a) = l_u(a)$ (see eqs. (17),(18))and $E_0 = 0$ in both limits $U/t \to 0$ and $U/t \to \infty$.

2.1 The FHNC scheme for $g(\mathbf{r}_{ij})$

The pair distribution function is calculated by first expressing G_J^2 as a sum of cluster terms[25]:

$$G_J^2 = \prod_{i<j} f^2(\mathbf{r}_{ij}) = X_2(\mathbf{r}_1, \mathbf{r}_2) + \sum_{i>2} X_3(\mathbf{r}_1, \mathbf{r}_2, \mathbf{r}_i) + ..., \tag{11}$$

$$X_2(\mathbf{r}_1, \mathbf{r}_2) = f^2(r_{12}) \tag{12}$$

$$X_3(\mathbf{r}_1, \mathbf{r}_2, \mathbf{r}_i) = f^2(r_{12})(h(\mathbf{r}_{1i}) + h(\mathbf{r}_{2i}) + h(\mathbf{r}_{1i})h(\mathbf{r}_{2i})), \tag{13}$$

where the dynamical correlation $h(\mathbf{r}_{ij}) = f^2(\mathbf{r}_{ij}) - 1$. It results that $g(\mathbf{r}_{ij})$ is given by

$$g(\mathbf{r}_{12}) = \frac{1}{\aleph}\rho^2 \sum_p \frac{\rho^p}{(p-2)!} \sum_{r_3,\ldots,r_p} X_p(\mathbf{r}_1, ..., \mathbf{r}_p)\Delta_p(\mathbf{r}-1, ..., \mathbf{r}_p), \tag{14}$$

where Δ_p are the p-body uncorrelated distribution functions:

$$\Delta_p(\mathbf{r}_1, ..., \mathbf{r}_p) = \frac{A(A-1)(A-p+1)}{\rho^p} \frac{\sum_{\mathbf{r}_{p+1}\ldots\mathbf{r}_N} |\Phi_{SDW}|^2}{\sum_{\mathbf{r}_1\ldots\mathbf{r}_N} |\Phi_{SDW}|^2}, \tag{15}$$

and $\aleph =< \Psi_{SDW}|\Psi_{SDW} >$ is the normalization. As in the ordinary paramagnetic case, Δ_p can be expressed in terms of the uncorrelated density matrix:

$$\begin{aligned} \hat{\rho}(1,2) &= \frac{1}{N}\sum_{\mathbf{k},\sigma} \phi^*_{\mathbf{k}\sigma}(1)\phi_{\mathbf{k}\sigma}(2) \\ &= \frac{1}{2}\sum_{\sigma} \xi^*_{\sigma}(1)\xi_{\sigma}(2)[l_u(\mathbf{r}_{12}) + e^{i\mathbf{Q}\cdot\mathbf{r}_1} l_v(\mathbf{r}_1) sgn(\sigma)] \\ &= \frac{1}{2}\sum_{\sigma} \xi^*_{\sigma}(1)\xi_{\sigma}(2)\rho_{\sigma}(1,2), \end{aligned} \tag{16}$$

where $sgn(\sigma)$ takes either $+1$ or -1 corresponding to $\xi_\sigma =\uparrow$ or $\downarrow$, and

$$l_u(\mathbf{r}_{12}) = \frac{2}{N}\sum_{\mathbf{k}\in F_1} \{(u^2(\mathbf{k})e^{-i\mathbf{k}\cdot\mathbf{r}_{12}} + v^2(\mathbf{k})e^{-i(\mathbf{k}+\mathbf{Q})\cdot\mathbf{r}_{12}}\}, \tag{17}$$

$$l_v(\mathbf{r}_{12}) = \frac{2}{N}\sum_{\mathbf{k}\in F_1} u(\mathbf{k})v(\mathbf{k})e^{-i\mathbf{k}\cdot\mathbf{r}_{12}}(1+\cos(\mathbf{Q}\cdot\mathbf{r}_{12})). \tag{18}$$

The density matrices $\hat{\rho}(i,j)$ appear in Δ_p in the form of closed separate loops, each with a factor $-(-1/2\rho)^m$, with m being the number of loop-points. The loops with $m > 2$ have an extra factor 2, since they are originated by two different cyclic permutations. Since the dynamical correlations are spin-independent they do not affect the spin traces and one has

$$\begin{aligned} \Delta_P &= 1 - (\frac{1}{2\rho})^2 \sum_{i<j}\sum_{\sigma_i\sigma_j} \hat{\rho}(i,j)\hat{\rho}(j,i) \\ &+ 2(\frac{1}{2\rho})^3 \sum_{i<j<k}\sum_{\sigma_i\sigma_j\sigma_k} \hat{\rho}(i,k)\hat{\rho}(k,j)\hat{\rho}(j,i) + \dots \end{aligned} \tag{19}$$

It follows that $\rho^2\aleph g(r_{12})$ is given by a series of cluster terms, each being a product of dynamical correlations and exchange operators, so that each particle variable of the cluster, except for $\mathbf{r}_1$ and $\mathbf{r}_2$ is argument of at least one dynamical correlation.

In eq. (14) for every p there are $(p-2)!/S$ cluster terms which differ only in the way of labelling the $p-2$ indices (S is the symmetry number). One can then remove the $(p-2)!$ factor in eq.(14) and sum over the topologically distinct cluster terms.

The normalization $\aleph$ has obviously the same structure as $\rho^2\aleph g(\mathbf{r}_{12})$. As a consequence (being the sum over p extended to infinity), $\rho^2\aleph g(\mathbf{r}_{12})$ can be factorized in a sum of linked cluster terms containing $f^2(r_{ij})$, times $\aleph$, with the result

$$g(r_{12}) = \frac{1}{\rho^2} f^2(r_{12}) \sum_p \rho^p \sum_{\mathbf{r}_3,\dots,\mathbf{r}_p} L_p(\mathbf{r}_3,\dots,\mathbf{r}_p), \tag{20}$$

where L_p represents the sum of all linked, topologically distinct cluster terms having the (external) variables $\mathbf{r}_1, \mathbf{r}_2$ and other $(p-2)$ internal particle variables.

Since $\hat{\rho}(i,j)$ satisfy the convolution property $\sum_{\mathbf{r}_k\sigma_k} \hat{\rho}(i,k)\,\hat{\rho}(k,j) = \hat{\rho}(i,j)$, then all the separable cluster terms of $g(\mathbf{r}_{12})$ cancel each other and one is finally left with the sum over the topologically distinct irreducible cluster terms I_p:

$$g(r_{12}) = \frac{1}{\rho^2} f^2(r_{12}) \sum_p \rho^p \sum_{\mathbf{r}_3,\dots\mathbf{r}_p} I_p(\mathbf{r}_1,\dots\mathbf{r}_p). \tag{21}$$

A few examples of cluster diagrams representing I_p are reported on Fig. 1.

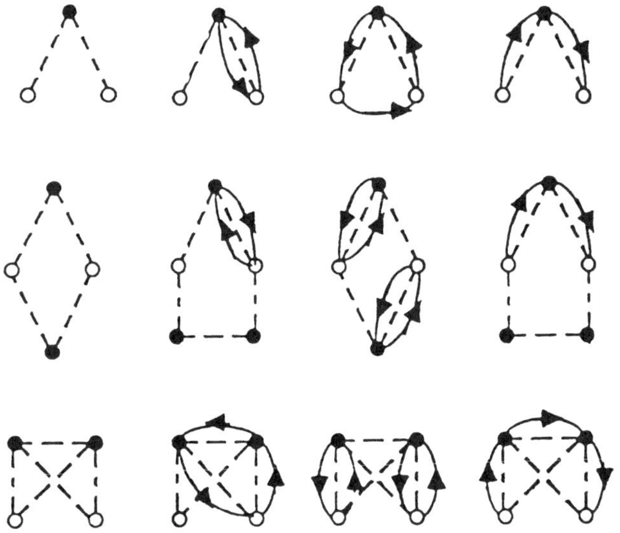

Fig.1 Examples of cluster diagrams contributing to $g(\mathbf{r}_{ij})$. The open dots represent the external variables i and j, whereas the solid dots denote internal variables, upon which summation is implied. Dashed and oriented solid lines represent dynamical correlation $h(r_{mn})$ and exchange operator $\hat{\rho}(m,n)$, respectively. The diagrams in the first row are chain (nodal), those in the second row are composite and the diagrams in the bottom row are elementary. Diagrams in the first column are of the $dd-$type; $de-$, $ee-$and $cc-$ diagrams are displayed in the subsequent columns.

The FHNC scheme to calculate $g(\mathbf{r}_{12})$ is very similar to that needed for the ordinary paramagnetic case[13]. The chain equations for N_{dd}, N_{de} and N_{ee} are exactly the same as in Ref. 13 and therefore will not be reported here. On the contrary, because of the two-component structure of $\hat{\rho}(i,j)$ and because of the presence of the phase factor $e^{i\mathbf{Q}\cdot\mathbf{r}_{ij}}$, the calculation of the (cyclic-cyclic) chain diagrams N_{cc} and of the composite diagrams X_{cc} and X_{ee} requires some more discussion. It is useful to represent the cyclic diagrams with two-component vector quantities as:

$$\begin{aligned}\hat{Z}(ij) &\equiv (Z_u(\mathbf{r}_{ij}), Z_v(\mathbf{r}_{ij})) \\ &= \sum_{\sigma} \xi^*_\sigma(i)\xi_\sigma(j)\{Z_u(\mathbf{r}_{ij}) + sign(\sigma)e^{i\mathbf{Q}\cdot\mathbf{r}_i} Z_v(\mathbf{r}_{ij})\}.\end{aligned} \tag{22}$$

The components of $\hat{\rho}(ij)$ are given by $l_u(\mathbf{r}_{ij})$ and $l_v(\mathbf{r}_{ij})$, respectively. The convolution integral is defined as

$$\hat{Z}(ij) = (\hat{F}(ik)|\hat{G}(kj)), \tag{23}$$

$$Z_u(\mathbf{r}_{ij}) = \rho\sum_{\mathbf{r}_k}\{F_u(r_{ik})G_u(r_{kj}) + F_v(r_{ik})G_v(r_{kj})\cos(\mathbf{Q}\cdot\mathbf{r}_{ik})\} \tag{24}$$

$$Z_v(\mathbf{r}_{ij}) = \rho\sum_{\mathbf{r}_k}\{F_v(r_{ik})G_u(r_{kj}) + F_u(r_{ik})G_v(r_{kj})\cos(\mathbf{Q}\cdot\mathbf{r}_{ik})\}. \tag{25}$$

The parallel connection between two vector quantities is a scalar given by:

$$\hat{F}(ij) * \hat{G}(ij) = 2F_u(r_{ij})G_u(r_{ij}) + 2\cos(\mathbf{Q}\cdot\mathbf{r}_{ij})F_v(r_{ij})G_v(r_{ij}), \tag{26}$$

In doing the chain summation one has to keep track of the ordering of the various cc-elements due to the asymmetrical structure of convolution integral (23-25), with the result

$$\begin{aligned}\hat{N}_{cc}(ij) &= (\hat{X}_{cc}(ik) + \hat{N}_{cc}(ik) - \frac{1}{2}\hat{\rho}(ik)|\hat{X}_{cc}(kj) \\ &- \frac{1}{2}(\hat{X}_{cc}(kl)|\hat{\rho}(lj))) - \frac{1}{2}(\hat{X}_{cc}(ik)|\hat{\rho}(kj)).\end{aligned} \tag{27}$$

This equation reduces to the conventional FHNC equation for $N_{cc}(r_{ij})$ when $l_v(r_{ij}) = 0$. The composite diagrams X_{cc} and X_{ee} are summed by

$$\hat{X}_{cc}(ij) = (g_{dd}(\mathbf{r}_{ij}) - 1)(\hat{N}_{cc}(ij) - \frac{1}{2}\hat{\rho}(ij)) + g_{dd}(\mathbf{r}_{ij})\hat{E}_{cc}(ij), \tag{28}$$

$$\begin{aligned}X_{ee}(ij) &= g_{dd}(\mathbf{r}_{ij})[N_{ee}(\mathbf{r}_{ij}) + E_{ee}(\mathbf{r}_{ij}) + (N_{de}(\mathbf{r}_{ij}) + E_{de}(\mathbf{r}_{ij}))^2 \\ &- 2(\hat{N}_{cc}(ij) + \hat{E}_{cc}(ij) - \frac{1}{2}\hat{\rho}(ij))^2] - N_{ee}(\mathbf{r}_{ij}),\end{aligned} \tag{29}$$

where $g_{dd}(\mathbf{r}_{ij}) = f^2(r_{ij})\exp\{N_{dd}(\mathbf{r}_{ij}) + E_{dd}(\mathbf{r}_{ij})\}$, and $E_{\alpha\beta}(ij)$ refers to the sum of bridge (elementary) diagrams of the $\alpha\beta$-type.

In terms of the FHNC-quantities the pair distribution function is given by:

$$\begin{aligned}g(\mathbf{r}_{12}) &= 1 + X_{dd}(\mathbf{r}_{12}) + N_{dd}(\mathbf{r}_{12}) + 2X_{de}(\mathbf{r}_{12}) + 2N_{de}(\mathbf{r}_{12}) \\ &+ N_{ee}(\mathbf{r}_{12}) + X_{ee}(\mathbf{r}_{12}),\end{aligned} \tag{30}$$

and the density-density correlation function $S(\mathbf{k}) = < \rho^{\dagger}_{\mathbf{k}}\rho_{\mathbf{k}} >$, with $\rho_{\mathbf{k}} = \frac{1}{\sqrt{N}}\sum_i exp(i\mathbf{k}\cdot\mathbf{r}_i)$, is obtained from $g(\mathbf{r}_{12})$ by means of the equation

$$S(k) = 1 + \rho\sum_{\mathbf{r}_i}(g(\mathbf{r}_i) - 1)\cos(\mathbf{k}\cdot\mathbf{r}_i). \tag{31}$$

Similarly, the spin-spin correlation function $S_\sigma(\mathbf{k}) = < \rho^{\dagger}_{\sigma,\mathbf{k}}\rho_{\sigma,\mathbf{k}} >$, with $\rho_{\sigma,\mathbf{k}} = \frac{1}{\sqrt{N}}\sum_i exp(i\mathbf{k}\cdot\mathbf{r}_i)\sigma_z(i)$, is given by

$$\begin{aligned}S_\sigma(\mathbf{k}) &= 1 + \rho\sum_i\cos(\mathbf{k}\cdot\mathbf{r}_i)\{X_{ee}(\mathbf{r}_i) + N_{ee}(\mathbf{r}_i)\} \\ &+ \delta(\mathbf{k} - \mathbf{Q})Nm^2,\end{aligned} \tag{32}$$

where the term which diverges at $\mathbf{k} = \mathbf{Q}$ arises from separable cluster diagrams in which the two external points are disconnected from each other and both are separability points. One can easily verify that the reducibility properties discussed in Ref. 25 do not apply here at $\mathbf{k} = \mathbf{Q}$, because of the spin operator $\sigma_z(i)$ for each external point. The resulting expression is constituted by the product of the two identical vertex corrections one for each external point and is given by m which coincides with the staggered magnetization discussed in the next subsection.

2.2 The FHNC scheme for $n(\mathbf{r}_{ij})$

The one-body density matrix is defined as

$$n(\mathbf{r}_{11'}) = \sum_\sigma \xi^*_\sigma(1')\hat{n}(1,1')\xi_\sigma(1), \tag{33}$$

$$\hat{n}(11') = \frac{A}{\aleph} \sum_{\mathbf{r}_2...\mathbf{r}_A,\sigma_2...\sigma_A} \Phi^*_{SDW}(12..A) \prod_{i,j\neq 1,1'} f(r_{1'i})f(r_{1j}) \times \prod_{i<j\neq 1,1'} f^2(r_{ij})\Phi_{SDW}(1'2...A). \quad (34)$$

If the correlation $f(\mathbf{r}_{ij})$ is set equal to 1, the uncorrelated one-body density matrix $n(\mathbf{r}_{11'}) = l_u(\mathbf{r}_{11'})$. More generally, the one-body density matrix is the u component of $\hat{n}(1,1')$. Its cluster expansion can be easily deduced from that given in Ref. 26 for a continuous system, following the procedure outlined in the previous subsection for $g(\mathbf{r}_{ij})$. One ends up with the result

$$n(11') = \sum_p \rho^p \sum_{\mathbf{r}_2,...,\mathbf{r}_p} \hat{L}_p(1,1',\mathbf{r}_3,...,\mathbf{r}_p), \quad (35)$$

where $\hat{L}_p(1,1',\mathbf{r}_3,...,\mathbf{r}_P)$ is the sum of the linked, topologically distinct terms having two external indices (1 and 1′) and $p-1$ internal indices. Each term is a product of dynamical and statistical correlations. The dynamical correlations having 1 or 1′ as particle variable must be of the type $\xi(r) = f(r) - 1$. As discussed in Ref. 26, the reducibilities in 1 and 1′ cannot be removed and they appear as vertex corrections. The density matrix is given by:

$$\hat{n}(11') = n_0\{\frac{1}{2}\hat{\rho}(11') - \hat{N}_{\xi c\xi c}(11') - \hat{E}_{\xi c\xi c}(11')\} \exp\{N_{\xi\xi}(\mathbf{r}_{11'}) + E_{\xi\xi}(\mathbf{r}_{11'})\}. \quad (36)$$

The FHNC equations for the vertex correction n_0, the chain functions $N_{\xi\xi}$, $N_{\xi d}$, $N_{\xi e}$ and the composite functions $X_{\xi\xi}$, $X_{\xi d}$, $X_{\xi e}$ coincide with those of the paramagnetic case[13] and will not be reported here. To calculate the chain-operator $\hat{N}_{\xi c\xi c}$,one has first to evaluate the chain operator $\hat{N}_{\rho a}(1,i)$ having ρ as the first element of the chain (that having 1 as particle variable) which is $\hat{\rho}$, provided by the integral equation:

$$\hat{N}_{\rho a}(1,i) = -\frac{1}{2}\hat{\rho}(1,i) + (\hat{N}_{\rho a}(1,k)|\hat{X}_{cc}(k,i) - \frac{1}{2}(\hat{X}_{cc}(k,l)|\hat{\rho}(l,i))), \quad (37)$$

and the chain operator $\hat{N}_{Xa}(1,i)$ having $\hat{X}_{\xi c}$ as first element:

$$\hat{N}_{Xa}(1,i) = \hat{X}_{\xi c}(1,i) - \frac{1}{2}(\hat{X}_{\xi c}(1,k)|\hat{\rho}(k,i)) + (\hat{N}_{Xa}(1,k)|\hat{X}_{cc}(k,i) - \frac{1}{2}(\hat{X}_{cc}(k,l)|\hat{\rho}(l,i))). \quad (38)$$

The chain operator $\hat{N}_{\xi c\xi c}(1,1')$ is given by:

$$\hat{N}_{\xi c\xi c}(1,1') - \frac{1}{2}\hat{\rho}(1,1') = (\hat{N}_{\rho a}(1,i)|\hat{X}_{c\xi}(i,1')) + (\hat{X}_{\xi c}(1,i)|\hat{N}_{a\rho}(i,1')) + (\hat{N}_{Xa}(1,i)|\hat{X}_{c\xi}(i,1')) + \hat{N}_{\rho a}(1,1') - (\hat{N}_{\rho a}(1,i)|\hat{X}_{cc}(i,1')), \quad (39)$$

where

$$\hat{X}_{\xi c}(1,i) = (g_{\xi d}(\mathbf{r}_{1i}) - 1)(\hat{N}_{\xi c}(1,i) - \frac{1}{2}\hat{\rho}(1,i)) + g_{\xi d}(\mathbf{r}_{1i})\hat{E}_{\xi c}(1,i), \quad (40)$$

$$\hat{N}_{\xi c}(1,i) = (\hat{X}_{\xi c}(1,k) + \hat{N}_{\xi c}(1,k) - \frac{1}{2}\hat{\rho}(1,k)|\hat{X}_{cc}(k,i) - \frac{1}{2}(\hat{X}_{cc}(k,l)|\hat{\rho}(l,i))) - \frac{1}{2}(\hat{X}_{\xi c}(1,k)|\hat{\rho}(k,i)). \quad (41)$$

The staggered magnetization m is defined as

$$m = \frac{1}{N} < S^z_{\mathbf{Q}} > = < \sum_{\mathbf{k}} c^\dagger_{\mathbf{k}+\mathbf{Q},\alpha} \sigma^z_{\alpha,\alpha'} c_{\mathbf{k},\alpha'} > . \tag{42}$$

It is related to $\hat{n}(11')$ by the equation

$$m = \frac{1}{N} \sum_{\mathbf{r}_1} e^{i\mathbf{Q}\cdot\mathbf{r}_1} (n_\uparrow(r_{11'} = 0) - n_\downarrow(r_{11'} = 0)), \tag{43}$$

where $n_\sigma(r_{11'}) = \xi^*_\sigma(1')\hat{n}(1,1')\xi_\sigma(1)$. In the case of ordinary paramagnetic wave function $m = 0$ since $n_\uparrow = n_\downarrow$.

The u-part of $n(1,1')$ does not contribute, because it does not depend upon σ. One gets the result

$$m = 2n_v(\mathbf{r}_{11'} = 0). \tag{44}$$

For the uncorrelated model, $n_v(0) = l_v(0)$ and one recovers eq. (9).

2.3 Excitation spectrum

As in the case of the paramagnetic model[13,27], the single particle excitation spectrum $e(\mathbf{p})$ can be calculated by modifying $\hat{\rho}(1,2)$:

$$\hat{\rho}_p(1,2) = \hat{\rho}(1,2) + \alpha \hat{f}_p(1,2), \tag{45}$$

where

$$\begin{aligned} f^u_p(\mathbf{r}_{12}) &= \frac{1}{2} \sum_{\mathbf{p}_i \in S_p} [v^2(\mathbf{p}_i)\cos(\mathbf{p}_i \cdot \mathbf{r}_{12}) + u^2(\mathbf{p}_i)\cos((\mathbf{p}_i + \mathbf{Q})\cdot\mathbf{r}_{12})] \\ &- \frac{1}{2} \sum_{\mathbf{k}_i \in S_F} [u^2(\mathbf{k}_i)\cos(\mathbf{k}_i \cdot \mathbf{r}_{12}) - v^2(\mathbf{k}_i)\cos((\mathbf{k}_i + \mathbf{Q})\cdot\mathbf{r}_{12})], \end{aligned} \tag{46}$$

$$\begin{aligned} f^v_p(\mathbf{r}_{12}) &= -\frac{1}{2} \sum_{\mathbf{p}_i \in S_p} u(\mathbf{p}_i)v(\mathbf{p}_i)[\cos(\mathbf{p}_i \cdot \mathbf{r}_{12}) + \cos((\mathbf{p}_i + \mathbf{Q})\cdot\mathbf{r}_{12})] \\ &+ \frac{1}{2} \sum_{\mathbf{k}_i \in S_F} u(\mathbf{k}_i)v(\mathbf{k}_i)[\cos(\mathbf{k}_i \cdot \mathbf{r}_{12}) + \cos((\mathbf{k}_i + \mathbf{Q})\cdot\mathbf{r}_{12})], \end{aligned} \tag{47}$$

where S_p is the set of momenta $\mathbf{k}_i$ characterizing the shell p. The terms linear in α in the expression $E_0(\hat{\rho}_p) - E_0(\hat{\rho})$, give $e_0(p) - e_0(k_F)$. A similar procedure can be used to evaluate the hole energies $e(h)$. The quantities $f^u_{k_F}$ and $f^v_{k_F}$ are different from zero, which implies the presence of a gap. It is easy to verify that, in the uncorrelated case, the above method leads to eq. (8).

The particle-hole correlated states Ψ_{ph} are not orthogonal to each other. However, it has been proved that, in the calculation of the diagonal matrix elements, the orthogonality corrections vanish in thermodynamic limit.

The behavior of the CBF theory based on $np - nh$ SDW-states as a function of the coupling constant U/t is an interesting issue. Particularly, in the limiting case $U/t \to \infty$ and $\nu = \frac{1}{2}$ the uncorrelated energy is already the best upperbound, therefore $f = 1$ and the *variational* single particle spectrum is given by $e_0(k)$ of eq. (8) which diverges. In fact, in this critical case, the number of states providing a finite energy expectation value is given by the square root of that allowed at any other finite value of U/t. As a consequence there are many states of the $np - nh$ basis which lead to unphysical energy expectation values. The understanding of the interplay between the $np - nh$ states and the magnon states[14] generated by the spin density fluctuation operator $\rho_{\sigma,k}$ could be relevant for this problem. Work in this direction is in progress.

3 Results and discussion

In this section, we review some numerical results obtained[13,29] for the $1D$ Hubbard model and present some preliminary results for the $2D$-SQL case[30].

The convergence of the FHNC scheme with respect to the inclusion of bridge diagrams has been studied for the $1D$-Gutzwiller ansatz

$$G_G = \prod_{i<j} f_G(\mathbf{r}_{ij}) = \prod_{i<j} (1 - (1-g)\delta_{r_i\uparrow r_j\downarrow}), \tag{48}$$

and the paramagnetic Slater determinant Φ_p, for which case exact results are available[31]. It has been found that FHNC/0 (no bridge diagrams) already provides an estimate of E/Nt for $U/t \leq 4$ and $\rho = 1$ which is accurate within less than 10%. The accuracy is somewhat worse for the sum rules $S(0) = S_\sigma(0) = 0$ and $n(0) = \rho$. The FHNC/4 approximation improves the results of E/Nt which are in agreement with the corresponding exact values for $U/t \leq 10$ at all values of ρ. For larger values of the coupling constant and, consequently smaller values of g ($g \leq 0.4$), the FHNC scheme has a low convergence and one needs to go beyond FHNC/4.

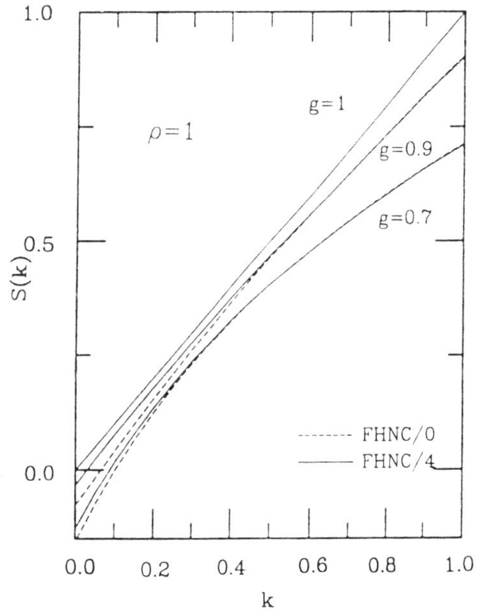

Fig.2 Taken from Ref. 13. $1D$-Hubbard model, $U/t = 4$. Density-density correlation functions for the Gutzwiller paramagnetic ansatz at $\rho = 1$. Solid and dashed lines correspond to calculations using FHNC/4 and FHNC/0, respectively. Results for values $g = 0.7$ and 0.9 of the variational parameter are compared with the Hartree-Fock ($g = 1$) results. k is in units of π/a.

Typical results of the density-density correlation function $S(k)$ are displayed in Fig.2 for $\rho = 1$. The effect of correlation is quite large even at $g = 0.9$. The contribution of bridge diagrams increases for larger values of the filling factor and smaller values of g. Their effect becomes negligible for $|\mathbf{k}| \geq \frac{1}{2}k_F$. $S(0)$ can be forced to be zero by redefining $S(k)$ in a small region of k around zero[32], with practically no change in E/Nt.

Table I Total energy E/Nt of the $1D$-Hubbard hamiltonian with $U/4t = 1$ calculated for various variational paramagnetic models, by using FHNC/4 (FHNC) or second order Power Series[25,13] (PS) approximation, and compared with the exact Gutzwiller model[31] (G_{exact}) and the exact results by Lieb and Wu[8]. G, J, S, Q stand for Gutzwiller, Jastrow, Gutzwiller-spin, Gutzwiller-quartet paramagnetic models, respectively. The bottom row displays the results of the one-body density matrix for the G-model at $r = 0$, calculated in FHNC/4 approximation.

ρ	1	0.8	0.6
G_{exact}	−0.518	−0.764	−0.791
$EXACT$	−0.573	−0.788	−0.821
G_{FHNC}	−0.519	−0.763	−0.792
J_{FHNC}	−0.520	/	/
J_{PS}	−0.522	/	/
S_{PS}	−0.528	/	/
Q_{FHNC}	−0.560	−0.779	−0.819
$n_{FHNC}(0)$	0.99	0.79	0.60

The results for E/Nt and $n(0)$ are summarized in Table I at $U/t = 4$. The second row of the Table displays the exact results for the $1D$-Hubbard hamiltonian by Lieb and Wu while the second row from the bottom shows the FHNC/4 results[13] for a correlation operator which includes four-particle correlations (quartets):

$$G = G_G \prod_{i<j<k<l} f^{(4)}(i,j,k,l), \tag{49}$$

where

$$f^{(4)}(i,j,k,l) = \begin{cases} \eta & if \;\; r_{ij} = 0, |\mathbf{r}_{ik}| = a, |\mathbf{r}_{il}| = 2a, |\mathbf{r}_{kl}| = 3a, \\ 1 & otherwise. \end{cases} \tag{50}$$

The quartet-correlations simulate short-range $e - d$ correlations, discussed originally by Kaplan, Horsch and Fulde[15] and later by others[15,33]. The estimate of E/Nt as a function of ρ at various values of $U/4t$ are shown in Fig. 3 and the resulting momentum distribution is displayed in Fig. 4. The following comments are in oder: (i) FHNC/0 and FHNC/4 approximations look reasonable for moderate values of the coupling constant ($U/t \leq 4$ for FHNC/0 and $U/t \leq 10$ for FHNC/4). The accuracy is expected to improve in more than one dimensional models; (ii) nearest-neighbor scalar ((J) in Table I) or spin-dependent ((S) in Table I) correlations do not lead to a significant lowering of the energy upperbound in the $1D$ paramagnetic case[13]; (iii)the quartet correlations account for 90% of the discrepancy between the results of the Gutzwiller model and the exact eigenfunctions[13,15]; (iv) the quartet correlations lead to a decrease of $n(k)$ for $k \leq k_F$, in accordance with the exact solution by Lieb and Wu; (v) the correlated paramagnetic wave function bear the Fermi liquid character which, strictly speaking, is incorrect at least in $1D$[34,7]; in spite of this, the energy upperbounds in the quartet-Gutzwiller model are very close to the exact results.

On the contrary, the correlated BCS[13,35] or the correlated SDW lead to a $n(k)$ which is continuous across k_F. In the following we discuss some preliminary results[29,30] obtained for the Jastrow correlated SDW wavefunction.

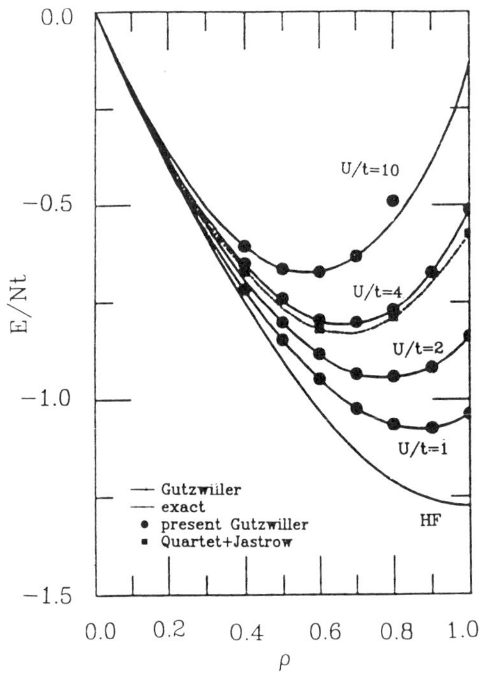

Fig.3 Taken from Ref. 13. The total energy for the $1D$ Hubbard model.The results obtained for the Gutzwiller paramagnetic wave function in FHNC/4 approximation are compared with the exact Gutzwiller results[31]. At $U/t = 4$, the estimates of the Quartet+Gutzwiller model which includes $e-d$ correlation are compared with the exact results of Lieb and Wu[8].

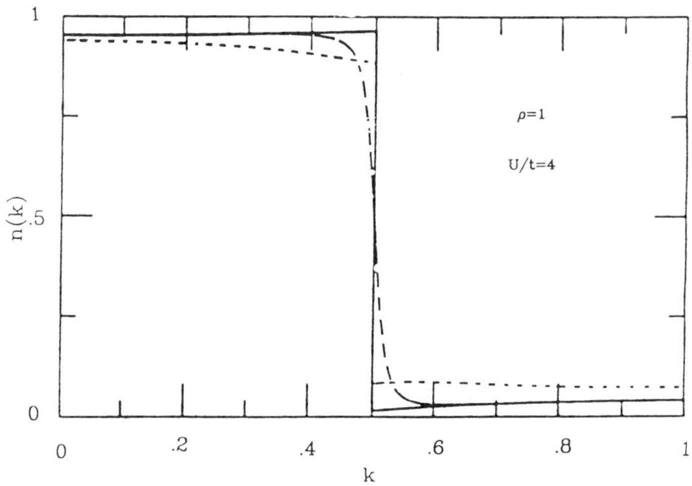

Fig.4 Momentum distribution functions for different correlated $1D$-models at $U/t = 4$ and $\rho = 1$. The $n(k)$ of the two paramagnetic wavefunctions ($\Delta = 0$) (solid: Gutzwiller; short-dashes: $e-d$) show the effect of quartet-correlations. The antiferromagnetic wave function (long-dashes) has $\Delta = 0.48$ and $g = 0.5$.

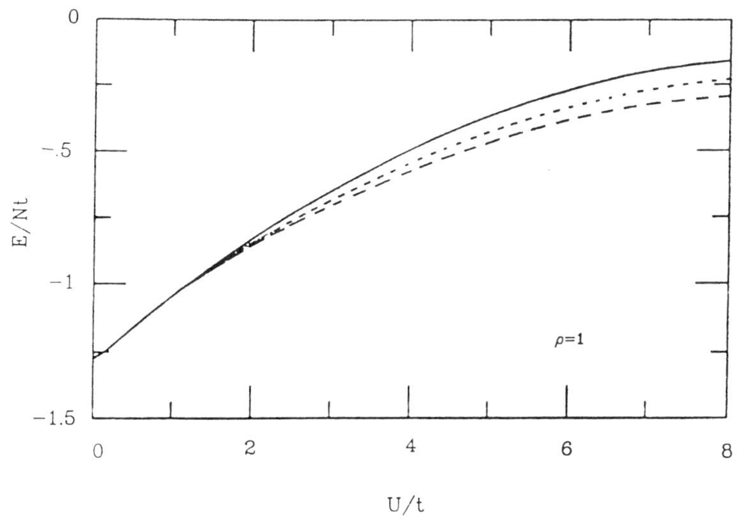

Fig.5 Energy expectation values of the $1D$-Hubbard model as a function of U/t. The results of the Gutzwiller correlated paramagnetic model [13,12] (solid line) and those of Gutzwiller correlated SDW model (short dashes)[29,12] are compared with the exact solution (long dashes)[8].

The $1D$-model results for momentum distribution and the energy upperbound are given in Figs. 4 and 5. The AF model is energetically preferred with respect to the paramagnetic one at all values of the coupling constant U/t. The FHNC/0 results agree with those of Ref. 12 within less than 5%. The smallest value of g occurs at $U/t = 4$ ($g = 0.6$) with $g = 1$ at both $U/t = 0$ and ∞. FHNC/4 is expected to reproduce the Monte Carlo results. The momentum distribution of the AF wave function differs from that of the paramagnetic one, mainly because of its continuous behavior across k_F. The features due to $e - d$ correlations, shown in Fig. 4, are not reproduced by the SDW correlations, which strongly indicates that $e - d$ correlations are important for the SDW model, too.

Preliminary results[30] obtained for the $2D$-SQL model at half filling are given in Figs. 6 – 8. As in the one dimensional case, the AF model is variationally prefered with respect to the paramagnetic one. The FHNC/0 evaluations are in good agreement with the Monte Carlo estimates of Ref. 12. For instance at $U/t = 4$ variational Monte Carlo gets $E/Nt = -0.84$ with $g = 0.65$, $\Delta = 0.45$ and $m = 0.58$ which compares reasonably well with the FHNC/0 results reported in Fig. 6 and Table II. In Fig. 6 the QMC estimates[9], the variationalMC calculation[12] and the results of 4×4 exact diagonalization[34] are reported for comparison.

The density-density correlation function $S(\mathbf{k})$ and the momentum distribution $n(\mathbf{k})$ are plotted in Figs. 7, 8 by following the path $\Gamma PQ\Gamma$ displayed in the insert Fig. 7. The correlated $S(\mathbf{k})$ should, in fact, heal to zero at Γ ($S(0) = 0$). The failure of this sum rule is entirely due to the lackness of elementary diagrams in FHNC/0 approximation. FHNC/4 approximation leads[30] indeed to a much better fulfillment of the sum rule, although the energy estimates change very little.

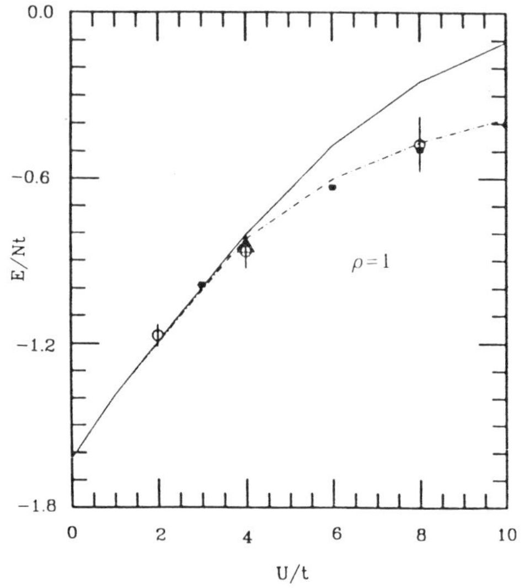

Fig.6 Energy expectation values of the $2D$-SQL model as a function of U/t. The results of the Gutzwiller correlated paramagnetic model (solid line) and SDW (dot-dashed line) models are compared with Hirsch's QMC results[9] (open circle and bars), that of Parola *et al.*[34] (triangle). and those of Yokoyama and Shiba[12] (solid square)

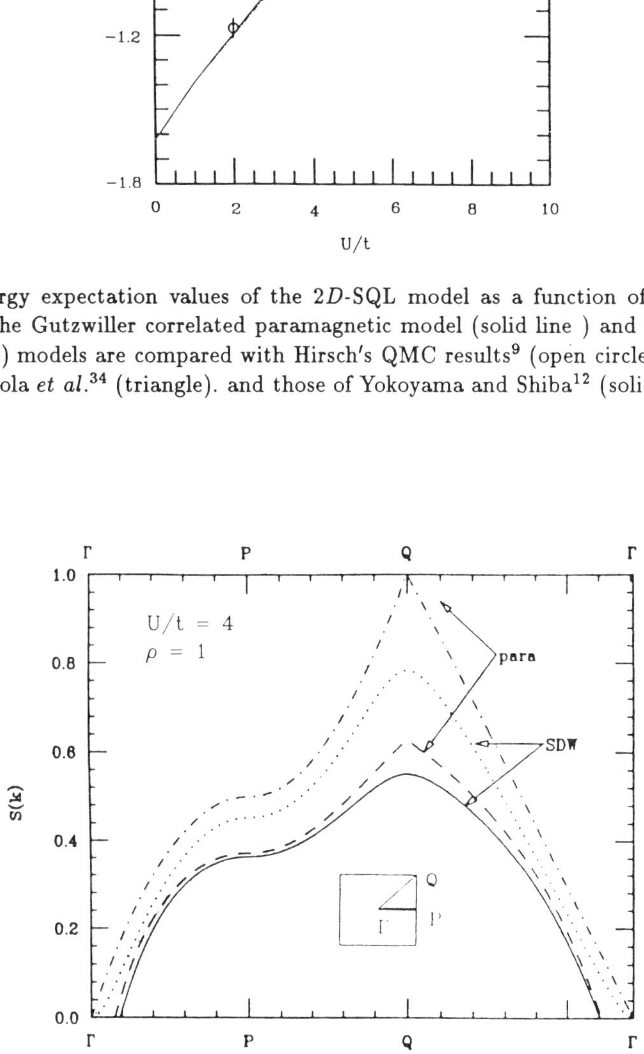

Fig.7 Density-density correlation functions $S(\mathbf{k})$ for the Gutzwiller correlated paramagnetic and AF models for the $2D$-SQL Hubbard hamiltonian at $\rho = 1$ and $U/t = 4$. The dotted and dot-dashed curves report the uncorrelated $S_{un}(\mathbf{k})$. All the results displayed, correspond to wave functions having the *optimal* values for the variational parameters g and Δ.

Table II. Breakdown of the energy and the staggered magnetization for the Gutzwiller correlated $2D$-SQL model calculated in FHNC/0 approximation.

ρ	1.0	1.0	1.0	0.946	0.946	0.90	0.90
U/t	1.0	4.0	6.0	1.0	4.0	4.0	6.0
g	0.85	0.62	0.98	0.86	0.60	0.61	0.50
Δ	0.02	0.43	2.40	0.02	0.30	0.06	0.39
$< T >$	-0.60	-1.48	-1.10	-1.61	-1.48	-1.49	-1.38
$< V >$	0.21	0.66	0.50	0.20	0.57	0.50	0.62
E/Nt	-1.39	-0.82	-0.60	-1.41	-0.91	-0.99	-0.76
m	0.20	0.41	0.81	0.02	0.30	0.08	0.34
$S(\pi,\pi)$	0.78	0.54	0.21	0.91	0.60	0.64	0.49
$S_\sigma(\pi,\pi) - m^2N$	1.19	1.21	0.25	1.11	1.34	1.46	1.19

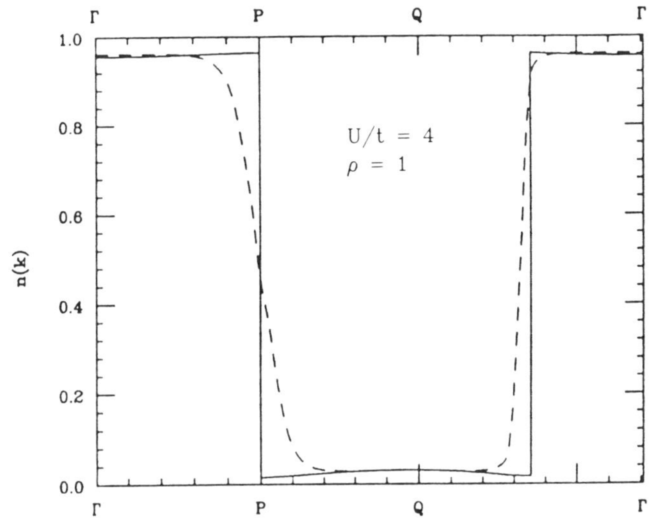

Fig.8 Momentum distribution of the Gutzwiller-correlated $2D$-SQL Hubbard model. The solid curve refers to the paramagnetic wave function and the dashed curve to the SDW wave function.

Some preliminary results obtained away from half filling ($\rho = 0.95$ and $\rho = 0.90$) are reported in Table II. They show that the Gutzwiller correlated models lead to AF-LRO away from half filling. Such interesting result persists in the Jastrow model, too[30], and it is in agreement with the 4×4 exact diagonalization calculation of Ref. 34. For a more realistic evaluation of the $U/t - \rho$ phase diagram, it is necessary to go beyond the Jastrow-type correlation operator. This can be done either by including triplet and quartet correlations in G or by employing CBF theory. Work in this direction is in progress.

References

[1] M.H.Kalos, *Phys. Rev.* **A2**, 250(1970); D.M.Ceperley and M.H. Kalos, in *Monte Carlo Method in Statistical Physics*, K. Binder ed. (Springer, Berlin 1978).

[2] E. Feenberg, *Theory of Quantum Liquids*, (Academic Press, New York 1969)

[3] *Recent Progress in Many-Body Theory*, Vol.I , A. Kallio, E. Pajanne and R. F. Bishop, eds., (Plenum Press, London, 1981).

[4] J. Hubbard, *Proc. Roy. Soc.* **A276**, 238(1964); **A285**, 542(1965).

[5] D. Vollhardt, *Rev. Mod. Phys.* **56**, 99(1984).

[6] W. F. Brinkman and T. M. Rice, *Phys. Rev.* **B 2**, 4302(1970).

[7] P. W. Anderson, *Science* **235**, 1196 (1987).

[8] E. H. Lieb and F. Y. Wu, *Phys. Rev. Lett.* **20**, 1445 (1968).

[9] J.E. Hirsch, *Phys. Rev.* **B31**, 4403 (1985); J.E. Hirsch and H.Q. Lin, *Phys. Rev. Lett.* **31**, 4403 (1989).

[10] S. Sorella, E. Tosatti, S. Baroni, R.Car, and M. Parrinello, *Intern. J. Mod. Phys.* **B1**, 993 (1988); S. Sorella, S. Baroni, R.Car, and M. Parrinello, *Europhys. Lett.* **8**, 663 (1989).

[11] C. Gros, R.Joynt, and T.M. Rice, *Phys. Rev.* **B36**, 381 (1987).

[12] H. Yokoyama and H. Shiba, *J. Phys. Soc. Jap.* **56**, 1490, 3570, 3582 (1987).

[13] S. Fantoni, X.Q. Wang, E. Tosatti, and Lu Yu, *Physica* **C153**, 1255 (1988); S. Fantoni, X.Q. Wang, E. Tosatti, Lu Yu and M. Viviani, submitted to *Phys. Rev.* **B**.

[14] E. Manousakis, Tallohssee preprint (1989) and references therein.

[15] T. A. Kaplan, P. Horsch and P. Fulde, *Phys. Rev. Lett.* **49**, 889(1982). P. Fazekas and K. Penc, *Int. J. Mod. Phys.* **B1** 1021(1988)

[16] E. Krotscheck, R. A. Smith, J. W. Clark and R. M. Panhoff, *Phys. Rev.* **B 24**, 6383(1981).

[17] V.R.Pandharipande and R.B. Wiringa, *Rev. Mod. Phys.* **51**, 821 (1979).

[18] M. Viviani, A. Buendia, S. Fantoni, and S. Rosati, *Phys. Rev.* **B 38**, 4523(1988).

[19] T. M. Rice, contribution in this volume.

[20] J. W. Clark, L. R. Mead, E. Krotscheck, K. E. Kurten and M.L. Ristig, *Nucl. Phys.* **A328**, 45(1979); E. Krotscheck and R. A. Smith, *Phys. Rev.* **B27**, 4222(1983); J. W. Clark, in *Progress in Particle and Nuclear Physics*, ed. D.H. Wilkinson (Pergamon, Oxford, 1979), Vol. 2.

[21] S. Fantoni, B.Friedman, and V. R. Pandharipande, *Nucl. Phys.* **A386**, 1(1982); **A339**, 1(1983); A. Fabrocini and S. Fantoni, *First Int. Course on Condensed Matter*, ACIF series Vol. 8, eds. D. Prosperi, S. Rosati, and G. Violini, eds. (World Sci., Singapore, 1986) and references therein.

[22] J. R. Schrieffer, X. G. Wen and S. C. Zhang, *Phys. Rev.* **B39**, 11663(1989).

[23] G. Shirane *et al. Phys. Rev. Lett.* **59**, 1613(1987). J. Tranrquada *et al. Phys. Rev. Lett* **60**, 156(1988).

[24] K. B. Lyons *et al. Phys. Rev.* **B37**, 2353(1988); *Phys. Rev. Lett* **60**, 732(1988).

[25] S. Fantoni and S. Rosati, *Nuovo Cim.* **20A**, 179 (1974); **25A**, 593(1975).

[26] S. Fantoni, *Nuovo Cim.* **44A**, 191 (1978).

[27] B. Friedman and V.R. Pandharipande, *Phys. Lett.* **B100**, 205 (1981).

[28] S. Fantoni and V. R. Pandharipande, *Phys. Rev.* **C37**, 1697 (1988).

[29] X. Q. Wang, S. Fantoni, E. Tosatti and Lu Yu, Perceedings of XIII International Workshop on Condensed Matter Theories, V. C. Angnilera-Navaro *et al.*, eds., to be published by Plenum Press.

[30] X. Q. Wang *et al.*, in preparation.

[31] W. Metzner and D. Vollhardt, *Phys. Rev. Lett.* **59**, 121 (1987); F. Gebhardt and D. Vollhardt, *Phys. Rev. Lett.* **59**, 1472 (1987).

[32] E. Krotscheck, *Nucl. Phys.* **A 317**, 149 (1979).

[33] D. Baeriswyl, in *Nonlinearity of Condensed Matter Physics*, A.R.Bishop *et al.*, eds. (Springer, Berlin 1987) *p*.183.

[34] A. Parola, S. Sorella, E. Tosatti, and M. Parrinello, to be published.

[35] S. Fantoni, *Nucl. Phys.* **A 363**, 381 (1981).

MANY BODY METHODS APPLIED TO SCATTERING OF COMPOSITE PARTICLES IN A GAUGE THEORY WITH CONFINEMENT

D. Bérubé[a], J.F. Brière[a], H. Kröger[a1], K.J.M. Moriarty[b,c,d] and J. Potvin[d,e]

[a] Département de Physique, Université Laval, Québec,
Prov. Qué., G1K 7P4, Canada
[b] Institute for Advanced Study, Princeton, NJ 08540
[c] John von Neumann National Super Computer Center, Princeton, NJ 08540
[d] Department of Mathematics, Statistics and Computer Science,
Dalhousie University, Halifax, Nova Scotia, Canada
[e] Department of Physics, Boston University, Boston, MA 02215

We present numerical results of a lattice calculation on scattering of two composite particles in QED_{1+1}, which confines single fermions. The composite particles are taken as neutral, fermion-antifermion, lowest mass eigen states of the Hamiltonian. We use the light-cone momentum representation on a lattice and compute the time evolution and S-matrix by diagonalizing the lattice Hamiltonian. Moreover, we discuss a new method based on the Hamiltonian formulation and on the Monte Carlo projector technique in order to compute the Minkowski time evolution and S-matrix. We present results for NN scattering in a non-relativistic potential model.

1. MANY-BODY ASPECTS IN COMPUTATION OF S-MATRIX FOR QUANTUM FIELDS

In many-body language, a Hamiltonian can be considered to act in occupation number space. If one goes over to discretized momenta (lattice) with a high momentum cut-off this reduces the number of degrees of freedom. Still there are infinitely many degrees of freedom , if its a bosonic system. In order to reduce the number of degrees of freedom to a manageable size, different routes can be taken: (a) The most naive way is to introduce a cut-off in the occupation number. This is useful in simple models in D=1 or D=2 dimensions. (b) There may be constraints from physics on the occupation number, e.g. as obtained from using light-cone momenta. This is applied in Sect.2. (c) The number of degrees of freedom , and in particular the occupation number, can be reduced by means of stochastic techniques . This has been shown to work successfully in Euclidean lattice field theory by evaluating Euclidean path integrals. It is also at the heart of the Monte Carlo

[1]Talk presented by H. Kröger at VI. International conference on recent progress in many-body theories, Arad, Israel 1989.

Université Laval preprint LAVAL-PHY-5/89
December 1989

Recent Progress in Many-Body Theories, Volume 2
Edited by Y. Avishai
Plenum Press, New York, 1990

Tab.[1]. S-matrix for elastic 2-boson scattering ($n_{init} = n_{final} = 2$) as function of virtual particle number cut-off n_{min} and n_{max}. Weak coupling $\lambda_{eff} = \lambda_{bare}/m^2_{bare} = 1/6$.

n_{min} n_{max} \ ν	10	20	30	40
2 2	0.8263	0.6801	0.5966	0.5150
	$-0.5261i$	$-0.7145i$	-0.7900	$-0.8481i$
2 4	0.8727	0.7444	0.6686	0.5928
	$-0.4615i$	$-0.6555i$	$-0.7353i$	$-0.7977i$
2 6	0.8744	0.7467	0.6712	0.5957
	$-0.4590i$	$-0.6531i$	$-0.7330i$	$-0.7956i$
2 8	0.8745	0.7469	0.6686	0.5959
	$-0.4589i$	$-0.6530i$	$-0.7355i$	$-0.7955i$

Tab.[2]. Same as Tab.[1], but $\lambda_{eff} = 1$.

n_{min} n_{max} \ ν	10	20	30	40
2 2	0.1016	0.4933	0.5152	0.3181
	$0.4008i$	$0.1014i$	$-0.2083i$	$-0.4199i$
2 4	-0.5054	-0.8075	-0.4372	-0.1587×10^{-1}
	$-0.6862i$	$0.3622i$	$0.7575i$	$0.7695i$
2 6	-0.1539	-0.8723	-0.7147	-0.3846
	$-0.9384i$	$-0.1322i$	$0.3368i$	$0.4716i$
2 8	0.3753×10^{-1}	-0.7771	-0.7100	-0.4361
	$-0.9672i$	$-0.3320i$	$0.1198i$	$0.2768i$
2 10	0.1188	-0.6757	-0.6535	-0.4241
	$-0.9568i$	$-0.4395i$	$-0.1031\times10^{-2}i$	$0.1351i$
2 12	0.1275	-0.6718	-0.6514	-0.4231
	$-0.9562i$	$-0.4426i$	$-0.5187\times10^{-2}i$	$0.1308i$

Greens Function method. It will be shown in Sect. 3, how Monte Carlo methods can be used to compute dynamical quantities like time evolution (real time). The crucial point of all these reduction techniques is that one has to establish a connection between the answer obtained with the reduced number of degrees of freedom and the continuum limit. In order to show that it is meaningful to use the occupation number in the context of an S-matrix, let us consider as example the Φ^4-model in 1+1 dimensions [1]. We have computed the S-matrix for elastic scattering of two bosons. In Tabs.[1,2] its dependence on the virtual particle number is shown. The parameter ν is related to the size of the lattice. Both cases, weak and intermediate coupling (the critical value for a phase transition is $\lambda_{eff}^{crit} \approx 1.5$), show convergence when increasing the upper limit of the virtual particle number.

2. SCATTERING IN QED_{1+1}

In this section we want to discuss scattering in QED in 1+1 dimensions using discretized light-cone momenta in order to reduce the number of degrees of freedom of the many-body system. We have chosen this model, because it is a gauge theory and it confines single fermions [2]. We use it to model meson-meson scattering in QCD. Besides there is new interest in QED, because Kogut et al. [3] have found evidence for new phenomena in strong coupling QED, i.e. a phase transition in the chiral condensate and a collapse of the e^+e^- wave function.

Eller, Pauli and Brodsky [4] have discussed this model in a Hamiltonian formulation, using light-cone (l.c.) momentum variables on the lattice and they have computed the mass spectrum. The method has been also applied by Hornbostel et al. [5] to $SU(2)$ gauge theory. The l.c. coordinates are $x^\pm = x^0 \pm x^1$. In order to quantize the corresponding l.c. Hamiltonian, one chooses the l.c. gauge $A^+ = 0$. Because there are no transverse photons in one space dimension, the Hamiltonian can be expressed in terms of fermionic fields only. One constructs the l.c. momentum operator P^+, the l.c. Hamiltonian P^- and the charge operator Q, which all commute. By a Fourier transformation one goes over to the variables k^+, k^-, i.e. l.c. momentum and energy. Then a l.c. momentum lattice is introduced via $k_n^+ = 2\pi n/L$, $n = 0, \ldots, \Lambda$. As a consequence of the l.c. formulation in 1+1 dimensions, the l.c. momentum operator is positive. This implies on the l.c. lattice a cut-off in the occupation number, if we consider a scattering process corresponding to a given total l.c. momentum. This leads to a sizeable reduction in the number of degrees of freedom compared with a formulation in terms of ordinary momenta, and is very convenient in numerical calculations. However, there is a singularity in the l.c. Hamiltonian for $k_0^+ = 0$, referred to as the zero-mode problem. This already occurs for the non-interacting Hamiltonian

$$H_0 = \sum_{n=0}^{\Lambda} \frac{1}{n} \left[b_n^\dagger b_n + d_n^\dagger d_n \right] . \tag{1}$$

Usually this is regularized by omitting the zero mode. However, one can show that this singularity does not occur in the l.c. time evolution $\exp[iH_0 x^+]$, after a regularization at $k^+ < \epsilon$ and letting $\epsilon \to 0$. No proof of this is available yet for the interacting case.

(a) FERMION ANTIFERMION SCATTERING IN REGULARIZED QED_{1+1}

As mentioned above this model confines fermions/antifermions. Fermion antifermion scattering can be studied when the interaction is modified. The interaction has terms which behave like

$1/(k_1^+ - k_2^+)^2$. By introducing a screening mass m_{scr}, the interaction is regularized. It behaves as $1/[(k_1^+ - k_2^+)^2 + m_{scr}^2]$ which is of short range. We compute the l.c. S-matrix

$$S_{fi}(L, x^+) = < \Phi_f | \exp[iP_{as}^-(L)x^+] \exp[-i2P^-(L)x^+] \exp[iP_{as}^-(L)x^+] | \Phi_i >, \tag{2}$$

by diagonalizing algebraically the l.c. Hamiltonians on the lattice. We also compute $\Delta_{<E>}(L, x^+)$, which is the violation of energy for scattering on the lattice (for more details see Ref. [6]). Results

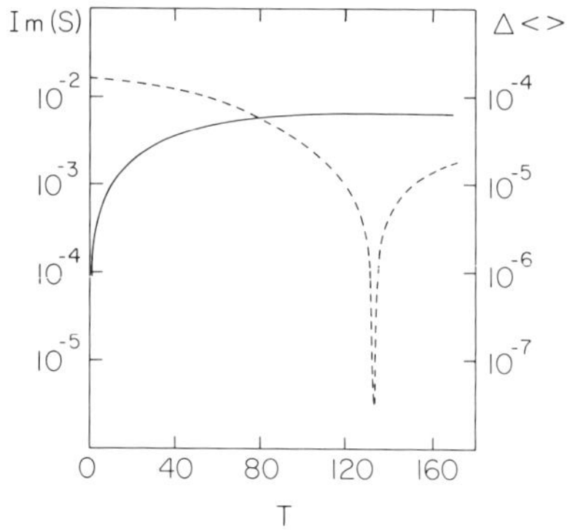

Fig.[1a]

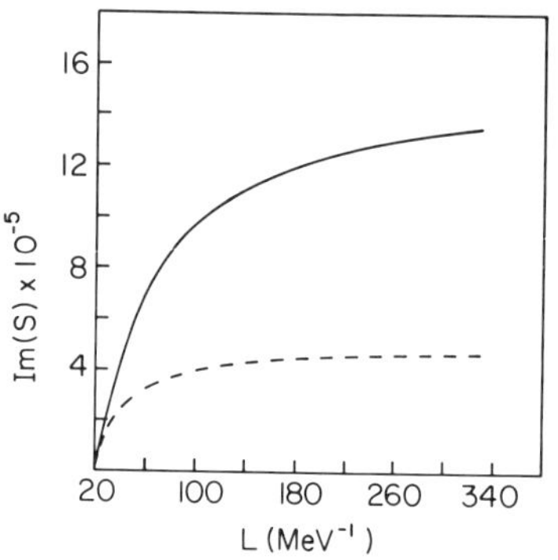

Fig.[1b]

Fig.[1a]. S-matrix (solid curve) and violation of energy (dashed curve) for elastic fermion antifermion scattering as a function of l.c. time $T \equiv x^+$. Fig.[1b]. S-matrix for elastic fermion antifermion scattering as a function of the lattice size L. The dashed curve corresponds to $x^+ = const$, while the solid curve corresponds to $x^+ = x^+_{scatt}$.

for fermion antifermion scattering are presented in Figs.[1a,b]. In Fig.[1a] one observes a region of stability for the imaginary part of the S-matrix as a function of the l.c. time, which coincides with the position of the minimum of the violation of energy. This calculation takes into account all virtual particle states admissible by the particle number cut-off implied by total momentum of the scattering process. In Fig.[1b] we show the behavior of the S-matrix when approaching the continuum limit $L \to \infty$, $\Lambda \to \infty$, but $L/\Lambda = const$. In this calculation the particle number has been constraint to the 1-fermion 1-antifermion sector. One observes convergence of the S-matrix computed with a fixed l.c. time. The more physical quantity is the S-matrix computed at the l.c. lattice scattering time x^+_{scatt}, defined as the value of x^+ associated with the minimum of energy violation. x^+_{scatt} itself varies with the lattice size L. It increases, when L increases and goes to infinity in the continuum limit. Based on this definition, convergence of the S-matrix with increasing the lattice is much harder to achieve. Some indication for convergence can be seen.

(b) COMPOSITE PARTICLE SCATTERING IN QED_{1+1}

We have taken as composite particle a fermion anti-fermion state, which forms a boson. It is a simultaneous eigenstate of the l.c. Hamiltonian, momentum, invariant mass and charge operator.

We have taken the lowest mass M (lowest eigenvalue of the mass spectrum) and charge $Q = 0$. This state $|B(k^+) >$ is then given by its l.c. momentum k^+ (or l.c. energy k^-, related by the mass shell condition $k^+k^- = M^2$). In order to compute scattering of two of those bosons we have taken as asymptotic state as a wave packet superposition of two bosons, to a given total momentum k^+_{tot}. The wave function of the boson state, its mass and the S-matrix for scattering of two bosons is computed numerically on the l.c. lattice. One should note that the l.c. Hamiltonian $P^-(L)$

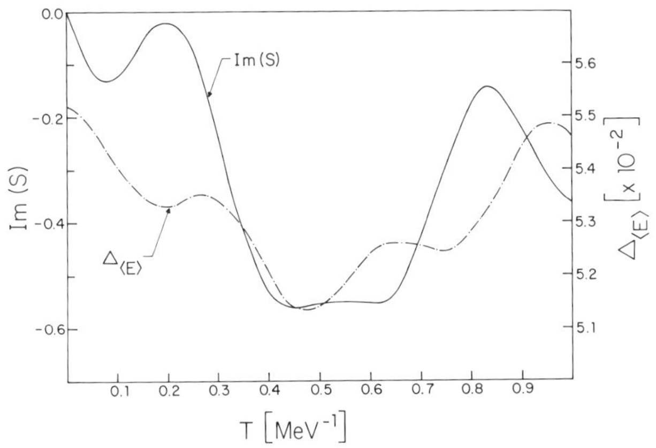

Fig.[2]. S-matrix (solid curve) and violation of energy (dashed dotted curve) for elastic boson boson scattering as a function of l.c. time $T \equiv x^+$.

corresponds to finite dimensional Hermitian matrix, which can be diagonalized numerically and yields a spectrum of energies and wave functions. In the numerical computation one diagonalizes the l.c. Hamiltonians $P^-(L)$, and computes the exponentials using the corresponding eigen representation. In order to obtain a physical S-matrix one has to carry out a mass and coupling renormalization (For details see Refs. [4, 6, 7]). In the numerical calculations we have taken as bare model parameters: $m = 3.14\ MeV$, $g/m = 0.5$. We have used the following lattice parameters: $L = 15\ MeV^{-1}$, $\Lambda = 18$. The bare coupling g has been chosen from a region which turned out in calculations of the mass spectrum to be a region of weak to medium coupling. We found for the invariant mass of the composite particle state $M = 6.02\ MeV$. Thus the difference with the bare mass of the constituents is $\delta m = 0.26\ MeV$. For comparison, one obtains from Eller et al. [4] the value $M = 5.99\ MeV$. Also in the region of weak to medium coupling, in particular for $g/m = 0.5$, the results of Ref. [4] agree well with those by Crewther and Hamer [8] from a standard space time lattice. In Fig.[2], which corresponds to $\Lambda = 18$, we have displayed the results for composite particle-composite particle (B-B) elastic scattering, using a reduced Fock space of 2 fermion- 2 antifermion states. We display the imaginary part of the S-matrix and $\Delta_{<E>}$, which is a measure for the violation of energy in the scattering process on the lattice, as a function of the l.c. time. One observes a region of stability of the S-matrix in the neighborhood of $x^+_{scatt} = 0.48\ MeV^{-1}$, i.e. where the violation of energy conservation is minimal. Also the real part of the S-matrix (not displayed) is stable in the same region. We have performed calculations for other values of the lattice parameter L in the neighbourhood of $L = 15\ MeV^{-1}$, which show a similar behavior.

Comparing boson-boson scattering in this framework with that which would have been obtained from non-relativistic boson-boson scattering with an effective short-range potential, one expects major differences in the following two situations: (a) If two fermions come closer than the size of a boson. The effective boson-boson potential is not valid at such a distance. The size of the boson can be estimated to be $0.417\ MeV^{-1}$ from the static fermion-fermion confining potential, given by Rothe et al. [9]. (b) If the energy is large enough to create another boson, i.e. $E > 2M_{boson}$. In the future, we hope to be able to extend these calculations to gauge theories in $2+1$ and $3+1$ dimensions, by using the Monte-Carlo technique applied in the next section to non-relativistic NN scattering.

3. S-MATRIX FROM MONTE CARLO PROJECTOR TECHNIQUE

The computation of the Minkowski time evolution, S-matrix or decay amplitudes is a numerical difficult problem if one starts from the Euclidean lattice formulation with a nonvanishing imaginary action. The Monte-Carlo projector method [10] based on a Hamiltonian formulation has been applied by Chin et al. [11] to compute ground state properties of glueballs. Here we suggest to use a Hamiltonian formulation, use renormalization group ideas and the Monte-Carlo projector method to compute the Minkowski S-matrix

$$< \Phi_{fin}|S|\Phi_{init} >= \lim_{t\to\infty} < \Phi_{fin}| \exp[iH^{as}t] \exp[-i2Ht] \exp[iH^{as}t]|\Phi_{init} >, \tag{3}$$

where H^{as} denotes an asymptotic Hamiltonian and the asymptotic states are assumed to have a wave packet distribution in energy in the interval $[E_{low}, E_{up}]$. In field theory, the Hamiltonian has infinitely many degress of freedom. In the scattering process described by the above matrix element, only the the degrees of freedom corresponding to energies in $[E_{low}, E_{up}]$ play a role. Thus we have the freedom to replace the Hamiltonian H by a new Hamiltonian $H(block)$, e.g. such that it has less degrees of freedom, but with the constraint to give the same physics for energies in the relevant energy interval. $H \to H(block)$ can be considered as a renormalization group transformation. We suggest to construct $H(block)$ as follows: One constructs a finite dimensional basis of so-called broad states $|\Phi(broad)>$. Then one constructs so-called zoomed states

$$|\Psi(zoom)>= \int_{E_{low}}^{E_{up}} dE \exp[-(H-E)^2\tau_{zoom}]|\Phi(block)>, \tag{4}$$

In the limit $\tau_{zoom} \to \infty$, the contributions of the broad state corresponding to energies E outside of $[E_{low}, E_{up}]$ are filtered out. The block Hamiltonian is constructed by taking matrix elements of the original Hamiltonian between the zoomed states. The projection operator is positive (τ_{zoom} is a real parameter) with a rapid exponential fall-off behavior for contributions corresponding to an energy outside of $[E_{low}, E_{up}]$. It is suitable for application of the Monte-Carlo projector method [10]. The S-matrix is replaced by a block S-matrix, by substituting $H \to H(block)$ in eq.(3), and replacing the time-limes by a finite scattering time. The numerical computation of the block S-matrix is straight forward after algebraic diagonalization of the block Hamiltonian.

We have tested this prescription for NN scattering in a non-relativistic potential model (Yamaguchi potential [12]). We consider s-wave scattering. The reference S-matrix is computed on a fine grid momentum lattice (40 nodes). Numerical results from the block method are presented in Figs.[3a,b]. A criterion for the usefulness of the block method is the propagation of statistical errors from the block Hamiltonian to the block S-matrix. We have simulated this by introducing a statistical error

$$H(block)_{i,j} \rightarrow H(block)_{i,j}(1 + \epsilon * P(x)), \tag{5}$$

where P has a Gaussian distribution. The results displayed in Fig.[3a] correspond to block dimension = 3 (small block size) and $\tau_{zoom} = 0.1$.

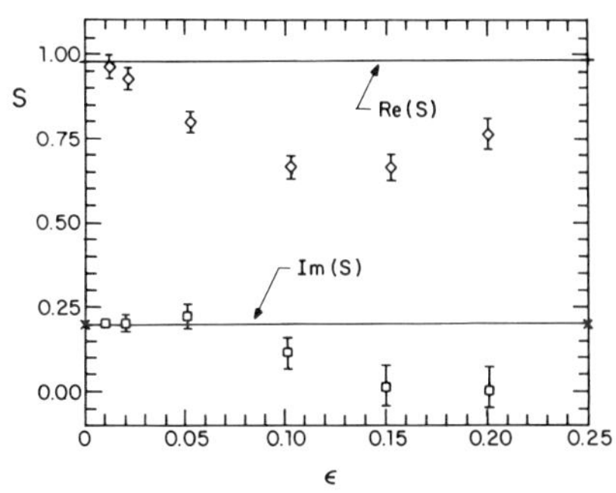

Fig.[3a]

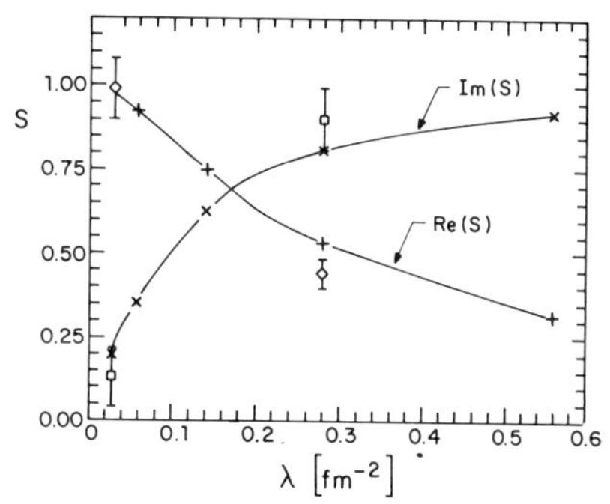

Fig.[3b]

Fig.[3a]. NN scattering. Dependence of S-matrix on statistical error of Hamiltonian (eq.(5)). Solid line corresponds to reference solution. Fig.[3b]. NN scattering. Dependence of S-matrix on coupling λ. Solid line corresponds to reference solution.

The important result is the linear behavior of the error in the S-matrix as a function of the error of the Hamiltonian in the region of small errors. For large errors, there is some saturation, probably due to the conservation of the unitarity of the S-matrix. The results for the S-matrix using the Monte-Carlo projector method to compute the block Hamiltonian are presented in Fig.[3b]. The S-matrix is displayed as a function of the coupling constant λ. The original Yamaguchi model (describing NN physics up to 100 MeV) corresponds to $\lambda = 0.0278811 fm^{-2}$.

ACKNOWLEDGEMENT

H. Kröger is grateful for computer time given at CRAY XMP 48 by HLRZ (Jülich), CRAY 1 S by CRAY Research (Montreal) and CRAY XMP 22 by CLSC (Toronto), and for support by NSERC of Canada. K.J.M. Moriarty and J. Potvin would like to thank John Buchanan, Terence R.B. Donahoe and Don Cameron of the Government of Nova Scotia for their continued interest, support and encouragement and grant support, James R. Berrett, Samuel W. Adams and Paul W. Crum for access to the NEC SX-2 in the Houston Area Research Center in The Woodlands, Texas, the National Allocation Committee for the John von Neumann National Supercomputer Center for access to the two CDC CYBER 205's and the two ETA-10's at the JVNC [Grant Nos. 110128, 171805, E171805, 171812, E171812, 171813, 551701-551706] where part of the calculations were performed. We acknowledge the Natural Sciences and Engineering Research Council of Canada [Grant Nos. NSERC A8420 and NSERC A9030] for financial support and the Canada/Nova Scotia Technology Transfer and Industrial Innovation Agreement [Grant Nos. 87II01, 88TTII01 and 89TTII01] for further financial support.

References

[1] H. Kröger, R. Girard and G. Dufour, Phys. Rev. D35 (1987) 3944.

[2] S. Coleman, R. Jackiw, and L. Susskind, Ann. Phys.(N.Y.) 93 (1975) 267; S. Coleman, Ann. Phys.(N.Y.) 101 (1976) 239.

[3] J.B. Kogut, E. Dagotto and A. Kocic, Phys. Rev. Lett. 60 (1988) 772; 61 (1988) 2410.

[4] T. Eller, H.C. Pauli and S.J. Brodsky, Phys. Rev. D35 (1987) 1493.

[5] K. Hornbostel, S.J. Brodsky and H.C. Pauli, Workshop on Relat. Nucl. Many-Body Phys., Ohio State Univ. 1988, eds. B.C. Clark et al., World Scientific, Singapore 1989, p. 573.

[6] J.F. Brière and H. Kröger, Phys. Rev. D41 (1990) 3197.

[7] J.F. Brière and H. Kröger, Phys. Rev. Lett. 63 (1989) 849.

[8] D.P. Crewther and C.J. Hamer, Nucl. Phys. B170 (1980) 353.

[9] H.J. Rothe, K.D. Rothe and J.A. Swieca, Phys. Rev. D19 (1979) 3020.

[10] J.E. Hirsch, D.J. Scalapino, R.L. Sugar, and R. Blankenbecler, Phys. Rev. Lett. 47 (1981) 1628.

[11] S.A. Chin, C. Long and D.Robson, Phys. Rev. Lett. 60 (1988) 1467.

[12] Y. Yamaguchi, Phys. Rev. 95 (1954) 1628.

TWO-BODY DENSITY MATRIX OF QUANTUM FLUIDS

M. L. Ristig

Institut für Theoretische Physik
der Universität zu Köln
D-5000 Köln 41, West-Germany

J. W. Clark

McDonnell Center for the Space Sciences
and Department of Physics
Washington University
St. Louis, Missouri 63130 USA

ABSTRACT

Final-state effects in deep-inelastic neutron scattering on liquid helium may be analyzed in terms of one- and two-body density matrix elements. We provide an analysis and discussion of the formal properties of the relevant two-body density matrix elements and consider their evaluation for a variational model within the correlated-basis framework. Numerical results have been obtained in the hypernetted-chain approximation ignoring elementary contributions. These results are compared with the predictions given by qualitative estimates of the two-body density matrix adopted in the theories of final-state interactions proposed by Silver and Gersch *et al.*

The dynamic response of a quantum fluid such as liquid helium at sufficiently large momentum transfer may be explored by deep-inelastic neutron scattering on the target helium fluid. Experimental results have conventionally been analyzed within the impulse approximation [1]. This approximation ignores completely the effects on the scattering due to correlations between the struck particle and other particles in the many-body system. However, in liquid ^{3}He and ^{4}He the constituents are strongly correlated even at high momentum transfer. The recent progress achieved in experimental techniques has led to an expanding collection of high-quality scattering data. These data allow one to make detailed tests of sophisticated theories of final-state effects arising from the dynamical correlations present in a dense quantum fluid such as liquid helium [2].

The novel theory of Refs. [3-5] suggests that the final-state effects in helium are dominated by the exchange of a virtual phonon (a density fluctuation) existing in the correlated ground state of the Bose (^{4}He) or Fermi (^{3}He) fluid. The proposed scattering mechanism in liquid ^{3}He is specified graphically in Figure 1. The probing neutron transfers momentum

Recent Progress in Many-Body Theories, Volume 2
Edited by Y. Avishai
Plenum Press, New York, 1990

$\hbar\underline{\mathbf{k}}$ and energy $\hbar\omega$ to a single ^{3}He atom, whereby it is ejected from the ground state with momentum $\hbar\underline{\mathbf{p}}$. Subsequently, the final-state interaction induces a scattering of the struck atom, shifting it from the momentum state $\hbar\underline{\mathbf{p}}$ to another final state of momentum $\hbar(\underline{\mathbf{p}} - \underline{\mathbf{q}})$ with the excitation of a virtual phonon. This momentum-conserving process is displayed in Figure 2.

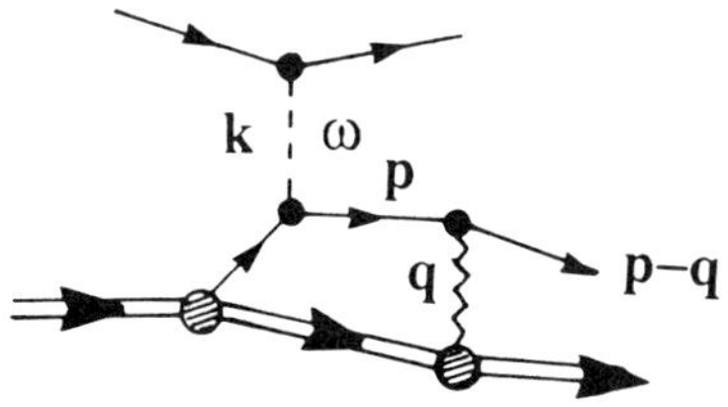

Fig. 1 Deep-inelastic scattering of neutrons from normal liquid ^{3}He, involving final-state interactions mediated by a virtual phonon of wave vector $\underline{\mathbf{q}}$.

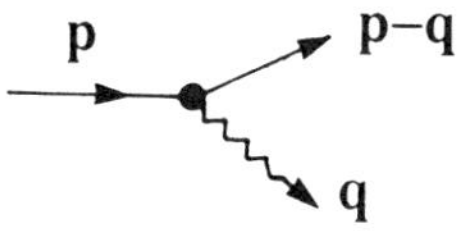

Fig. 2 Graphical representation of the scattering of a fermion from orbital $\underline{\mathbf{p}}$ into another orbital $\underline{\mathbf{p}} - \underline{\mathbf{q}}$, with creation of a phonon.

Fig. 3 Graphical representation of the creation (absorption) of a boson out of (into) the zero-momentum condensate with the intervention of a phonon.

The deep-inelastic scattering of neutrons on a liquid ^{4}He target involves the same scattering process (Figs. 1 and 2). However, due to the presence of a zero-momentum condensate in the Bose fluid the final-state interaction mechanism displayed in Fig. 2 is accompanied by processes which involve the creation (absorption) of a ^{4}He atom out of (into) the condensate (Figure 3).

The form factors associated with the scattering processes implicated by Silver's theory [2-5] are determined (i) by the single-particle momentum distribution [6]

$$n(p) = \langle \Psi | a^{\dagger}_{\hat{\mathbf{p}}} a_{\hat{\mathbf{p}}} | \Psi \rangle / \langle \Psi | \Psi \rangle \tag{1}$$

and (ii) by the generalized momentum distribution [7-10]

$$n(\underline{\mathbf{p}},\underline{\mathbf{q}}) = \sum_{\widehat{\mathbf{k}}} \langle \Psi | a^{\dagger}_{\widehat{\mathbf{k}+\mathbf{q}}} a^{\dagger}_{\widehat{\mathbf{p}-\mathbf{q}}} a_{\widehat{\mathbf{p}}} a_{\widehat{\mathbf{k}}} | \Psi \rangle / \langle \Psi | \Psi \rangle \quad . \tag{2}$$

For a boson fluid $\widehat{\mathbf{p}}$ denotes the single-particle orbital of momentum $\hbar\underline{\mathbf{p}}$. For a fermion fluid the orbital $\widehat{\mathbf{p}}$ is characterized by the momentum $\hbar\underline{\mathbf{p}}$ and by the spin projection σ associated with the ν-fold spin degeneracy of the momentum state.

The basic quantities (1) and (2) may be most efficiently analyzed in terms of one- and two-body density matrix elements $\rho_1(\underline{\mathbf{r}}_1,\underline{\mathbf{r}}_1')$ and $\rho_2(\underline{\mathbf{r}}_1,\underline{\mathbf{r}}_2,\underline{\mathbf{r}}_1',\underline{\mathbf{r}}_2)$ corresponding to the correlated target ground state Ψ. For N atoms in the homogeneous fluid at particle density ρ we have [6-8]

$$n(p) = \frac{1}{N}\frac{\rho}{\nu}\int \rho_1(\underline{\mathbf{r}}_1,\underline{\mathbf{r}}_1')e^{-i\underline{\mathbf{p}}(\underline{\mathbf{r}}_1'-\underline{\mathbf{r}}_1)}d\underline{\mathbf{r}}_1\,d\underline{\mathbf{r}}_1' \quad , \tag{3}$$

$$n(\underline{\mathbf{p}},\underline{\mathbf{q}}) = \frac{1}{N}\frac{\rho}{\nu}\int \rho_2(\underline{\mathbf{r}}_1,\underline{\mathbf{r}}_2,\underline{\mathbf{r}}_1')e^{-i\underline{\mathbf{p}}(\underline{\mathbf{r}}_1'-\underline{\mathbf{r}}_1)}e^{-i\underline{\mathbf{q}}(\underline{\mathbf{r}}_1-\underline{\mathbf{r}}_2)}d\underline{\mathbf{r}}_1\,d\underline{\mathbf{r}}_2\,d\underline{\mathbf{r}}_1' \quad , \tag{4}$$

where $\rho_2(\underline{\mathbf{r}}_1,\underline{\mathbf{r}}_2,\underline{\mathbf{r}}_1') \equiv \rho_2(\underline{\mathbf{r}}_1,\underline{\mathbf{r}}_2,\underline{\mathbf{r}}_1',\underline{\mathbf{r}}_2)$.

The matrices ρ_1 and ρ_2 possess certain symmetry properties and asymptotic features depending on the statistics. They also fulfill important sum rules and a sequential relation, as well as boundary conditions imposed by the hard-core character of the interatomic potential. A detailed analysis of these formal properties of the density matrices has been presented in Ref. [6-8]. We list some of the structural results on Bose fluids in the following paragraphs.

For an isotropic homogeneous many-boson system such as liquid ^{4}He, the one-body density matrix (a) depends only on the relative distance, i.e., $\rho_1(\underline{\mathbf{r}}_1,\underline{\mathbf{r}}_1') = \rho_1(r)$ with $r = |\underline{\mathbf{r}}_1 - \underline{\mathbf{r}}_1'|$, (b) is non-negative, (c) may be suitably normalized, with $\rho_1(0) = \rho$, and (d) exhibits long-range order, i.e., $\rho_1(\infty) = \rho\, n_0$, where the zero-momentum condensate is characterized by the occupancy fraction n_0, $0 \le n_0 \le 1$.

The diagonal elements $\rho_2(\underline{\mathbf{r}}_1,\underline{\mathbf{r}}_2,\underline{\mathbf{r}}_1)$ which contribute to the generalized momentum distribution (4) are proportional to the radial distribution function,

$$\rho_2(\underline{\mathbf{r}}_1,\underline{\mathbf{r}}_2,\underline{\mathbf{r}}_1) = \rho^2 g(|\underline{\mathbf{r}}_1 - \underline{\mathbf{r}}_2|) \quad . \tag{5}$$

Further, the elements $\rho_2(\underline{\mathbf{r}}_1,\underline{\mathbf{r}}_2,\underline{\mathbf{r}}_1')$ fulfill the sequential relation [11]

$$\int \rho_2(\underline{\mathbf{r}}_1,\underline{\mathbf{r}}_2,\underline{\mathbf{r}}_1')d\underline{\mathbf{r}}_2 = (N-1)\,\rho_1(\underline{\mathbf{r}}_1,\underline{\mathbf{r}}_1') \tag{6}$$

and decouple in the asymptotic region, with

$$\lim \rho_2(\underline{\mathbf{r}}_1,\underline{\mathbf{r}}_2,\underline{\mathbf{r}}_1') = \rho\,\rho_1(\underline{\mathbf{r}}_1,\underline{\mathbf{r}}_1') \tag{7}$$

as $\underline{\mathbf{r}}_2 \to \infty$.

The existence of spatial correlations between and among the Bose particles is apparent in the relation

$$\lim \rho_2(\underline{\mathbf{r}}_1,\underline{\mathbf{r}}_2,\underline{\mathbf{r}}_1') = n_0\rho^2 g_1(\underline{\mathbf{r}}_1,\underline{\mathbf{r}}_2) = n_0\rho^2\{1 + F_1(r)\} \tag{8}$$

as $\underline{\mathbf{r}}_1' \to \infty$, with $r = |\underline{\mathbf{r}}_1 - \underline{\mathbf{r}}_2|$ and $F_1(\infty) = 0$.

Finally, the strong repulsion between the interacting bosons requires the boundary condition $\Psi(\underline{\mathbf{r}}_1, \underline{\mathbf{r}}_1, \underline{\mathbf{r}}_3, ..., \underline{\mathbf{r}}_N) = 0$ and, consequently, $g(0) = 0$ and

$$\rho_2(\underline{\mathbf{r}}_1, \underline{\mathbf{r}}_1, \underline{\mathbf{r}}'_1) = 0 \quad . \tag{9}$$

The structural results (5)-(9) permit a decomposition of the two-body density matrix elements into compact two- and three-point portions [7],

$$\rho_2(\underline{\mathbf{r}}_1, \underline{\mathbf{r}}_2, \underline{\mathbf{r}}'_1) = \rho\, \rho_1(\underline{\mathbf{r}}_1, \underline{\mathbf{r}}'_1) \{1 + F_1(\underline{\mathbf{r}}_1, \underline{\mathbf{r}}_2) + F_1(\underline{\mathbf{r}}'_1, \underline{\mathbf{r}}_2) + F_1(\underline{\mathbf{r}}_1, \underline{\mathbf{r}}_2, \underline{\mathbf{r}}'_1)\} \quad . \tag{10}$$

The function $F_1(\underline{\mathbf{r}}_1, \underline{\mathbf{r}}_2)$ depends only on the relative distance $r = |\underline{\mathbf{r}}_1 - \underline{\mathbf{r}}_2|$, and the three-point compact portion $F_1(\underline{\mathbf{r}}_1, \underline{\mathbf{r}}_2, \underline{\mathbf{r}}'_1)$ vanishes whenever $\underline{\mathbf{r}}_1$, $\underline{\mathbf{r}}_2$ or $\underline{\mathbf{r}}'_1$ recedes to infinity. The quantities $\rho_1(\underline{\mathbf{r}}_1, \underline{\mathbf{r}}'_1)$, $F_1(\underline{\mathbf{r}}_1, \underline{\mathbf{r}}_2)$ and $F_1(\underline{\mathbf{r}}_1, \underline{\mathbf{r}}_2, \underline{\mathbf{r}}'_1)$ completely determine the form factors for the scattering processes of Figures 2 and 3 which contribute to the generalized momentum distribution function (2) or (4) for a Bose fluid. They may be calculated for a given correlated wave function by employing stochastic methods, perturbative approaches or correlated-basis techniques [6-8,12].

Within the correlated-basis framework one begins with a variationally determined ansatz for the ground-state wave function of the many-body system [7, 8]. For an initial study of the ^{4}He-boson fluid it is appropriate to choose a simple Jastrow form for the trial ground-state boson wave function,

$$\Psi(\underline{\mathbf{r}}_1, \underline{\mathbf{r}}_2, \underline{\mathbf{r}}_3, ..., \underline{\mathbf{r}}_N) = \prod_{i<j}^{N} f(\underline{\mathbf{r}}_i, \underline{\mathbf{r}}_j) \quad , \tag{11}$$

with $f(\underline{\mathbf{r}}_i, \underline{\mathbf{r}}_j) = f(|\underline{\mathbf{r}}_i - \underline{\mathbf{r}}_j|)$. The associated one- and two-body density matrix elements of interest here may be represented as follows in terms of irreducible (non-factorizable) functions,

$$\rho_1(\underline{\mathbf{r}}_1, \underline{\mathbf{r}}'_1) = \rho\, n_0 \exp\{-Q(\underline{\mathbf{r}}_1, \underline{\mathbf{r}}'_1)\} \quad , \quad n_0 = \exp\{Q(\underline{\mathbf{r}}_1, \underline{\mathbf{r}}_1)\} \quad , \tag{12}$$

and

$$\rho_2(\underline{\mathbf{r}}_1, \underline{\mathbf{r}}_2, \underline{\mathbf{r}}'_1) = \rho\, \rho_1(\underline{\mathbf{r}}_1, \underline{\mathbf{r}}'_1)\, f(\underline{\mathbf{r}}_1, \underline{\mathbf{r}}_2)\, f(\underline{\mathbf{r}}'_1, \underline{\mathbf{r}}_2) \exp\{-R(\underline{\mathbf{r}}_1, \underline{\mathbf{r}}_2, \underline{\mathbf{r}}'_1)\} \quad , \tag{13}$$

where $R(\underline{\mathbf{r}}_1, \underline{\mathbf{r}}_2, \underline{\mathbf{r}}'_1) = P(\underline{\mathbf{r}}_1, \underline{\mathbf{r}}_2) + P(\underline{\mathbf{r}}'_1, \underline{\mathbf{r}}_2) + P(\underline{\mathbf{r}}_1, \underline{\mathbf{r}}_2, \underline{\mathbf{r}}'_1)$. The compact quantities $Q(\underline{\mathbf{r}}_1, \underline{\mathbf{r}}'_1)$, $P(\underline{\mathbf{r}}_1, \underline{\mathbf{r}}_2)$ and $P(\underline{\mathbf{r}}_1, \underline{\mathbf{r}}_2, \underline{\mathbf{r}}'_1)$ have been explicitly analyzed in Ref. [7] employing the familiar hypernetted-chain classification scheme [13, 14]. Hypernetted-chain techniques may also be exploited for a numerical calculation of these quantities and therewith of the elements (12), (13).

An analogous study of the one- and two-body density matrix elements associated with the correlated ground state of a Fermi fluid such as liquid ^{3}He has recently been completed [8]. In this case the development is based on a completely antisymmetric trial ground state of the Jastrow-Slater type. The corresponding decomposition of the associated density matrix elements obtains upon proper generalization of results (12) and (13) to incorporate the exchange effects arising from the Pauli exclusion principle. A detailed analysis leads to the formal results [6, 8]

$$\rho_1(\underline{\mathbf{r}}_1, \underline{\mathbf{r}}'_1) = \rho_{1D}(\underline{\mathbf{r}}_1, \underline{\mathbf{r}}'_1)\, L(\underline{\mathbf{r}}_1, \underline{\mathbf{r}}'_1) \quad , \tag{14}$$

$$\rho_2(\underline{\mathbf{r}}_1, \underline{\mathbf{r}}_2, \underline{\mathbf{r}}'_1) = \rho_{2D}(\underline{\mathbf{r}}_1, \underline{\mathbf{r}}_2, \underline{\mathbf{r}}'_1) \{L(\underline{\mathbf{r}}_1, \underline{\mathbf{r}}'_1) + L(\underline{\mathbf{r}}_1, \underline{\mathbf{r}}_2, \underline{\mathbf{r}}'_1)\} \quad . \tag{15}$$

The factors ρ_{1D} and ρ_{2D} in Eqs. (14), (15) collect the direct-direct portions which contribute to the density matrix elements. The irreducible representations of ρ_{1D} and ρ_{2D} take the forms (12) and (13), respectively. The two- and three-point exchange factors $L(\underline{\mathbf{r}}_1, \underline{\mathbf{r}}_1')$ and $L(\underline{\mathbf{r}}_1, \underline{\mathbf{r}}_2, \underline{\mathbf{r}}_1')$ may be analyzed within the Fermi hypernetted-chain framework [13, 14] in terms of the associated irreducible components $P_{\alpha\beta}(\underline{\mathbf{r}}_1, \underline{\mathbf{r}}_1')$ and $P_{\alpha\beta\gamma}(\underline{\mathbf{r}}_1, \underline{\mathbf{r}}_2, \underline{\mathbf{r}}_1')$. A detailed study of these quantities is given in Ref. [8]. The subscripts $(\alpha\beta)$ indicate the circular (cc) and direct-exchange (de) portions of the two-point quantity $P(\underline{\mathbf{r}}_1, \underline{\mathbf{r}}_1')$, while the subscripts $(\alpha\beta\gamma)$ characterize the direct-circular (dcc), the direct-exchange-direct (ded), the circular-direct-circular (cdc) and the circular-exchange-circular (cec) three-point elements $P_{\alpha\beta\gamma}$. Ref. [8] provides explicit expressions for these irreducible functions.

We may recast expression (15) by employing the standard Fermi hypernetted-chain equations [13, 14]. The result reads

$$\rho_2(\underline{\mathbf{r}}_1, \underline{\mathbf{r}}_2, \underline{\mathbf{r}}_1') = \rho_2^{(2)}(\underline{\mathbf{r}}_1, \underline{\mathbf{r}}_2, \underline{\mathbf{r}}_1') + \rho_2^{(3)}(\underline{\mathbf{r}}_1, \underline{\mathbf{r}}_2, \underline{\mathbf{r}}_1') \quad . \tag{16}$$

The first component may be expressed entirely in terms of irreducible two-point functions [8],

$$\begin{aligned} \rho_2^{(2)}(\underline{\mathbf{r}}_1, \underline{\mathbf{r}}_2, \underline{\mathbf{r}}_1') = \; & \rho\, \rho_1(\underline{\mathbf{r}}_1, \underline{\mathbf{r}}_1')\, g_{Qdd}(r)\, g_{Qdd}(r') \\ & + \rho\, \rho_{1D}(\underline{\mathbf{r}}_1, \underline{\mathbf{r}}_1')\, l(\underline{\mathbf{r}}_1, \underline{\mathbf{r}}_1')\, \{ g_{Qdd}(r)\, F_{de}(r') + g_{Qdd}(r')\, F_{de}(r) \} \\ & - \rho\, \rho_{1D}(\underline{\mathbf{r}}_1, \underline{\mathbf{r}}_1')\, \nu\, \{ \nu^{-1} l(r) - F_{cc}(r) \} \{ \nu^{-1} l(r') - F_{cc}(r') \} \quad , \end{aligned} \tag{17}$$

where, as before, $r = |\underline{\mathbf{r}}_1 - \underline{\mathbf{r}}_2|$, while $r' = |\underline{\mathbf{r}}_1' - \underline{\mathbf{r}}_2|$. The second term in Eq. (16) collects the contributions which are generated by irreducible three-point functions $P_{\alpha\beta\gamma}$ (see Ref. [8] for the explicit expression). Eq. (17) contains the Slater exchange function $l(r)$ and the form factors $F_{\alpha\beta}(r) = g_{\alpha\beta}(r) - 1$ for the differing scattering processes which contribute to the generalized momentum distribution function (2) or (4).

To date only rough estimates for the two-body density matrix elements have been adopted in numerical calculations of the final-state corrections to the dynamic response functions [5, 15-17]. These calculations are based on the approximants

$$\rho_2(\underline{\mathbf{r}}_1, \underline{\mathbf{r}}_2, \underline{\mathbf{r}}_1') \approx \rho\, \rho_1(\underline{\mathbf{r}}_1, \underline{\mathbf{r}}_1')\, g(|\underline{\mathbf{r}}_1 - \underline{\mathbf{r}}_2|) \tag{18}$$

(Silver's estimate [5, 17]), or

$$\rho_2(\underline{\mathbf{r}}_1, \underline{\mathbf{r}}_2, \underline{\mathbf{r}}_1') \approx \rho\, \rho_1(\underline{\mathbf{r}}_1, \underline{\mathbf{r}}_1')\, \{ g(|\underline{\mathbf{r}}_1 - \underline{\mathbf{r}}_2|)\, g(|\underline{\mathbf{r}}_1' - \underline{\mathbf{r}}_2|) \}^{\frac{1}{2}} \tag{19}$$

(Gersch, Ref. [15]), or

$$\rho_2(\underline{\mathbf{r}}_1, \underline{\mathbf{r}}_2, \underline{\mathbf{r}}_1') \approx \rho\, \rho_1(\underline{\mathbf{r}}_1, \underline{\mathbf{r}}_1')\, g\left(|\tfrac{1}{2}(\underline{\mathbf{r}}_1 + \underline{\mathbf{r}}_1') - \underline{\mathbf{r}}_2|\right) \tag{20}$$

(Rinat, Ref. [16]).

Using estimate (18), Silver found spectacular agreement with experiment within his version of final-state theory [5]. The present study makes possible an improved calculation of the two-body density matrix elements and the dynamic response functions, based on the structural results reported above in conjunction with hypernetted-chain approximation. An initial numerical comparison uncovers serious deficiencies of the numerical results based on the estimates (18)-(20). For example, Figures 4 and 5 depict results on the two-point form factor $F_1(r)$ for liquid ^{4}He and on the two-point function $F_{dd}(r) + F_{de}(r) = g_{Qdd}(r) + g_{Qde}(r) - 2$ for liquid ^{3}He, corresponding to the estimates (18) and (19), in comparison with results calculated more precisely within hypernetted-chain approximations. The figures clearly reveal the rather poor quality of the simple estimates (18) and (19). More detailed comparisons and discussions of these approximations are contained in Refs. [7] and [8]. Our findings call for a reexamination of the final-state correction to the impulse approximation within Silver's hard-core perturbation scheme, using an improved approximation for the generalized momentum distribution function (4) derived from variational theory or the more sophisticated correlated-basis approach.

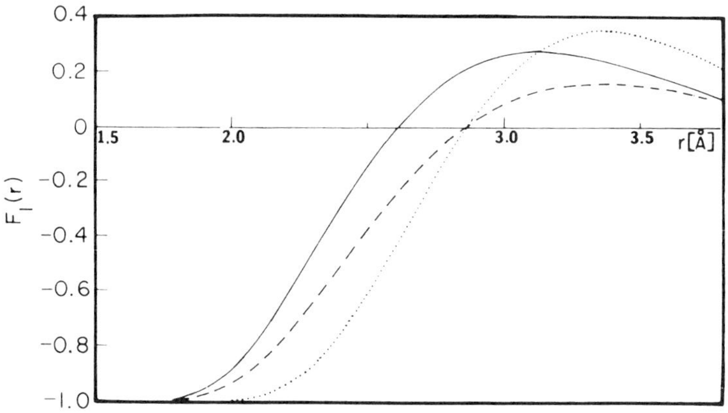

Fig. 4 Numerical results on the function $F_1(r)$ of decomposition (10). Solid line: optimized hypernetted-chain approximation, neglecting elementary diagrams. Dotted: approximation based on Silver's estimate (18). Dashed: based on Gersch's approximation (19). Required inputs correspond to liquid ^{4}He at density $\rho = 0.0218\text{Å}^{-3}$.

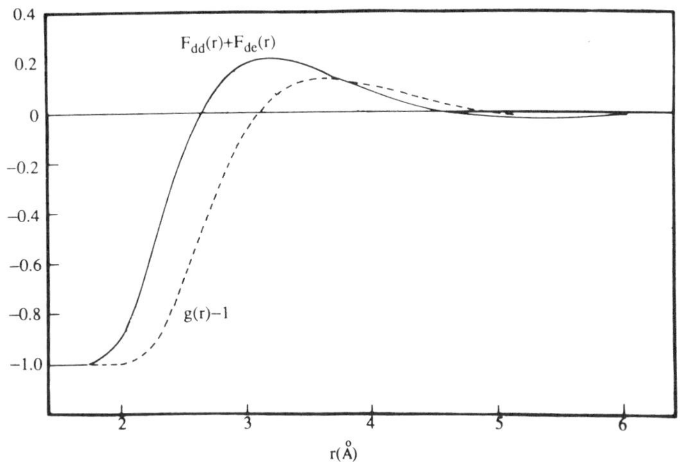

Fig. 5 Numerical results on function $F_{dd}(r) + F_{de}(r)$ for normal liquid ^{3}He at density $\rho = 0.01658\text{Å}^{-3}$. The calculation is based on a Jastrow-Slater ansatz with a correlation factor $f(r)$ of Schiff-Verlet type. Solid curve: result in Fermi-hypernetted-chain approximation, neglecting elementary diagrams. Dashed curve: result based on Silver's estimate (18).

We take this opportunity to correct a statement made in Ref. [7] to the effect that the Rinat estimate (20) fulfills the sequential relation (6) whereas Silver's estimate (18) does not. In fact, an integration over $\underline{\mathbf{r}}_2$, followed by a change of integration variable and invocation of the condition $\rho \int [g(r) - 1] d\underline{\mathbf{r}} = -1$, suffices to demonstrate *formal* satisfaction of the sequential relation by either approximant. However, Rinat's approximant meets the relation at the expense of imposing an unphysical asymptotic behavior on the function $F_1(\underline{\mathbf{r}}_1, \underline{\mathbf{r}}_2, \underline{\mathbf{r}}_1')$ entering decomposition (10), while the Silver approximant does so only by compensation of incorrect, unsymmetrical behaviors of the functions $F_1(\underline{\mathbf{r}}_1, \underline{\mathbf{r}}_2)$ and $F_1(\underline{\mathbf{r}}_1', \underline{\mathbf{r}}_2)$.

This research was supported in part by the Condensed Matter Theory Program of the Division of Materials Research of the U. S. National Science Foundation under Grant No. DMR85-19077, by the Deutsche Forschungsgemeinschaft under Grant Ri 267 and by the DAAD, Acciones Integradas. We benefited from a bitnet exchange with S. Stringari.

REFERENCES

[1] P.C.Hohenberg and P.M.Platzman, Phys. Rev. 152, 198 (1966).
[2] R.N.Silver, in *Momentum Distributions*, ed. R.N.Silver (Plenum, New York, 1989).
[3] R.N.Silver, in *Condensed Matter Theories*, Vol. 3, eds. J.S.Arponen, R.F.Bishop and M.Manninen (Plenum, New York, 1988); ibid. Vol. 4, ed. J.Keller (Plenum, New York, 1989).
[4] R.N.Silver, Phys. Rev. B 37, 3794 (1988).
[5] R.N.Silver, Phys. Rev. B 38, 2283 (1988).
[6] J.W.Clark and M.L.Ristig, in *Momentum Distributions*, ed. R.N.Silver (Plenum, New York, 1989).
[7] M.L.Ristig and J.W.Clark, Phys. Rev. B 40, 4355 (1989).
[8] M.L.Ristig and J.W.Clark, submitted Phys. Rev. B.
[9] J.W.Clark and M.L.Ristig, in *Condensed Matter Theories*, Vol. 5, ed. V.C.Aguilera-Navarro (Plenum, New York, 1990), in press.
[10] M.L.Ristig, Foundations of Physics 19, 1041 (1989).
[11] E.Feenberg, *Theory of Quantum Fluids* (Academic, New York, 1969).
[12] D.Ceperley, in *Momentum Distributions*, ed. R.N.Silver (Plenum, New York, 1989).
[13] S.Fantoni, Nuovo Cimento A 44, 191 (1978).
[14] M.L.Ristig, in *From Nuclei to Particles*, Proceedings of the International School of Physics "Enrico Fermi", Course LVII, Varenna 1981, ed. A.Molinari (North Holland, Amsterdam, 1982).
[15] H.A.Gersch, L.J.Rodriguez and P.N.Smith, Phys. Rev. A 5, 1547 (1972).
[16] A.S.Rinat, Phys. Rev. B 40, 6625 (1989).
[17] R.N.Silver, these proceedings.

NEW MICROSCOPIC DESCRIPTION OF LIQUID HELIUM 3

J.P. Bouchaud

Laboratoire de Physique Statistique, Ecole Normale Superieure
24 rue Lhomond, 75231 Paris Cedex 05

1. INTRODUCTION. LANDAU THEORY AND JASTROW-SLATER WAVE FUNCTIONS

In his seminal attempt to describe liquid He^3, Landau [1] made the rather bold hypothesis that no `catastrophy' is met if one starts from the non-interacting Fermi gas and progressively switches on the interaction. In particular, if one writes the Hamiltonian as:

$$H = H_0 + \varepsilon H_{int},$$

the low energy excitations smoothly evolve with ε. Since generically energy levels do not cross, their hierarchy is not modified. Hence the *label* adequate to identify the energy levels in the free gas is also a faithful label for the interacting liquid. This label is simply the momentum distribution function $N°(\mathbf{k})$; and thus the low energy levels of the interacting liquid are determined by the (small) difference $\delta N(\mathbf{k})$ between $N°(\mathbf{k})$ and the step function $\theta(k_f-k)$. The assumed regular expansion of the energy in powers of $\delta N(\mathbf{k})$ then leads to the standard predictions of Landau's theory (linear specific heat, Pauli susceptibility, etc..), which describe correctly the low (< 0.1 K) temperature properties of liquid He^3.
After this strikingly simple phenomenological approach came more microscopic justifications of the whole scenario. Many body theory showed in particular: [1,2]
a) That Landau's theory was indeed exact if perturbation theory (in ε) converges[3].
b) That although if $N°(\mathbf{k})$ is only a label for the interacting liquid's excitations, the *true momentum distribution* $N(\mathbf{k})$ of the liquid is also discontinuous right at the Fermi momentum (Migdal)
c) The Fermi momentum is independent of ε and sticks to its free value $k_f^\circ=(3\pi^2\rho)^{1/3}$, again if perturbation theory converges (Landau, Luttinger).
Now, one might take another point of view and ask how to describe in terms of (variational) wave functions (w.f.) the properties of liquid He^3, as Feenberg and Wu [4] initiated. (And Feynman and Cohen [4] for liquid He^4) Points b) and c) show that Landau's apparently innocent hypothesis on the low energy *excitations* has an important consequence on the *ground state* of the liquid. Of course, this is not such a stringent constraint because many wave functions can be devised to have the required discontinuity of $N(\mathbf{k})$. But it is clear that the philosophy behind Landau's theory strongly suggests that the antisymmetric part of the ground state w.f. should be a Slater determinant of single particle states labeled by $\mathbf{k}$, $k \le k_f$- so that a quasi-particle excitation corresponds to replacing a state with $k \le k_f$ by a state with $k \ge k_f$. Hence the most natural trial is the Jastrow-Slater w.f. Ψ_{JS}, which is the product of a determinant of plane waves by a product of two body functions going to zero rapidly when hard cores start to overlap. The properties of this class of w.f. has been intensely investigated

Recent Progress in Many-Body Theories, Volume 2
Edited by Y. Avishai
Plenum Press, New York, 1990

analytically and numerically, both for He^3, and for Hubbard's model (it is rather called Gutzwiller's w.f. in this context). The main results of variational Monte-Carlo calculations are the following [5]:

i) With only a two-body Jastrow part and plane waves as single particle states, the best energy per particle is found to be -1.5 K, while the experimental value (at s.v.p) is -2.5 K. Note that the binding energy is 5 times smaller than both the kinetic and potential energy. This shows at once that He^3 is a strongly correlated system: interaction effects are large.

ii) Inclusion of three-body correlations and ``backflow" in the plane waves only lowers, with 6 variational parameters, the binding energy to• about -2.0 K. The missing 0.5 K can be compared e.g. to the temperature for which the specific heat quite abruptly changes from a γT behavior to a plateau, which gives a scale of the correlation energy. This occurs around 300 mK.

iii) The (coherent) structure factor (density fluctuations) is dominated by the Jastrow part of the w.f., and agrees well with the experimental X-ray data of B. Hallock [6]

iv) The spin-spin correlations (or the incoherent structure factor, probed by neutrons) shows quite a large nearest neighbor antiferromagnetic (AF) correlation, which translates to a peak in the incoherent factor, showed in figure 1. We have also given the neutron scattering results of Sköld et al.[7] , which, although rather imprecise, seem to favor less local AF correlations

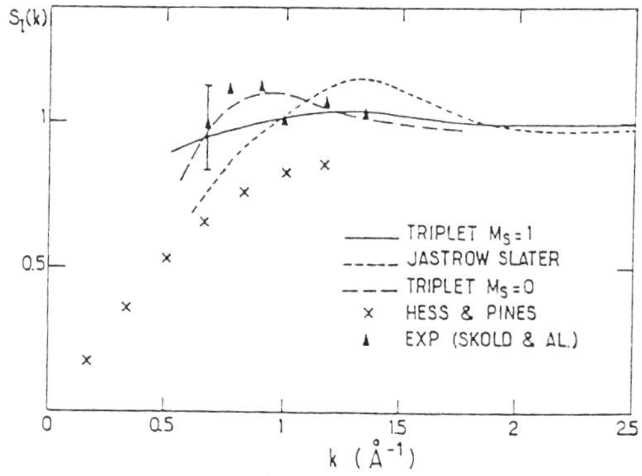

Fig 1. Incoherent structure factor, probing the Spin-Spin correlation function. "Triplet" denotes the pairing scheme investigated within a Jastrow-BCS approach (see Below). The calculation of Hess and Pines may be found in J. Low. Temp. Phys 72 (1988) 247

v) The momentum distribution is -as expected- discontinuous at k_f, with $Z=N(k_{f+})- N(k_{f-})$ ~ 0.3 (instead of 1 in the free gas). Note that Z compares relatively well with the 'local' prediction Z= m/m*.

Hence, the `nearly free' description *à la Landau* gives an unsatisfactory binding energy and a rather strong local AF tendency** .This has motivated us to propose a completely different antisymmetrization scheme, *à la BCS*, which sees He^3 as an assembly of small resonating pairs- in fact much in the same spirit as the "RVB" description of Hubbard's model in the strong coupling regime. This description is quite successful (see below), but raises a theoretical problem, which can be summarized as "Is liquid He^3 a standard Landau Fermi liquid? What is the microscopic nature of its low lying excitations ?"

2. NEW DESCRIPTION: STRONG CORRELATIONS AND RESONANT PAIRING

It is of course well known that even a very weak attraction between fermions causes an instability of the Fermi surface, through Cooper pairing. Usually, however, the coherence length of those pairs is large, and thus the momentum distribution, although smooth, varies rapidly around k_f

In the case of He^3, the bare interaction between two atoms is quite strongly attractive, since the potential well is -10 K deep•••. Abandoning all prejudices about He^3, we assume that, in a first stage, the Fermi surface is completely destructed by the attraction: the building blocks of our description are taken to be <u>pair</u> wavefunctions $\varphi(\mathbf{r}_i-\mathbf{r}_j)$. | *Spin Ket* >, characterized by a rather short 'molecular' size ξ (ξ ~ 5-15 A°). This pairing is supposed to handle more adequately the strong correlations between atoms. The complete w.f. then reads:

$$\Psi_{JBCS} = (\text{Jastrow}) \,.\, \mathcal{A}\{\varphi(12)\varphi(34)\varphi(56) \ldots\} \quad (1)$$

where $\mathcal{A}$ denotes the antisymmetrizer.

We have considered [9] three different pairing scheme, depending upon the total spin **S** of the pairs: Triplet (S=1), with M_S = 0 or ±1, or Singlet (S=0). Those correspond either to an odd or an even $\varphi(\mathbf{r}_i-\mathbf{r}_j)$. Taking into account (rather large) size effects, we find that:

* The Singlet or Triplet (M_S = 0) already yield an energy comparable to the simple Jastrow-Slater wave function.

** The Triplet M_S = 1 pairing, with only 2 variational parameters, does as well (-2.05 K per particle) as the best result based on a Slater determinant description of the fermionic correlations (see however [10], and •• above). This is remarkable since our Jastrow part is not optimised. This Jastrow factor is basically devised to treat the hard-core interactions and does not really care about the antisymmetric part of Ψ. Hence, by analogy with Jastrow-Slater wave functions, we strongly believe that 0.3 K can be gained by taking an optimised Jastrow part. This would lead to a variational energy of -2.35 K, only 5% above the true ground state energy.

Within this pairing scheme, the Spin-Spin correlation function may be computed. We find that local AF correlations are much smaller in this description than in the previous one. The incoherent structure factor is shown in figure 1, compared to the Jastrow-Slater prediction and to experiments.

The momentum distribution may only be computed (for technical reasons explained in [9,11]) for M_S = 0 pairing schemes. We do not however expect drastic differences with the

- - - - - - - - - - - - - - - - -

• *Finite size effects are not very easy to assess. For the simple Jastrow Slater w.f. we have estimated the difference between a N=54 particles calculation and the infinite volume limit to be as large as 150 or 200mK, assuming a $N^{-1/3}$ size dependence.*

•• *It must however be mentionned that recently, Viviani and Fantoni [10] claimed that a Jastrow Slater w.f. with an explicit Spin Spin correlation of the form $\exp(\Sigma f_{ij}\, \sigma_i\, \sigma_j)$ did bring the energy very close to the experimental value. Their w.f. however leads to an enhanced AF local correlation, and would lead to quite a large bump in the incoherent structure factor. More experimental results are obviously needed to see if this is indeed the case, or if the AF local correlation is in fact small, as we suggest below (see fig 1)*

••• *Although not deep enough for a bound state to appear in three dimensions.*

$M_S = 1$ case; and the results are shown in figure 2. As expected, the momentum distribution is smooth and nothing special occurs near the Fermi surface. Again, our numerical prediction is quite different from that obtained within a Jastrow-Slater description; the difference could perhaps be seen (although the experiments are most difficult [12]) by neutron scattering, or analysis of the γ-rays emitted by neutron capture.

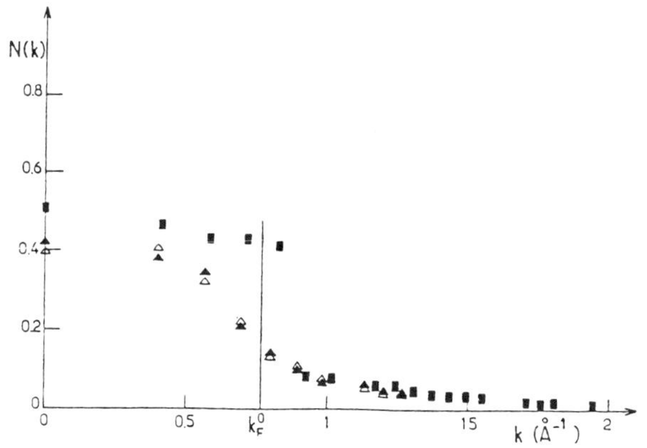

Figure 2. The momentum distribution, calculated within a JS description (squares) exhibiting the familiar jump at k_f , and calculated for JBCS wave function (triangles).

3. BACK TO FERMI LIQUID BEHAVIOUR?

From a physical point of view, however, the situation is odd. Nothing experimentally urges one to think that liquid He^3 might not be *the* model Landau Fermi liquid- even if some of its properties are intriguiging, such as for example the specific heat plateau for intermediate temperatures. It is also *a priori* difficult to admit that a perturbation theory can be stretched so far out without any harm (The Landau parameters are large by any standards). It is of course conceivable that a Landau like theory without Fermi surface can be constructed [see e.g. 13 for a 'localized' approach] to account for all experimental results (including the superfluid properties), but even in that case Occam's razor would probably compell us to stick to Landau's simple picture. Hence we have to figure out how a Fermi surface might reappear *after all* , and insure quasi-particle like excitations.

Two possibilities may be proposed: since our w.f. (1) is of course not the true ground state, it could be improved

-either by adding to the pair wave function φ a small correction, containing oscillations of period $2\pi/k_f$, in such a way to make N(k) discontinuous. The physical meaning of such a procedure is unclear.

-or by trying a 'mixed' w.f., where a fraction f of the atoms are paired, and the remaining (1-f) left in single particle orbitals (the whole w.f. being of course correctly antisymmetrized [11]). This would in a sense describe a resonating mixture of Bosons (pairs) and Fermions••••. The latter should possess a Fermi surface and thus be describable

- - - - - - - - - - - - - - - - -

•••• *A phenomenological approach of He^3 and of Heavy fermions, based on the idea of a decoupled mixture of 'Heavy' fermions and bosons, was in fact quite sucessfully developed by Ranninger et al. [14]. In particular, they found the fraction 1-f of fermions to be around 0.5* as a Fermi liquid, although the Fermi momentum is a priori no longer fixed by the free gas value, but could be for example related to the 'density of Fermions' through

$$k_f = [3\pi^2 \rho(1-f)]^{1/3} \leq k_f^\circ$$

For 1-f=0.5 [14], one would thus obtain $k_f=0.8\ k_f^\circ$.

This scenario would make liquid He^3 a 'non standard' Fermi liquid, and raises the following question: how precisely is k_f known from experimental results ? Bearing in mind that a shift in k_f could be absorbed in a redefinition of the effective mass to reproduce thermodynamical quantities, only transport coefficients give constraints on k_f. Unfortunately, those also depend on scattering amplitudes, for which only plausible models (such as the s-p approximation, or the polarisation potential theory [15]) are available.

It would be interesting to construct a model Hamiltonian for which the above Boson-Fermion mixture is the ground state, to see if our proposal is at least internally consistent (i.e. that the Fermion part decouples from the Boson part and indeed behaves as a Fermi liquid), if at all useful to describe He^3. It could nevertheless be a way to marry satisfactorily strong correlations and Fermi liquid theory, and as such be helpful to understand strongly correlated electronic systems [see e.g. 16].

ACKNOWLEDGMENTS

I am deeply indebted to Claire Lhuillier, with whom the work presented here was developed. Many discussions with P. Nozieres and A. Georges have been of great help.

REFERENCES

[1] L. LANDAU, E. LIFSHITZ Statistical Mechanics vol.2, Pergamon Press (New York) 1981
[2] D. PINES, P. NOZIERES The Theory of quantum liquids (I) Benjamin, New York, 1966
[3] R. BALIAN, C. DE DOMENICIS Physica 30 (1964) 1927
[4] F.Y. WU, E. FEENBERG, Phys Rev 128 (1962) 141; R. FEYNMAN, M. COHEN, Phys Rev 102 (1956) 1189
[5] see e.g R.M. PANOFF, J. CARLSON, Phys Rev Lett 62 (1989) 1130 and references therein; see also R.M. PANOFF, this volume.
[6]R.B HALLOCK, J. Low Temp. Phys. 9 (1972) 109
[7] K. SKOLD, C. PELIZZARI, R.KLEB, G. OSTROWSKI, Phys Rev Lett 27 (1976) 842
[8]R.M. PANOFF, P.WHITLOCK, Can. J. Phys. 65 (1987) 1409
[9]J.P. BOUCHAUD, C. LHUILLIER, Z. Phys B 75 (1989) 283
[10] S. FANTONI, M.VIVIANI , in Spin Polarized Quantum Systems, S. Stringari Edt, World Scientific Publishers, 1989
[11] J.P. BOUCHAUD, A. GEORGES, C. LHUILLIER, J. Physique (Paris) 49 (1988) 553
[12]R. N. SILVER This volume, and Phys Rev. B39, (1989) 4022 .
[13]B. CASTAING, J Physique (Paris) Lett 41 (1980) 333, J.P. BOUCHAUD, C. LHUILLIER, Physica 148B, (1987) 87
[14]J. RANNINGER, S. ROBASZKIEWICZ, A. SULPICE, R. TOURNIER, Europhys Lett 3 (1987) 347
[15] for a review, see K. LEVIN, O. VALLS, Phys. Rep. 98(1983)1
[16] Y. BAR-YAM `Two fluid superconductivity', Weizmann Institute preprint.

EXCITATION SPECTRUM OF A ^{3}He IMPURITY IN ^{4}He

M. Saarela
Department of Theoretical Physics
University of Oulu, SF-90570 Oulu 57, Finland

INTRODUCTION

Recent measurements of the effective mass of a ^{3}He impurity on ^{4}He films by Wang and Gasparini[1] and Valles et al.[2] have provided a challenging problem for many-body theorists. In the case of a very thin ^{4}He film - less than a monolayer - one is studying a homogeneous, quasi-two-dimensional fluid. When the film thickness is increased the impurity begins to float on top of the film and the shape of the surface profile and layered structure of the film become important.[3] The effective mass of the impurity depends on the density of surrounding particles and thus can be used as a sensitive test to our understanding of the behaviour of ^{4}He fluid. Recent neutron scattering experiments[4] on dilute mixtures of ^{3}He (1% and 5%) in ^{4}He measured the elementary excitation spectrum of ^{3}He for wave vectors between 0.4 and 2.2 $\mathring{A}^{-1}$. They tested the effective mass approximation at finite momentum in the three-dimensional, homogeneous ^{4}He -^{3}He mixture. Deviations from the quadratic Landau-Pomeranchuck spectrum with increasing momentum were found in agreement with earlier thermodynamical measurements.[5]

A lot of work has been done to understand both the effective mass and the excitation spectrum of ^{3}He in ^{4}He using variational methods[6] and different approximations for the dynamic density-density response function[7]. In this work we shall use the variational approach to calculate the proper self energy for the impurity which then determines its effective mass and excitation spectrum. We divide the calculation into three parts. First we minimize the t tal energy of the pure ^{4}He fluid and calculate its structure function S(k). Secondly we minimize the chemical potential of the impurity to obtain the impurity-background structure function S^I(k), and finally we minimize the action integral which determines continuity equations and the proper self energy. In the three-dimensional case we can compare our results with experiments at each step, but in two-dimensions we must rely on recent Monte Carlo analysis.[8]

It is well known that in a quantitative calculation of the ground state of ^{4}He one has to include three-body correlations and elementary diagrams. This is especially important when one wants to calculate pressure and chemical potential. At the experimental saturation density the chemical potential in the hypernetted chain (HNC) approximation is +2K whereas the experimental result is -7.18K. The chemical potential of the impurity suffers from the same problem because the main diffence between ^{4}He and ^{3}He chemical potentials comes from the mass difference in the kinetic energy. Using this simple approximation one gets an estimate for the impurity ^{3}He chemical potential, μ^I = 9K. It means that ^{3}He is unbound whereas experimentally it should be bound by 2.78K.

In the evaluation of the three-body correlations we use the Euler equation derived by Chang and Campbell[9]. It turns out to be quite important to include the three-body correlation as well as elementary diagram contibutions also to the induced interaction in the iteration of the Euler equations.[10] This puts some constraints to the numerical calculation of elementary diagrams because it becomes very time consuming when six or more particles are included. A carefull analysis by Smith et al.[11] fortunately shows that the structure of the higher order diagrams is similar to the lowest order ones. The approximation where only the 4-body diagram is evaluated and multiplyed with a scaling factor, called scaling approximation, has been used[12]. On the other hand the evaluation of the two simplest five-body diagrams does not require much extra efford after the calculation of the 4-body diagram. Thus we follow the suggestion[10] that the contribution of the higher order diagrams can be estimated by multiplying the sum of 4- and 5-body diagrams by an enhancement factor in order to get the correct saturation energy.

A moving ^{3}He impurity couples strongly with low lying phonon and roton excitations of ^{4}He yielding the large effective mass of ^{3}He . This mechanism, suggested by Landau and Pomeranchuck, has been studied carefully over the decades, but still it seems to be difficult to derive a consistent microscopic theory which gives the correct effective mass as a function of density[6]. The main reason is the fact that a good account of the excitation spectrum of pure ^{4}He already requires backflow contribution and this mechanism is complicated to implement consistently into the calculation of the impurity spectrum.

The excitations of the impurity deviate, with increasing momentum, slightly from the effective mass approximation and earlier it was suggested that the spectrum might even have a roton like minimum.[13,14] More recent analysis[5,4] does not show that behavior. The impurity spectrum increases monotonically and merges with the ^{4}He spectrum near roton minimum with the possibility of roton emition. In the case of a single impurity the pure ^{4}He spectrum remains uneffected whereas a finite concentration mixture of ^{3}He in ^{4}He would result a mode-mode coupling which shows up as a shift in the ^{4}He spectrum. The behavior of the mixture differs also at small momenta where it has two linear eigenmodes corresponding to the superfluid (^{4}He) zero sound and normal fluid (^{3}He) second sound.

THE VARIATIONAL THEORY

The groundstate of ^{4}He is calculated using the extented Jastrow ansatz for the A particle wave function which includes both two- and three-body correlations

$$\Psi_B(\vec{r}_1,\vec{r}_2,...,\vec{r}_A) = exp\frac{1}{2}\Big[\Big(\sum_{\substack{i,j=1\\ i<j}}^{A} u_2(\vec{r}_i,\vec{r}_j)\Big) + \sum_{\substack{i,j,k=1\\ i<j<k}}^{A} u_3(\vec{r}_i,\vec{r}_j,\vec{r}_k)\Big]. \tag{1}$$

The method we use here is well described in the literature[10] and thus we summarize only the main ideas. The correlation functions are determined variationally by minimizing the total energy of the system which can be devited into three parts, (1) contribution E_{HNC} coming from the two particle quantities using HNC - summation, (2) contribution ΔE_e from elementary diagrams, and (3) contribution ΔE_3 from three-body correlations.

$$E/A \equiv E_{HNC} + \Delta E_e + \Delta E_3, \tag{2}$$

$$E_{HNC} = \frac{\rho}{2}\int d^3r g(r)\Big[v(r) - \frac{\hbar^2}{4m}\vec{\nabla}^2(lng(r) - N(r))\Big], \tag{3}$$

$$\Delta E_e = \frac{\hbar^2\rho}{8m}\int d^3r g(r)\nabla^2 E(r), \tag{4}$$

$$\Delta E_3 = \frac{\hbar^2\rho^3}{8m}\int d^3r_1 d^3r_2 d^3r_3\Big[\kappa_3(\vec{r}_1,\vec{r}_2,\vec{r}_3) - \frac{1}{3!}K_3(\vec{r}_1,\vec{r}_2,\vec{r}_3)\Big] \tag{5}$$

with radial distribution function $g(r)$, two-particle potential $v(r)$, sum of nodal diagrams $N(r)$ and kernels

$$\kappa_3(\vec{r}_1,\vec{r}_2,\vec{r}_3) = g(\vec{r}_1,\vec{r}_2)\vec{\nabla}_1^2\Big[(g(\vec{r}_1,\vec{r}_3)g(\vec{r}_2,\vec{r}_3) - 1)u_3(\vec{r}_1,\vec{r}_2,\vec{r}_3)\Big],$$
$$K_3(\vec{r}_1,\vec{r}_2,\vec{r}_3) = [g_3(\vec{r}_1,\vec{r}_2,\vec{r}_3) - g(\vec{r}_1,\vec{r}_2) - g(\vec{r}_1,\vec{r}_3) - g(\vec{r}_2,\vec{r}_3) + 2] \times (\vec{\nabla}_1^2 + \vec{\nabla}_2^2 + \vec{\nabla}_3^2)u_3(\vec{r}_1,\vec{r}_2,\vec{r}_3). \tag{6}$$

The above variational problem leads to two Euler equations. The two-body equation is traditionally written for the square root of the radial distribution function $\sqrt{g(r)}$ or for the structure function $S(k)$. The latter version,

$$S(k) = [1 + \frac{4m}{\hbar^2 k^2}\tilde{V}_{p-h}(k)]^{-\frac{1}{2}}, \tag{7}$$

can be solved numerically very efficiently and extented to inhomogeneous systems as well as many component mixtures. The key quantity in the expression is the Fourier transform of the particle -hole potential,

$$V_{p-h}(r) = g(r)[v(r) + \Delta v(r)] + \frac{\hbar^2}{m}|\vec{\nabla}\sqrt{g(r)}|^2 + [g(r) - 1]w(r), \tag{8}$$

where $w(r)$ is the induced potential from the HNC approximation. It is more conveniently defined in momentum space,

$$\tilde{w}(k) = -\frac{\hbar^2 k^2}{4m}[2S(k) + 1][1 - S^{-1}(k)]^2, \tag{9}$$

The induced potential from three-body correlations and elementary diagrams is determined through functional differentiation with respect to the radial distribution function,

$$\Delta v(r) = \frac{2}{\rho}\frac{\delta(\Delta E_3 + \Delta E_e)}{\delta g(r)}. \tag{10}$$

The three-body Euler equation is quite easily solved in the convolution approximation as suggested by Chang and Campbell[9],

$$\tilde{u}_3(\vec{k}_1,\vec{k}_2,\vec{k}_3) = -\delta_{\vec{k}_1+\vec{k}_2+\vec{k}_3,0}\frac{\vec{k}_1\cdot\vec{k}_2\tilde{X}(k_1)\tilde{X}(k_2) + c.p.}{k_1^2/S(k_1) + k_2^2/S(k_2) + k_3^2/S(k_3)}, \tag{11}$$

where $\tilde{X}(k)$ is the direct correlation function and the notation c.p. refers to the cyclic permutation of indeces 1, 2 and 3. The graphical analysis of the corresponding energy expression for ΔE_3 [10],

$$\Delta E_3 = \frac{1}{24}\frac{\hbar^2}{2m}\int\frac{d^3k_1 d^3k_2 d^3k_3}{(2\pi)^9\rho^3}\tilde{u}_3(\vec{k}_1,\vec{k}_2,\vec{k}_3) \times \Big[[2\vec{k}_1\cdot\vec{k}_2\tilde{X}(k_1)\tilde{X}(k_2) + c.p.] + [\frac{k_1^2}{S(k_1)} + \frac{k_2^2}{S(k_2)} + \frac{k_3^2}{S(k_3)}]\tilde{u}_3(\vec{k}_1,\vec{k}_2,\vec{k}_3)\Big] \times S(k_1)S(k_2)S(k_3)\delta(\vec{k}_1 + \vec{k}_2 + \vec{k}_3), \tag{12}$$

shows that already in the convolution approximation a number of elementary diagrams involving three-particle triagles up to six-particle diagrams are included in the summation.

Since we are also interested in the chemical potential,

$$\mu = \frac{E}{A} + \frac{P}{\rho}, \tag{13}$$

we have to calculate the pressure. The elementary particle contribution into it is obtained by straight forward differentiation with respect to the density, but the three-body term is complicated by functional differentiation of the nodal diagrams and has the following expression

$$\begin{aligned} \frac{P_3}{\rho} &= 2\Delta E_3 \\ &+ \frac{\hbar^2}{48m}\int \frac{d^3k_1 d^3k_2 d^3k_3}{(2\pi)^9\rho^3}\Big[-(k_1^2+k_2^2+k_3^2)s(k_1)s(k_2)s(k_3)\tilde{u}_3(\vec{k}_1,\vec{k}_2,\vec{k}_3) \\ &+ 2[k_1^2 s(k_2)S(k_3) + c.p.]\tilde{u}_3^2(\vec{k}_1,\vec{k}_2,\vec{k}_3)\Big]\delta(\vec{k}_1+\vec{k}_2+\vec{k}_3), \end{aligned} \tag{14}$$

where we have used the notation $s(k) \equiv S(k) - 1$.

The impurity particle can be added into the system by giving it a coordinate index $\vec{r}_0$ and starting the summations in the wave function of eq.(1) from zero. Then

$$\Psi_I(\vec{r}_0,\vec{r}_1,...,\vec{r}_A) = exp\frac{1}{2}\Big[\sum_{0\le i<j\le A} u_2(\vec{r}_i,\vec{r}_j) + \sum_{\substack{i,j,k=0 \\ i<j<k}}^{A} u_3(\vec{r}_1,\vec{r}_j,\vec{r}_k)\Big]. \tag{15}$$

The variational problem is turned into minimizing the chemical potential of the impurity particle or, in other words, minimizing the energy required or gained when one impurity particle is added into the background fluid,

$$\mu^I \equiv E_A^I - E_A, \tag{16}$$

where E_A^I is the energy of the system with the impurity and E_A without it. This determines correlations between the impurity and background particles and leaves correlations between background particles unaltered.

It is usefull to devide μ^I into the same three contributions as in eq.(2),

$$\mu^I \equiv \mu_{HNC}^I + \mu_e^I + \mu_3^I \tag{17}.$$

The expressions for these quantities are very similar those of eqs.(3), (4) and (5), but in that comparison one should remember that the pressure terms must also be included. The HNC part is well studied and we refer to the appendix of ref. 15 for details. The number of elementary diagrams to be calculated increases when one of the particles can be an impurity but the generalization is straight forward.

The three-body correlation function between one impurity and two background particles can again be solved analytically in the convolution approximation yielding the following expression,

$$\tilde{u}_3^I(\vec{k}_0,\vec{k}_1,\vec{k}_2) = -\frac{N(\vec{k}_0,\vec{k}_1,\vec{k}_2)}{t(k_0)+\varepsilon(k_1)+\varepsilon(k_2)} \tag{18}$$

with

$$\begin{aligned} N(\vec{k}_0,\vec{k}_1,\vec{k}_2) &= S^I(k_0)[\varepsilon(k_1)+\varepsilon(k_2)]\tilde{u}_3(\vec{k}_0,\vec{k}_1,\vec{k}_2) \\ &+ \frac{\hbar^2}{2m_I}\vec{k}_1\cdot\vec{k}_2\tilde{X}^I(k_1)\tilde{X}^I(k_2) \\ &+ \frac{\hbar^2}{2m}S^I(k_0)[\vec{k}_0\cdot\vec{k}_1\tilde{X}(k_1) + \vec{k}_0\cdot\vec{k}_2\tilde{X}(k_2) + \vec{k}_1\cdot\vec{k}_2\tilde{X}(k_1)\tilde{X}(k_2)]. \end{aligned} \tag{19}$$

The quatities with index I refer to impurity-background functions and those without it are background-background functions. We have introduced here the notation,

$$t(k) = \frac{\hbar^2 k^2}{2m_I}, \tag{20}$$

for the kinetic energy of the impurity particle and

$$\varepsilon(k) = \frac{\hbar^2 k^2}{2mS(k)} \tag{21}$$

for the elementary excitations in the background fluid. Using the solution of the Euler equation (18) we get the following expession for the three-body part of the chemical potential,

$$\begin{aligned} \mu_3^I = &-\frac{1}{16} \int \frac{d^3k_0 d^3k_1 d^3k_2}{(2\pi)^9 \rho^3} \delta(\vec{k}_0 + \vec{k}_1 + \vec{k}_2) S(k_1) S(k_2) \\ &\times \Big[[t(k_0) + \varepsilon(k_1) + \varepsilon(k_2)] (\tilde{u}_3^I(\vec{k}_0, \vec{k}_1, \vec{k}_2))^2 \\ &+ \frac{1}{3} [t(k_0) + t(k_1) + t(k_2)] S^I(k_0) \tilde{X}^I(k_1) \tilde{X}^I(k_2) \tilde{u}_3(\vec{k}_0, \vec{k}_1, \vec{k}_2) \Big]. \end{aligned} \tag{22}$$

RESULTS FOR THE GROUNDSTATES

The iteration of the Euler equation (7) starts from a guess for the particle-hole potential. Equation (7) gives then S(k) and its Fourier transform g(r). This g(r) can not be used as such because it has wrong small r behavior and one has to impose the condition g(r)=0, when r< r_c, the core radius of the potential. When S(k) and g(r) are known one can calculate the new particle-hole potential and repeat the same procedure again.

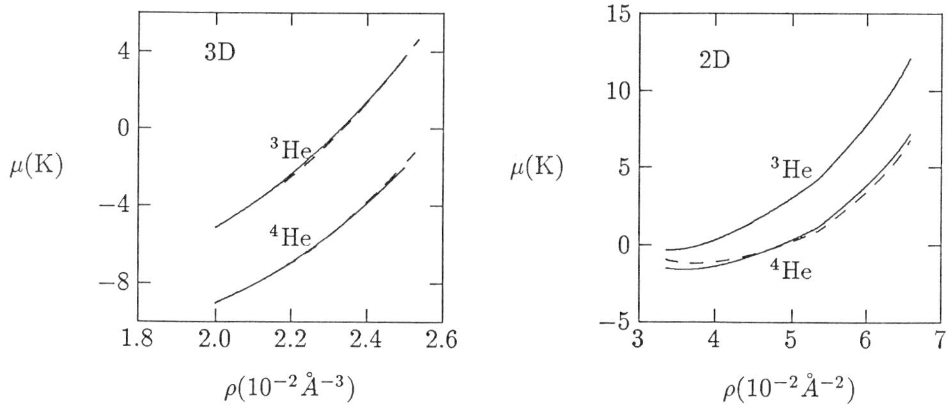

Fig. 1. Chemical potentials of the pure ^{4}He and the ^{3}He impurity in ^{4}He liquid. The solid lines give present results. In the three dimensional case (3D) the dotted lines give fits to experiments from ref. 16 and in the two dimensional case (2D) the dotted line is a fit to Monte Carlo results of ref. 8.

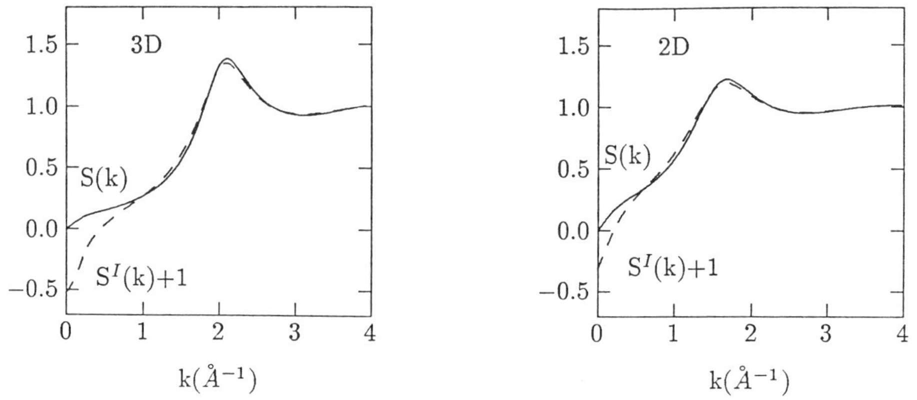

Fig. 2. The structure functions of the pure ^{4}He , S(k), and the ^{3}He impurity, $S^I(k)+1$. The three dimensional case is labeled with (3D) and the two dimensional one with (2D).

The bare two-body potential used in these calculations is the Aziz potential[17]. As discussed earlier we have included elementary diagrams upto the two simples five-body diagrams and estimate the higher order diagrams by multiplying the sum of four- and five-body diagrams by an enhancment factor. This factor is chosen in such a way that the saturation properties (energy/particle and pressure) fit the experimental values. The same factor is then used in the calculation of the impurity chemical potential. In three-dimensional case the factor is 1.4, and it gives a very good fit over the whole density range for both the pure ^{4}He and the impurity ^{3}He as shown in Fig.1. In the two-dimensional case where the comparison is done with the recent Monte Carlo results[8] the enhancement factor is 1.2. The smaller value is due to the fact that the number of near by particles is reduced in two dimensional system. Here we also get a good fit over a large density range around the saturation density, $\rho = 0.043\text{Å}^{-2}$, for pure ^{4}He . Reliable Monte Carlo results for the ^{3}He impurity in 2D system do not exist and they would be very welcome.

Corresponding structure functions are shown in Fig.2. The ^{3}He -^{4}He structure function plus a unity, $S^I(k)+1$, is very similar to the pure ^{4}He structure function except at small momenta. In the limit k=0 $S^I(0)$=-α -1 where α is the volume coefficient related to the excess volume occupied by the lighter mass ^{3}He [18,19]. The experimental value[20] at the saturation density for α is 0.29 whereas our calculation gives 0.55. This number is, however, very sensitive to the speed of the zero sound c, because the impurity-background particle-hole potential is proportional to the square of the zero sound, $V^I_{p-h}(0) = m\alpha\, c^2$. Thus the typical 25 % underestimate in c when it is calculated from the slop of S(k) gives a factor of two too large volume coefficient. The structure functions in two-dimensional case behave qualitatively in the same way, but show less pronounced oscillations.

EXITATIONS

In the variational approach to a dynamical system one searches for the extremum of the action integral

$$L = \int_{t_0}^{t} \frac{< \varphi(t)|H - i\hbar\frac{\partial}{\partial t}|\varphi(t) >}{< \varphi(t)|\varphi(t) >} \tag{23}$$

where H is the Hamiltonian of the system, and the time dependent wave function can be written as follows,

$$\varphi(t) = e^{-i(A\mu+\mu^I)t/\hbar}\psi_I(\vec{r}_0, \vec{r}_1, ...\vec{r}_A; t), \tag{24}$$

with chemical potentials μ and μ^I. We start with the approximation where the time dependence is included only in the one-body impurity correlation function and two-body impurity-background correlations

$$\psi_I(\vec{r}_0,\vec{r}_1,...\vec{r}_A;t) = exp\frac{1}{2}\Big(\delta u_1^I(\vec{r}_0;t) + \sum_{i=1}^{A}\delta u_2^I(\vec{r}_0,\vec{r}_i;t)\Big)\psi_I(\vec{r}_0,\vec{r}_1,...,\vec{r}_A). \tag{25}$$

The time independent part given in eq.(15) remains uneffected. The variation of eq.(23) with respect to $\delta u_1^I(\vec{r}_0;t)$ and $\delta u_2^I(\vec{r}_0,\vec{r}_i;t)$ leads to the following continuity equations[21],

$$\begin{aligned}&\vec{\nabla}_0\cdot\vec{j}^{(1)}(\vec{r}_o;t) + i\delta\dot{\rho}_1(\vec{r}_0;t) = 0\\ &\vec{\nabla}_0\cdot\vec{j}^{(2)}(\vec{r}_0,\vec{r}_1;t) + \vec{\nabla}_1\cdot\vec{J}^{(2)}(\vec{r}_0,\vec{r}_1;t) + i\delta\dot{\rho}_2(\vec{r}_0,\vec{r}_1;t) = 0\end{aligned} \tag{26}$$

where the currents and densities are defined in terms of δu_1^I and δu_2^I.

$$\begin{aligned}\vec{j}^{(1)}(\vec{r}_o;t) &= \frac{\hbar}{2m_I}\Big[\rho_1(\vec{r}_0)\vec{\nabla}_0\delta u_1^I(\vec{r}_0;t) + \int d^3r_1\rho_2(\vec{r}_0,\vec{r}_1)\vec{\nabla}_0\delta u_2^I(\vec{r}_0,\vec{r}_1;t)\Big]\\ \vec{j}^{(2)}(\vec{r}_0,\vec{r}_1;t) &= \frac{\hbar}{2m_I}\Big[\rho_2(\vec{r}_0,\vec{r}_1)\Big(\vec{\nabla}_0\delta u_1^I(\vec{r}_0;t) + \vec{\nabla}_0\delta u_2^I(\vec{r}_0,\vec{r}_1;t)\Big)\\ &\quad + \int d^3r_2\rho_3(\vec{r}_0,\vec{r}_1,\vec{r}_2)\vec{\nabla}_2\delta u_2^I(\vec{r}_0,\vec{r}_2;t)\Big]\\ \vec{J}^{(2)}(\vec{r}_0,\vec{r}_1;t) &= \frac{\hbar}{2m}\rho_2(\vec{r}_0,\vec{r}_1)\vec{\nabla}_1\delta u_2^I(\vec{r}_0,\vec{r}_1;t),\end{aligned} \tag{27}$$

and

$$\begin{aligned}\delta\rho_1(\vec{r}_0;t) &= \rho_1(\vec{r}_0)\delta u_1^I(\vec{r}_0;t) + \int d^3r_1\rho_2(\vec{r}_0,\vec{r}_1)\delta u_2^I(\vec{r}_0,\vec{r}_1;t)\\ \delta\rho_2(\vec{r}_0,\vec{r}_1;t) &= \rho_2(\vec{r}_0,\vec{r}_1)(\delta u_1^I(\vec{r}_0;t) + \delta u_2^I(\vec{r}_0,\vec{r}_1;t))\\ &\quad + \int d^3r_2\rho_3(\vec{r}_0,\vec{r}_1,\vec{r}_2)\delta u_2^I(\vec{r}_0,\vec{r}_2;t).\end{aligned} \tag{28}$$

The first continuity equation is easily solved in the Fourier space,

$$\hbar w = \frac{\hbar^2k^2}{2m_I} + \Sigma(k,w). \tag{29}$$

It includes the proper self energy $\Sigma(k,w)$,

$$\Sigma(k,w) = \frac{\hbar^2}{2m_I}\int\frac{d^3p}{(2\pi)^3\rho}\vec{k}\cdot\vec{p}S^I(p)\alpha_{k,w}(\vec{p}), \tag{30}$$

which in turn depends on the Fourier transform of the fluctuating part of the two-body correlation function,

$$\alpha_{k,w}(\vec{p}) = \frac{\rho_0F\{\delta u_2^I(\vec{r}_0,\vec{r}_1;t)\}}{\delta\rho(k,w)}. \tag{31}$$

In the convolution approximation the second continuity equation becomes an integral equation,

$$\begin{aligned}(\hbar w - t(k+p) - \varepsilon(p))\alpha_{k,w}(\vec{p}) &= \hbar w\frac{\vec{k}\cdot\vec{p}}{k^2}\frac{S^I(p)}{S(p)} - I_{k,w}(\vec{p}),\\ I_{k,w}(\vec{p}) &= \int\frac{d^3q}{(2\pi)^3\rho}\alpha_{k,w}(\vec{q})S^I(|\vec{p}-\vec{q}|)K_{k,w}(\vec{p},\vec{q})\end{aligned} \tag{32}$$

where the kernel is of the form

$$K_{k,w}(\vec{p},\vec{q}) = \hbar w - S(q)\frac{(\vec{k}+\vec{p})\cdot(\vec{k}+\vec{q})}{p^2}t(p) - \frac{\vec{p}\cdot\vec{q}}{p^2}\varepsilon(p). \tag{33}$$

If we ignore the integral term we get an analytic expression for the two-particle correlation function. When this is inserted into the self energy of eq.(30) we have discovered a similar formula as the one from the Green's function approach by Bhatt[7], but with a well defined form for the coupling to the phonon-roton spectrum,

$$\Sigma(k,w) = \frac{\hbar^2}{2m_I}\int \frac{d^3p}{(2\pi)^3\rho}\frac{(\vec{k}\cdot\vec{p})^2}{k^2}\frac{S^I(p)^2}{S(p)}\frac{\hbar\omega}{(\hbar w - t(k+p) - \varepsilon(p))}. \tag{34}$$

The integral term in eq.(32) takes care of the correct short range behavior in coordinate space, and thus the iteration of eq.(32) leads to a ladder-type summation in the diagramatic expansion.

In the limit where the impurity is a static spectator, i.e. its mass is infinite and its structure function $S^I(k)=0$ we find that the above formulation simply treats the phonon-roton spectrum as a Feynman-type spectrum of eq.(21) which is known to lie in the three-dimensional system by a factor of two too high in the roton region. In order to include the backflow correction consistently one should add the fluctuating impurity-background-background three-body correlation factor into the wave function and solve the three-body continuity equation which leads to a rather complicated integral equation. In this work we have chosen to take the ^{4}He excitation spectrum directly from experiments and use that for $\varepsilon(p)$ in eqs.(32),(33) and (34). Similarily we replace the impurity mass in the definition of t(p) in eq.(20) with its effective mass.

The results for the effective mass are shown in Fig. 3. as a function of density. In the three-dimensional case the calculated effective mass is some 15% below the experimental value. The approximation we have used ignores three-body corrections in the impurity-background coupling and that would enhance the effective mass.[6] In the two dimensional case the results agree surprisingly well with experiments even when the basic definitions of eq.(20) and (21) are used.

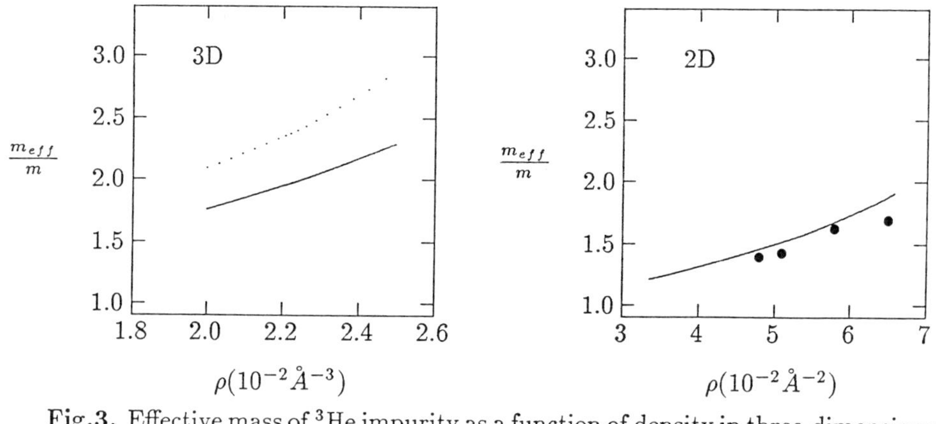

Fig.3. Effective mass of ^{3}He impurity as a function of density in three-dimensional and two-dimensional ^{4}He backgrounds. Black dots are measured values. In the three-dimensional case we have use the the formula $m_{eff}/m = (1-26.1\rho)^{-1}$ which fits experimental values very well[22]. In the two-dimensional case the measurements are taken from ref. 1.

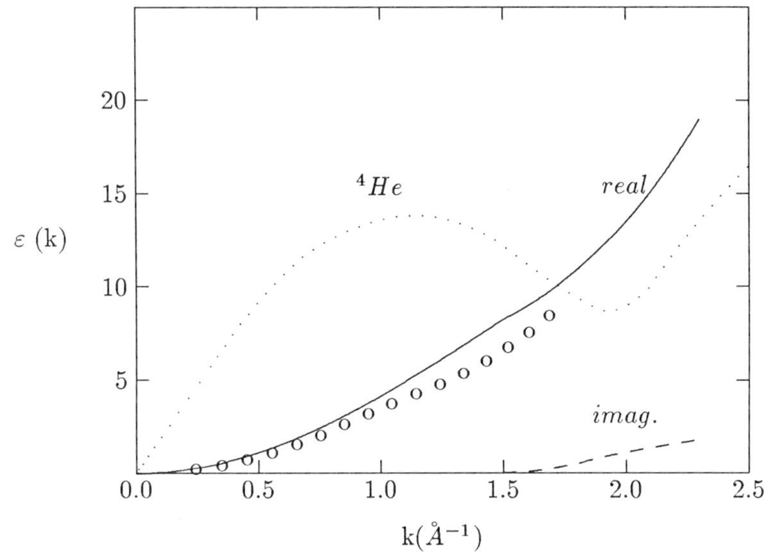

Fig.4. Excitation spectrum of the ^{3}He impurity. Solid curve is the real part and dashed line the imaginary part of the spectrum. Dotted line is the phonon-roton spectrum of the background ^{4}He and open circles give a fit to experiments, called FIT4 in ref. 22.

Finally in Fig.4 we show the excitation spectrum of the ^{3}He impurity. The spectrum does not show any roton-like minimum in agreement with recent neutron scattering experiments[4]. The crossing of the ^{4}He spectrum makes roton emition possible and that gives rise to the imaginary part of the ^{3}He spectrum. Numerically the ^{3}He spectrum terminates at $k \sim 2.3 Å^{-1}$ indicating that all the additional energy would be lost to roton excitations. We also find increasing deviation from the purely quadratic momentum dependence with increasing momentum. This is in agreement with both neutron scattering experiments and the analysis using specific heat and second sound measurements[22].

ACKNOWLEDGEMENTS

I like to thank Eckhard Krotscheck and Janice Epstein for collaboration, and Roger Smith, Alpo Kallio and Pekka Pietiläinen for many valuable discussions. I also like to thank Texas A&M University for pleasent working conditions during my visit.

REFERENCES

1. Xingwu Wang and F. M. Gasparini Phys. Rev. **B38**, 11245, (1988).
2. J. M. Valles, Jr., R. H. Higley, R. B. Johnson, and R. B. Hallock, Phys. Rev. Lett. **60**, 428, (1988).
3. E. Krotscheck, M. Saarela, and J. L. Epstein, Phys. Rev. Lett. **61**, 1728 (1988), and J. Epstein, E. Krotscheck, and M. Saarela, to be published in Phys. Rev. Lett.
4. B. Fåk, K. Guckelsberger, M. Körfer, R. Scherm, and A.J. Dianoux, submitted to Phys. Rev. B.
5. W. Greywall Phys. Rev. Lett. **41**, 177 (1978).
6. See
J.C. Owen, Phys. Rev. **B23**, 5815 (1981),
A. Fabrocini, S. Fantoni, S. Rosati, and A. Polls, Phys. Rev. **B33**, 6057 (1986) and references therein.

7. See
R.N. Bhatt, Phys. Rev. **B18**,2108 (1978),
M. Lücke, A. Szprynger, Phys. Rev. **B26**, 1374 (1982),
W. Hsu, and D. Pines, Phys. Rev. **B32**, 7179 (1985), and references therein.
8. P. A. Whitlock, G. V. Chester M. H. Kalos , Phys. Rev. **B38**, 2418, (1988).
9. C. C. Chang and C. E. Campbell, Phys. Rev. **B15**, 4238 (1977).
10. E. Krotscheck, Phys. Rev. **B33**, 3158 (1986).
11. R.A. Smith, A. Kallio, M. Puoskari, and P Toropainen, Nucl. Phys. **A328**, 186 (1979), and,
R.A. Smith, Phys. Lett. **85B**, 183 (1979).
12. Q.N. Usmani, B. Friedman, and V.R. Pandharipande, Phys. Rev. **B25**, 4502 (1982).
13. P. A. Hilton, R. Scherm, and W. G. Stirling, J. Low Temp. Phys. **27**, 851 (1977)).
14. L. Pitaevskii, unpublished.
15. E. Krotscheck, M. Saarela, and J.L. Epstein, Phys. Rev. **B38**, 111 (1988).
16. C. Ebner, D.O. Edwards, Phys. Rep. **2**, 77 (1970).
17. R.A. Aziz, V.P.S. Nain, J.C. Carley, W.L. Taylor, and G.T. McColville, J. Chem. Phys. **70**, 4330 (1979).
18. J. Bardeen, G. Baym, and D. Pines, Phys. Rev. **156**, 207 (1967).
19. J.C. Owen, Phys. Rev. Lett. **47**, 586 (1981).
20. A.E. Watson, J.D. Reppy, and R.C. Richardson, Phys. Rev. **188**, 384 (1969)
21. M. Saarela, Phys. Rev. **B33**, 4596 (1986), and M. Saarela and J. Suominen, in *Condenced Matter Theories*, edited by J.S. Arponen, R.F. Bishop, and M. Manninen, vol. 3, 157 (Plenum Press, New York, 1988).
22 D. Greywall, Phys. Rev. **B20**, 2643 (1979).

ACHIEVEMENTS IN QUANTUM MONTE CARLO

Presentation of the third Eugene Feenberg Memorial Medal in Many-Body Physics to MALVIN H.KALOS

M. L. Ristig

Institut für Theoretische Physik
der Universität zu Köln
D-5000 Köln 41, West-Germany

I have the great honour to present the Eugene Feenberg Memorial Medal on the occasion of the sixth International Conference on Recent Progress in Many-Body Theories. The Feenberg Medal was created by the International Advisory Committee of the third conference of this kind held in 1983 in Altenberg, West-Germany. It has been established – in memoriam of our esteemed colleague and teacher Eugene Feenberg – for keeping the spirit of his approach to physics alive and flowering, and for honoring outstanding contributions to the advancement of many-body theories from first principles.

Eugene Feenberg, who was born in 1906 and died in 1977, was one of the great pioneers in the application of non-relativistic quantum mechanics to realistic microscopic systems. His original and very important contributions to nuclear theory, to approximation methods and to the theory of strongly correlated quantum fluids have had a profound and lasting impact on the diverse subfields of many-body physics [1]. His comprehensive studies in nuclear physics were fundamental for the development of modern nuclear shell theory. The work culminated in the publication of his book in 1955 on the 'Shell Theory of the Nucleus'. His later, eminently fruitful, and highly original studies on the microscopic theory of the ground and low excited states of quantum many-particle systems reflect the characteristics of Feenberg's independent, analytic and farseeing mind. However, for a long time this part of his work was not fully appreciated by the majority of his contemporaries who went along following the dominant scientific spirit of that period. Within almost two decades we have now experienced an explosive burst of activities in which the methods introduced or highly regarded by Feenberg have been developed into powerful tools for analyzing quantum many-body systems of strongly coupled particles on a quantitatively accurate level. These developments have brought and continue to bring to fruition Feenberg's strong conviction that the outstanding achievements of physics at the end of this century will be the remarkable accuracy of the quantitative results we may derive from first principles.

The Feenberg Medal was awarded for the first time at the San Francisco meeting in 1985 to David Pines from the University of Illinois. The second medallist has been John Walter Clark, Washington University, who received the prize in 1987 at the fifth many-body conference held in Oulu, Finland [2]. The present choice has been – similar as in the preceding years – far from simple since there are many excellent physicists in our international many-body physics community who are regarded as distinct candidates for the award.

Recent Progress in Many-Body Theories, Volume 2
Edited by Y. Avishai
Plenum Press, New York, 1990

Based on the collected proposals which the acting committee had requested and received the members of the search group have been able to reach a convincing and unanimous decision on the third recipient of the Feenberg award.

The third Eugene Feenberg Memorial Medal is awarded to Malvin H.Kalos. He receives it for his pioneering highly original and profound corpus of work on stochastic methods in quantum many-body theories specifically on his invention of the Green's function Monte Carlo method and for his early recognition of the importance of computational physics and high performance computing to meet its needs.

In a 1929 paper Paul Dirac observed that the physical laws necessary for the mathematical theory of a large part of physics are completely known and that it is only necessary to find practical methods for solving the equations governing the behavior of complex systems [3].The first direct suggestion of developing and employing stochastic methods for solving the Schrödinger equation came from Fermi, Metropolis and Ulam [4]. However, the early attempts to solve this equation using Monte Carlo methods contained finite difference approximations and were done for the most part as feasibility studies rather than as serious attempts for analyzing physical quantum many-body problems [5]. An important step forward was made by McMillan who employed a variational quantum Monte Carlo (VMC) method to simulate the ground state properties of liquid helium [6]. The first successful Green's function Monte Carlo (GFMC) calculation was designed in 1962 in a Physical Review paper [7] where Malvin Kalos reported on his formulation of the problem and on his calculations on the binding energy of the alpha particle and the triton. In the following years this method developed into one of the essential keys for the dramatic progress in computational physics.

Malvin Kalos received his PhD-degree from the University of Illinois where he was one of John Blatt's students. As a post-doctoral fellow he then went to Cornell University to work with Hans Bethe and later accepted a position with United Nuclear Corporation in White Plains, New York. It was there that he became deeply involved in Monte Carlo methods primarily to solve neutron transport problems. Kalos' review article together with G.Goertzel in 1958 on this topic contains one of the earliest and clearest expositions of the Monte Carlo method and of statistical sampling [8]. In fact, his studies on neutron transport problems laid the foundation for Kalos' profound GFMC work developing from the 1962 paper on the few-body nuclear ground state, Ref.[7]. It is typical of his philosophy that the method he used in this early calculation did not contain any unnecessary approximations. Kalos sampled the Green's function exactly and the numerical results were exact except for statistical and machine precision errors. These calculations were done at the Courant Institute of Mathematical Sciences at New York University on the IBM 7090 computer. Later on Kalos accepted a permanent appointment with this Institute. It is noteworthy that his results on the alpha particle and on the triton were directly comparable to experiment and that Faddeev's treatment of the three-body problem was published just the year before, in 1961.

Malvin Kalos has been motivated throughout his career by the importance of the physical problem rather than the novelty of his methods. His study of the electronic structure of the helium atom [9] – probably the first serious Monte Carlo calculation of an atomic system – was done in 1966 in response to discrepancies found between variational estimates for the electronic density in the helium atom. He continued improving the GFMC method for the next several years [10,11] and greatly benefited by the facilities accessible at the Courant Computer Laboratory.

In 1970 he published his first GFMC calculation on liquid ^{4}He at equilibrium density [12]. In this paper the method had been extended so that it could calculate the properties of the ground state of hard-core systems such as liquid helium. These calculations simulated a system of thirty-two ^{4}He atoms interacting via a Lennard-Jones potential. The result showed clearly that the true ground state energy was well below the available variational results based on Jastrow-type wave functions. We may remark that the total computer time to run these calculations at that time did amount to perhaps ten hours on a CDC-6600.

It was the problem that arose when Kalos attempted to extend his calculations to larger systems to eliminate any size dependence that led him to include importance sampling and to improve efficiently the Green's function expansion that he used. These calculations were carried out with D.Levesque and L.Verlet [13] and published in 1974. Because it was clear that the major numerical problems were caused by the hard core they concentrated on the Bose hard-sphere system. They included an attractive well by perturbation theory and used these results to compare to experimental data on liquid ^{4}He. Their numerical studies included the first GFMC calculations on the solid phase and gave the phase diagram for the system considered. These calculations were also important as benchmark calculations for other methods including variational approaches, integral equation methods etc. Their work showed that the optimal importance function is the exact ground state wave function, that a variational trial function is a good choice for an importance function, and that if the optimal importance function is used the energy is obtained with zero variance. Algorithms were given for sampling the Green's function with importance sampling. This and the preceding paper give all the important ingredients needed to solve the many-boson Schrödinger equation yielding exact results except only for statistical sampling errors.

In 1979 Kalos and Chester together with Whitlock and Ceperley published a complete study [14] of the bulk ground state properties of liquid and solid ^{4}He over the range of experimentally accessible densities employing the interatomic Lennard-Jones potential. They were able to demonstrate in this work that the GFMC method ist accurate enough to show the discrepancies between experimental results on the equation of state and theoretical results based on the Lennard-Jones potential. The authors attributed these differences to the inadequacy of the Lennard-Jones potential. Two years later Kalos, Lee, Whitlock and Chester re-examined the helium boson system with the GFMC method employing several different interatomic interactions and also explored the influence of the three-body Axilrod-Teller potential [15]. It was the stunning agreement of these calculations using the HFDHE2 potential by Aziz et al. with experiment that convinced the many-body physics community that this potential gives a better description of the ^{4}He – ^{4}He interaction than the hitherto adopted Lennard-Jones potential. At this point Feenberg's early attempts [16] to correct the helium pair potential by exploiting accurate results of many-body calculations were brought to full fruition. In the same period Kalos, Chester, and collaborators applied the GFMC method to other boson systems such as a system of bosons interacting via Yukawa potentials [17] and a system of hard spheres in a channel [18].

The great success of the GFMC approach led to attempts for generalizing and employing these Monte Carlo procedures to deal with fermionic many-body systems, in particular, to explore the properties of liquid ^{3}He. These attempts developed into the fixed-node approximation [11] and the procedure of transient estimates [19]. A brief outline of these random walk computational methods for solving the many-fermion Schrödinger equation may be found in a recent article by D.Ceperley and B.Alder [20]. The authors pay also tribute to Malvin Kalos' important share in developing computational physics to the present status.

We may refer to the work with Ceperley and Chester where Malvin Kalos developed several methods to calculate variationally – employing Monte Carlo procedures – the ground state properties of Fermi systems [21]. It was shown that linear algebra techniques could be used for rapidly calculating, for example, energy results based on fermionic variational wave functions and to update them after a Monte Carlo step. In 1981 Schmidt, Kalos, Lee and Chester reported on a GFMC calculation on liquid ^{3}He in its ground state [22]. The method used the fact that the GFMC approach could be constructed such that the calculated energy at each step represents an upper bound to the exact ground state energy. While the fermionic character of the problem causes the variance of the energy to diverge exponentially it was possible to arrive, in particular, at a series of upper bounds to the ground state energy before the variance became unmanagable. This calculation gave an upper bound of − 2.20 ± 0.05 K at the equilibrium density of liquid ^{3}He which should be compared to the experimental value of − 2.47 ± 0.01 K. Since then, the bounds have been successively lowered. Panoff and Carlson, two of Kalos' coworkers, employed improved variational trial functions as initial input data and could use more computer time. Their result on the ground state energy [23], the best theoretical data to date, is − 2.44 ± 0.04 K.

Throughout this period Malvin Kalos has been actively involved in intense efforts for including more and more characteristic physical features into the variationally determined trial wave functions to both lower the GFMC error estimates and to improve our detailed understanding of the underlying physical effects. In this he maintained that while the GFMC method could give precise answers on the equilibrium properties of quantum matter and guide theoreticians to develop a correct physical picture of the phenomena a variational analysis involving a few physically motivated parameters are valuable in their own right since they can give more and deeper insights. Some of the VMC calculations Kalos has collaborated on include the work with Chester, Lee and Schmidt [24] on liquid ^{4}He employing pair plus triplet correlations and on Fermi fluids including the effects of triplet and backflow correlations [25] as well as his recent studies − together with Vitiello and Runge − of 'shadow' wave functions to produce translationally invariant wave functions for the solid phase [26,27].

During his scientific career Malvin Kalos has made important contributions to theoretical developments in other fields of research. Like Feenberg, Kalos began his studies in nuclear physics [28,29,7]. He has collaborated with dinstinguished scientists on various diverse subjects in statistical physics such as phase separations in binary alloys [30-32], percolation problems [33], liquid-vapor interfaces [34] critical clusters [35,36], polymer physics [37-40] and spin systems [41,42]. His talk at Brookhaven [43] in 1982 inspired several researchers to study the phenomenon of critical slowing down near second- order phase transitions [44,45].

To date the quantum Monte Carlo methods which Malvin Kalos helped to develop so much have a prominent place in all areas of physics from the unifying theories of many-body systems, condensed matter and solid state physics to lattice and field theories. Today we recognize finally that his pioneering work laid the fundament for this explosion of interest and outburst of scientific activities and successes.

Although Kalos' position at the Courant Institute did not formally allow him to teach physics classes he is a great and stimulating teacher. Among the physicists who have much benefited from his teaching are J.Carlson, D.Ceperley, J.M.C.Chen, K.E.Kürten, M.A.Lee, R.M.Panoff, K.E.Schmidt, P.A.Whitlock, J.Zabolitzky. Recently, Kalos has written with Whitlock an introductory text on Monte Carlo methods [46]. Until last year he has been the head of the New York University Ultra computer project before leaving to head the Cornell Theory Center which runs the National Supercomputer Center. Yet, he is still very active in physics research and is currently working on adapting Monte Carlo methods to 'real time' and Fermi systems as well as excitations in quantum fluids.

ACKNOWLEDGEMENT

The author acknowledges the substantial assistance of K.E.Schmidt in preparing this article.

REFERENCES

[1] Eugene Feenberg obituary, Nucl. Phys. A 317(1979).
[2] R.F.Bishop, in *Recent Progress in Many-Body Theories*, eds. A.J. Kallio, E.Pajanne and R.F.Bishop, pages 385-393 (Plenum, New York, 1988) Vol. I.

[3] P.A.M.Dirac, Proc.Roy.Soc.London Ser. A 123,714(1929).
[4] N.Metropolis and S.Ulam, J.Am.Stat.Assoc. 44,247(1949).
[5] M.D.Donsker and M.Kac, J.Res.Natl.Bur.Stand. 44,551(1950).
[6] W.L.McMillan, Phys.Rev. 138, A442(1965).
[7] M.H.Kalos, Phys.Rev. 128,1791(1962).
[8] G.Goertzel and M.H.Kalos, Progress in Nuclear Energy 2,315(1958).
[9] M.H.Kalos, J. Comput. Phys. 1,257(1966).
[10] M.H.Kalos, Nucl. Phys. A 328,153(1979).
[11] D.M.Ceperley and M.H.Kalos, in *Monte Carlo Methods in Statistical Physics*, Topics in Current Physics, Vol. 7, ed. K.Binder (Springer, Berlin, 1979).
[12] M.H.Kalos, Phys.Rev. A 2,250(1970).
[13] M.H.Kalos, D.Levesque, and L.Verlet, Phys.Rev. A 9,2178(1974).
[14] P.A.Whitlock, D.M.Ceperley, G.V.Chester, and M.H.Kalos, Phys.Rev.B19,5598(1979).
[15] M.H.Kalos, M.A.Lee, P.A.Whitlock, and G.V.Chester, Phys.Rev. B 24,115(1981).
[16] E.Feenberg, *Theory of Quantum Fluids* (Academic, New York, 1969).
[17] D.Ceperley, G.V.Chester, and M.H.Kalos, Phys.Rev. B 17,1070(1978).
[18] K.S.Liu, M.H.Kalos, and G.V.Chester, Phys.Rev. A 10,303(1974).
[19] K.E.Schmidt and M.H.Kalos, in *Applications of the Monte Carlo Method in Statistical Physics*, ed. K.Binder (Springer, Berlin 1984).
[20] D.Ceperley and B.Alder, Science, Vol. 231, 555(1986).
[21] D.Ceperley, G.V.Chester, and M.H.Kalos, Phys.Rev. B 16,3081(1977).
[22] M.A.Lee, K.E.Schmidt, M.H.Kalos, and G.V.Chester, Phys.Rev.Lett. 46,728(1981).
[23] R.M.Panoff and J.Carlson, Phys.Rev.Lett. 62,1130(1989).
[24] K.Schmidt, M.H.Kalos, M.A.Lee and G.V.Chester, Phys.Rev.Lett. 45,573,(1980).
[25] K.E.Schmidt, M.A.Lee, M.H.Kalos, and G.V.Chester, Phys.Rev.Lett. 47,807,(1981).
[26] S.Vitiello, K.Runge and M.H.Kalos, Phys.Rev.Lett. 60,1970(1988).
[27] S.Vitiello, K.J.Runge and M.H.Kalos, in *Computer Simulation Studies in Condensed Matter Physics: Recent Developments*, Proceedings in Physics,Vol.33, eds. D.P.Landau, K.K.Mon and H.B.Schüttler (Springer, Berlin, 1988).
[28] J.M.Blatt and M.H.Kalos, Phys.Rev. 92,1563(1953).
[29] M.H.Kalos, L.C.Biedenharn, and J.M.Blatt, Nucl.Phys. 1, 233(1956).
[30] J.L.Lebowitz and M.H.Kalos, Scr.Metall. 10,9(1976).
[31] K.Binder, M.H.Kalos, J.L.Lebowitz and J.Marro, Adv. Colloid and Interface Sci. 10,173(1979).
[32] K.Binder, J.L.Lebowitz, M.K.Phani, and M.H.Kalos, Acta Metall. 29,1655(1981).
[33] A.Sur, J.L.Lebowitz, J.Marro, M.H.Kalos and S.Kirkpatrick, J.Stat.Phys.15,345(1976).
[34] M.H.Kalos, J.K.Percus, and M.Rao, J.Stat.Phys. 17,111(1977).
[35] M.H.Kalos, J.L.Lebowitz, O.Penrose, and A.Sur, J.Stat.Phys. 18,39(1978).
[36] K.Binder and M.H.Kalos, J.Stat.Phys. 22,363(1980).
[37] D.Ceperley, M.H.Kalos, and J.L.Lebowitz, Phys.Rev.Lett. 41,313(1978).
[38] I.Webman, J.L.Lebowitz, and M.H.Kalos, Phys.Rev. A 23,316(1981).
[39] M.Bishop, D.Ceperley, H.L.Frisch, and M.H.Kalos, J.Chem.Phys. 72,3228(1980).
[40] M.Bishop, D.Ceperley, H.L.Frisch, and M.H.Kalos, J.Chem.Phys. 75,5538(1981).
[41] A.B.Bortz, J.L.Lebowitz, and M.H.Kalos, J.Comput.Phys. 17,10(1975).
[42] M.K.Phani, J.L.Lebowitz, and M.H.Kalos, Phys.Rev. B21,4027(1980).
[43] M.H.Kalos, Brookhaven Conference on Monte Carlo Methods and Future Computer Architectures, 1983.
[44] K.E.Schmidt, Phys.Rev.Lett. 51,2175(1983).
[45] J.Goodman and A.D.Sokal, Phys.Rev.Lett. 56,1015(1986).
[46] M.H.Kalos and P.A.Whitlock, *Monte Carlo Methods*, Vol I: Basics (Wiley, New York, 1986).

SUMMARY OF THE VI INTERNATIONAL CONFERENCE ON RECENT PROGRESS IN MANY BODY THEORIES, ARAD (ISRAEL)

Hermann G. Kümmel*

Institut für theoretische Physik II
Ruhr Universität, Bochum (Germany)

1. Prologue

This conference suffers to some extent from what we may consider as the very strength of many body theory, namely the wide applicability to almost all branches of physics. Indeed, this is clear from the breadth of the topics presented at this meeting. We heard talks about biophysics, solid state, atomic, molecular and nuclear physics, about fluids and field theory. Already at the last conferences of this series this has lead to the problem that it became increasingly difficult to bridge the gap between so widely different fields. After all, slang and even mentality of the people involved very much depend on the fields they are working in. This was clearly observable in the talks presented and the discussions I had with colleagues at this conference. Several times I have asked groups of participants whether they ever heard about some objects and even words which were presented by speakers as a matter which everybody in the audience should know. In many cases the answer was "no". Inevitably this situation will show up in the present review of the conference by the somewhat subjective selection of topics and talks, although I have tried hard to obtain the missing informations afterwards. Anyway, something has to be done with the future conferences of this series, reconciling specialization in the applications with the generality of methods and ideas of many body theory. Of course, because methods without applications and/or tests are of academic interest only, this Gordian knot cannot be cut by the exclusion of applications from the program.

2. High T_c Superconductors

Let me start with this technologically most important, yet intellectually appealing (and least understood) topic dealt with at this conference. The present situation can best be characterized by the following statement: the number of people working on the theory of high T_c superconductors (HT_cSC) - and thus the number of different theories and models- is steadily increasing. This leaves a marginal chance for each individual to hit upon the right theory.

It is a remarkable fact that - in contrast to elementary particle theory - there is a well defined hamiltonian in solid state physics. But, it is horribly complicated, especially for the geometry of the HT_cSC's. Thus one is forced to replace the exact Hamiltonian by a model. Finding out that this model still is not solvable, make a model of the model, then a model of the model of the model etc. until something solvable is obtained:

Recent Progress in Many-Body Theories, Volume 2
Edited by Y. Avishai
Plenum Press, New York, 1990

$$H_{exact} = H_{electron} + H_{ion} + H_{ion-electron\ interaction}$$
$$\rightarrow H_{electron} + H_{phonon} + H_{phonon-electron\ interaction}$$
$$\rightarrow H_{model\ 1} = H_{model\ 2} \rightarrow H_{model\ 3} \ldots.$$

The models in question are the Anderson, Hubbard, t-J models and others, as discussed in this conference. The opening talk by *Rice* gave an example of the reduction of one model (Anderson's) to the next one (the t-J model) as outlined above. The latter just incorporates a Heisenberg type spin-spin interaction plus site occupancy by three states (spin up and down and empty). Mean field theory then eventually leads to flux states with symmetry breaking and all that. These features are built in from the beginning into the so called anyon model as presented by *Wiegmann*. Although this model - related to gauge field theories - is very sophisticated and deep, I have trouble to see convincing arguments that it has something to do with reality. Probably some experts will disagree with me if I say that from this point of view the approach presented by *Aharony* is nearer to the special Cu-O layer structure of the actual HT_cSC: he step by step explains the complicated phase structure of these substances and manages to construct hole-hole pairs from holes hopping from O to O. Of course this is an a posteriori explanation of experimental facts and it reduces the full hamiltonian to the bare necessities - and this leaves one a bit dissatisfied. A completely different approach was suggested by *Weger*. He observed oscillating solutions of the BCS equations and has arguments for their existence in the Eliashberg equations which after all in principle, not in practice, are exact. I like this idea since it means that all solids have a chance for high T_c: it merely depends on the parameters of a given solid whether or not it will show up in the end. Whatever this means: the mere fact that there are such strange solutions is interesting enough and their meaning has to be explored. I also mention the electron liquid approach by *Kallio* exploiting the similarity to superfluidity and two studies of the Hubbard model by *Shmeltzer* and *Fantoni* and the coupled cluster like incorporation of correlations in the t-J model by *Oles*. As seen from this short overview there indeed are quite a few approaches to HT_cSC's competing with each other - and what has been presented at this conference is only a very small fraction of what exists in this field. It also is seen that there two major questions for which one has to find answers:

i) what is the gross structure of the superconducting phase ?
ii) what do electron correlations do on top of them ?

The first one one tries to answer by (mostly) guessing zero order solutions in the framework of mean field theories, Hartree-Fock-Bogoliubov approaches etc. There is no "safe" systematic method. For the correlations all standard methods fail if the correlations are strong. These problems are not new. They are typical for most many body systems, of course. But for HT_cSc's they are more serious than for any other many body system I know.

3. How to Deal with Systems with Strong Correlations

There were two talks presenting methods to get out of this dilemma: *Bishop* gave an overview of the extended coupled cluster method (ECCM) as compared to the good old configuration interaction (CI) method and the "normal" coupled cluster method (NCCM). I like very much his stringent test using the anharmonic oscillator, since I once did play with the same model and found - using the standard truncation scheme - that the correlation amplitudes eventually did *in*crease with the degree of correlations. This problem is settled now: the Manchester group has completely understood the underlying mathematics in terms of the Bargmann space and found that the amplitudes in fact *de*crease. I also believe that the ECCM (invented by Arponen from Helsinki) is an extremely powerful method - even for the gross structure (phase transitions, spontaneous symmetry breaking and the like), since there is a mapping to classical multivariable hamiltonian equations. I recommend this method very much to solid state theorists since already the NCCM has turned out as the most universal method in quantum chemistry, where after all also Coulomb forces are governing everything and also the geometry may be complicated. Another paper - presented by *Fulde* -attacks the same kind of problems from a different angle: using the super operator formalism allowing a cumulant form of energy and one particle Green's function he even managed to use a part of the *hamiltonian containing interactions as a starting point*. Of course, older methods as used in quantum chemistry are embedded in this theory. I feel that the connection with the NCCM and ECCM should be explored. (Certainly the latter methods normally do not start from an interacting hamiltonian, but this can be implemented).

4. Quantum Fluids

Before talking about theory let me start with the question how to extract from experiments the structure of quantum fluids. There was a contribution by *Silver* on this topic: momentum distributions can be measured by deep inelastic scattering of neutrons. One easily sees that the impulse approximation is not sufficient and that at least double scattering must be taken into account. This implies that some information about the two body density matrix is needed - and thus this cannot be disentangled from theory. But if one does use results as obtained from microscopic calculations like the ones reported by *Ristig* at this conference (based on Jastrow type wave functions) one obtains good results and a consistent picture with about 9 % condensate for the ^{4}He Bose fluid. Finally there were two talks by *Krotschek* and *Saarela* on inhomogeneous fluids, e.g. layers of ^{4}He fluid with and without ^{3}He impurities, both for the ground state as well as excited states. It is quite remarkable to which high degree of sophistication the standard methods have been pushed. This is also true for the Green's function (GFMC) method: *Panoff* presented impressive results which have the advantage of being final. Fortunately one is moving away from the once rather disturbing fixed node approximation for Fermions, using one more of the simple - yet brilliant -tricks typical for Monte Carlo methods. The talk by *Bouchaud* was quite a surprise to me: he presented a new BCS type approach to quantum fluids which apparently works very well, yet uses less CPU time. Even a mixture of BCS and Jastrow type wave functions is feasible, opening the avenue for new classes of (bosonic plus fermionic) excited states.

5. Nuclear Physics

Nuclei are somewhat "dirty" objects for many body theorists since the forces are not known. A long time ago - using various many body methods - all work in this field lead to the conclusion that two body forces - however sophisticated - will not suffice: it was not possible to obtain binding energy and density (radius) at the same time in agreement with experiments. Now from the theory of strong interactions it is clear that there are three, four,.. body forces between nucleons. Unfortunately, at present it is not possible to reliably construct such potentials. All one can do is to try out some model forces. An example of this was presented by *Pieper*: the people from the Argonne-Illinois collaboration are using sophisticated forms of the wave function with free parameters, model potentials and Monte Carlo variational methods to compute properties of the ^{16}O nucleus. Although they could slightly improve the situation, there is still a deviation from experiment. In this context, the fresh look at the old many body calculations taken by *Dickhoff* came as a surprise: he uses the Green's function many body method. This has the feature that particles and holes are treated on the same footing already in low order, something Goldstone perturbation theory cannot do. In NCCM such a treatment has been done iteratively only and in this way pairing effects could not be reached (except by using a BCS type starting wave function). I think that the recently observed strong spread of single particle states Dickhoff has discussed is not surprising. After all, the strong nucleon-nucleon interaction favours (Landau) Fermi liquids. It always was some kind of a wonder that the shell model did work so well. However, he also presented the related result that the binding energy has to be increased in such a way that one comes near to the experiments. This indeed is exciting. He claims that there is a pairing effect producing this increase. Certainly this result has to be checked carefully, just because it differs from anything believed so far. In the talk by *Zuker* the assumed fact that two body forces will not suffice was used to correct the nucleon-nucleon force for spectrocopy. The monopole part is the main "sinner" and has been corrected accordingly with quite convincing results. There also was a talk by *Giraud* offering a new approach to the shell model description of scattering of many body systems, with the advantage that one can save computer time and obtain correct absolute values - a notoriously difficult problem. The idea behind this method is the use of optimized channel wave packets as inhomogenities.

6. Random Systems and Methods

There are many kinds of random systems and many reasons for their existence. Some of them are produced as disordered systems, for instance by doping solids with some "dirt". There were three talks (by *Kaveh, Stephen* and *Shapiro*) on how to obtain information about t hem from the "outside", by letting electromagnetic waves pass through them or putting them into a magnetic field. Actually, this is not many body theory and thus I may be permitted just to quote them. There was a study of random polymers and their localization and folding by *Orland*. I found it interesting how this classical approach could be put into the mathematical form of path integrals and thus, naturally, of a Schrödinger equation. It is an example for the old wisdom that physicists are "girls for everything": give them problem, physics or not, with some luck they will find a solution - often remembering the tricks they have learned at college.

More interesting and more fundamental are signatures of "chaos" like those observed in fluctuations. The typical many body theorist - and many other physicists - tend to ignore the fact that "order" is the exception rather than the rule. Even a known hamiltonian becomes useless once one wants to learn something about higher excited states typically showing fluctuations. Yet, even in this situation one may extract physical information as we have learned from the talk given by *Weidenmüller*. First one has to simulate the hamiltonian in terms of random matrices. For certain observables they then may give important informations. One example is the extraction of parity violating forces and effects. These forces are "weak" forces, typically producing effects of the order of 10^{-6}. But their random effects may be many orders of magnitude larger. This enhancement can be studied with random matrices because the (simulated) symmetry violating part of the hamiltonian has a structure quite different from the symmetrical one, and this difference is well known. These techniques by now have achieved a high level of sophistication. Not all the tricks and ideas could be described in this talk, of course. But I think it served well the purpose of making many body theorists aware of this new field, which after all mostly is concerned with many body systems, too. It also is useful for "mesoscopic" systems, where there is genuine disorder. We have learned from *Pepper*'s talk about such a system. Due to the increasingly refined techniques in producing smaller and smaller electronic devices it now is possible to enclose electrons in a small one dimensional box the size of which can be varied by electric fields. One finds quantization of resistivity, fluctuations and a whole bunch of interesting and amusing things, like some consequences of the Bohm-Aharanov effect.

7.Miscellaneous Topics

I also want to mention the new approach to critical phenomena by *Suzuki* which somehow is meant to replace the standard renormalization group methods. The idea is to start with small clusters, each described by effective mean fields ,and then to increase the cluster size until convergence is achieved. The convergence is extremely rapid - and this is the strength of the method.

Finally I want to mention the paper by *Kröger* on the scattering of composite particles. He studies very explicitly quantum electrodynamics in 1 + 1 dimension. The method is going back to the original definition of the S-matrix as a time limit of exponentials with hamiltonians. To keep the size of the problem small, he uses discretization (i.e. a lattice form) and light front dynamics. The reason for the last step is that the light front momenta K_+ always are positive and as a consequence the Hilbert space is greatly reduced in size. Doing all this one obtains trustworthy results. One has to assume, however, that the singularities inherent in the light front formulation of quantum field theories do not hurt. The numerical results indicate that this is so. This may be due to the fact that for this model in the mass operator the singularities cancel. Kröger also presented ideas how one could proceed in 3 + 1 dimensions by mixing this method with Monte Carlo procedures.

I enjoyed very much the lecture on quasicrystals by *Levin*. Although certainly outside the topics of this conference, this must have been of interest to anybody looking beyond his own horizon. Historically, it all started with a general disbelief that such a thing could exist in nature - with rejections of papers as a natural consequence. This is just one more example for the fact that one should not trust the experts. Of course, one should not trust new ideas either - but nevertheless one should study them carefully. After all, one new idea out of several hundred others may be a good one.

8. Epilogue

After such a conference probably every participant asks himself the question: was it worth it ? Did I learn something useful ? Did I hear about some new ideas - or even got a new idea of my own ? My own answer to these questions is "yes". I have indicated in this summary where I was surprised and where something unexpected did show up. These are some of the points convincing me to answer "yes". I also trace some of my satisfaction to the hospitality and the friendliness we all encountered during this conference. The athmosphere was as relaxed as it could be with a conference as densily packed with talks as this one. For this achievement I would like to thank the organizers, especially Yishai Avishai. I know that it was a hard job done under not very favourable circumstances. For us everything went along quite smoothly, and if there were any problems - the participants did not notice anyone of them. This is how it should be, since there were and still are enough physical problems to bother us (and to amuse us nevertheless).

---

* Supported in part by the Deutsche Forschungsgemeinschaft

INDEX